Therapeutic Potential of Natural Products in Neurodegenerative Disorders

This book reviews the potential of natural products in the treatment and management of neurodegenerative disorders. The introductory chapter provides fundamental insights into neurodegenerative disorders. Subsequently, the book covers the therapeutic potential of polyphenols and other agents for amyotrophic lateral sclerosis, the application of natural ingredients for Friedreich's ataxia, and the combined effects of natural ingredients in addressing Huntington's disease. The book also explores conventional therapeutic regimens for Lewy body disease and examines the synergistic effects of both natural and synthetic agents. Furthermore, the chapters discuss the utilization of natural extracts in the management of spinal muscular atrophy while examining broader synthetic management approaches. The second section of the book explores nanodelivery systems for managing neurodegenerative disorders. It discusses synergistic treatment approaches, followed by an analysis of advanced techniques such as electron spinning and electron spraying, and their application in neurodegenerative disorders. The book also reviews the therapeutic potential of flavonoids in managing neurodegenerative disorders. Toward the end, the book also examines the ethical framework of natural remedies used in the treatment of neurodegenerative disorders. This book is intended for neuroscientists and researchers in the fields of neurology, pharmacology, and nanotechnology.

Therapeutic Potential of Natural Products in Neurodegenerative Disorders

Edited by
Ghulam Md Ashraf, Md. Habibur Rahman,
Karri V. V. S. Narayana Reddy, Chenmala Karthika,
Ali T. Zari, and Gowthamarajan Kuppusamy

CRC Press
Taylor & Francis Group
Boca Raton London New York

CRC Press is an imprint of the
Taylor & Francis Group, an **informa** business

Designed cover image: Shutterstock

First edition published 2026
by CRC Press
2385 NW Executive Center Drive, Suite 320, Boca Raton FL 33431

and by CRC Press
4 Park Square, Milton Park, Abingdon, Oxon, OX14 4RN

CRC Press is an imprint of Taylor & Francis Group, LLC

ISBN: 978-1-032-76483-2 (hbk)
ISBN: 978-1-032-76484-9 (pbk)
ISBN: 978-1-003-47865-2 (ebk)

DOI: 10.1201/9781003478652

Typeset in Times
by Apex CoVantage, LLC

Contents

About the Editors ix

PART I Neurodegenerative Disorder and Correlation with Natural Ingredients

Chapter 1 General Introduction of Neurodegenerative Disorders 3

Nayudu Teja, Sounderrajan Vijayakhiruba, Vadabam Valasa Reddy, Neelufar Shaik, Aparna Baliwada, and Amol S. Rakte

Chapter 2 Role of Natural Herbs in Alzheimer's Disease and Their Future Prospects 15

Gifty M. Jojo, Mr. Jobin Johnson, Gowthamarajan Kuppusamy, and Karri V. V. S. Narayana Reddy

Chapter 3 Polyphenols and Other Agents with Their Uses for Amyotrophic Lateral Sclerosis 27

Subhasri Bogadi, Gowthamarajan Kuppusamy, and Karri V. V. S. Narayana Reddy

Chapter 4 Natural Ingredients for Friedreich's Ataxia 45

Madhu Tanya Singh, Sai Varshini Magham, Raja Shekhar Nunavath, Sai Kiran S. S. Pindiprolu, and Praveen Thaggikuppe Krishnamurthy

Chapter 5 The Effect of Natural Ingredients and Their Combination for Huntington's Disease 56

Nagasen Dasari, Sankhanil Panda, Sunanda Kaligithi, and Sai Kiran S.S. Pindiprolu

Chapter 6 Lewy Body Disease and Its Therapeutic Regimen Using Conventional Methods 73

Sagarkanta Mohanty, Srikanth Jupudi, Neeru Dugar, Swati Swagatika Swain, and Nayudu Teja, and Karri V. V. S. Narayana Reddy

Chapter 7 Parkinson's Disease and Synergistic Effect of Natural and Synthetic Agents 87

K.M. Muhasina, Mehrukh Zehravi, N.P. Deepika, Krishnaveni Nagappan, Md Ekhtiar Uddin, and Dhanabal S. Palaniswamy

Chapter 8 Natural Extracts for Spinal Muscular Atrophy 97

Swati Swagatika Swain, Karri V. V. S. Narayana Reddy, Gowthamarajan Kuppusamy, Subhasri Bogadi, Sagarkanta Mohanty, Ashish Sriram Mishra, and Vetriselvan Subramaniyan

Chapter 9 Synthetic Way of Management in General 114

Magham Sai Varshini, Madhu Tanya Singh, Sai Kiran S.S. Pindiprolu, Praveen Thaggikuppe Krishnamurthy, and R. Aishwarya Reddy

Chapter 10 Credentials of Natural Therapy over Synthetic Treatment 123

Vetriselvan Subramaniyan, Muralikrishnan Dhanasekaran, Mukesh Kumar, Sivaraman Dhanasekaran, Srikanth Jeyabalan, Gowthamarajan Kuppusamy, and Karri V. V. S. Narayana Reddy

Chapter 11 Nano-Delivery System for the Management of Neurodegenerative Disorder 135

Sai Kiran S.S. Pindiprolu, Ramakrishna Gummadi, Nagasen Dasari, and Chirravuri Satyanarayana Phani Kumar

PART II *Neurodegenerative Disease and Its Therapeutic Approaches*

Chapter 12 Synergistic Approach and the Treatment of Neurodegenerative Disorder 155

Nusrat K. Shaikh

Chapter 13 Electron Spinning and Electron Spraying Techniques and Their Correlation with Neurodegenerative Disorders 170

Subhasri Bogadi, Gowthamarajan Kuppusamy, Karri V. V. S. Narayana Reddy, and Swati Swagatika Swain

Chapter 14 Neurodegenerative Disease and Its Therapeutic Approaches: Formulation Approach for Overcoming the Barriers during the Treatment of Neurodegenerative Disorder 191

Koushik Yetukuri, M.S. Umashankar, Sandeep Kanna, and Rama Rao Nadendla

Chapter 15 Food and Nutrition as a Medicine Class for the Prevention and Treatment of Neurodegenerative Disorder 211

Vetriselvan Subramaniyan, Muralikrishnan Dhanasekaran, Mukesh Kumar, Mahendran Sekar, Neeraj Kumar Fuloria, Shivkanya Fuloria, Gowthamarajan Kuppusamy, and Karri V. V. S. Narayana Reddy

Chapter 16 Flavonoids: Role in Neurodegenerative Studies 224

Swathi Nalla, Nagaraju Bandaru, Naidu Teja, and Vadabam Valasa Reddy

Chapter 17 Side Effects during Treatment of Neurodegenerative Disorder 243

Swathi Nalla, Nagaraju Bandaru, Nayudu Teja, and Vadabam Valasa Reddy

Chapter 18 Differentiation between Synthetic and Natural Treatment Modalities 261

Deepak Vasudevan Sajini and Amritha Chakkitukandiyil

Chapter 19 Challenges during the Use of Natural Remedies for Neurodegenerative Disorders 271

K.M. Muhasina, Puja Ghosh, Mehrukh Zehravi, N.P. Deepika, Krishna Veni Nagappan, Md Ekhtiar Uddin, and Dhanabal S. Palaniswamy

Chapter 20 Conclusion and Future Aspects during Treatment of Neurodegenerative Disorders 279

Mirunalini Gobinath, L. Priyanka Dwarampudi, R. Shanmugam, Parikshit Roychowdhury, and Subhajit Dutta

Index 291

About the Editors

Ghulam Md Ashraf is an accomplished biochemist and molecular neuroscientist with 15 years of experience in research, teaching, and innovation across the biological, health, and medical sciences. He holds a dual degree—Ph.D. in Biochemistry (Aligarh Muslim University, India) and PGDBA in Operations Management (Symbiosis Center for Distance Learning, India)—and completed the prestigious Instructional Skills Workshop (ISW) in collaboration with the University of British Columbia, Canada, enhancing his pedagogical expertise. His research focuses on molecular mechanisms and therapeutic strategies for neurodegenerative disorders, including multiple sclerosis and Alzheimer's disease. His lab investigates serotonin receptor signaling, develops selective modulators to counteract pathological protein aggregation, and advances next-generation Multiple Sclerosis therapies, including engineered interferons and nanomedicine-based neuroprotective delivery. Dr. Ashraf is recognized among the Top 2% of Scientists Worldwide since 2021 (Stanford–PLOS–Elsevier), the leading Biochemistry Scientist in the UAE (AD Scientific Index), and a World Expert in Neurodegenerative Diseases (Expertscape). He is listed in Marquis Who's Who in the World (since 2020), has received the Publons/Web of Science Top Reviewer Awards in Cross-Field and Biology & Biochemistry (2019), holds 4 international patents, and has authored nearly 450 publications with ~20,000 citations (H-index ~70). Dr. Ashraf has held leadership roles in research administration and laboratory management and has extensive editorial experience with journals including *Journal of Advanced Research, Scientific Reports, and Plos One*. He is a Fellow of the Royal Society of Medicine (FRSM), a Member of the Royal Society of Biology (MRSB), and an active member of international scientific societies such as the Canadian Academy of Neuroscience and the American Society for Biochemistry and Molecular Biology. Dr. Ashraf's work exemplifies the integration of advanced research, innovation, and global collaboration to advance biomedical science and improve human health.

Md. Habibur Rahman is Registered Pharmacist in Bangladesh and a member of the Novel Global Community Education Foundation (Australia). He earned his Ph.D. in Medical Science from Yonsei University, South Korea, and has contributed his expertise to several leading pharmaceutical organizations. His research is primarily focused on neuroscience, particularly the molecular mechanisms underlying clinical dementia, with emphasis on biomarker discovery and targeted diagnostic strategies. In addition to his neuroscience work, Dr. Rahman is an expert in molecular biology, with extensive experience in gene expression analysis, oxidative stress pathways, and neuroinflammation. He has been recognized among the top 2% of scientists worldwide (Stanford University, citation-based global ranking) for his impactful contributions. Dr. Rahman is also an active scientific communicator and public speaker, frequently presenting at international conferences and academic forums. He serves as a peer reviewer for several high-impact journals and is involved in multiple ongoing research collaborations in neuroscience and molecular medicine.

Karri V. V. S. Narayana Reddy is Associate Professor in the Department of Pharmaceutics at JSS College of Pharmacy, Ooty, India. A proud alumnus of the same institution, he completed his B.Pharm, M.Pharm, and Ph.D. in Pharmaceutics from JSS College of Pharmacy, Ooty. Recognized among the Top 2% of Scientists in the World by Stanford University for three consecutive years (2022, 2023, and 2024), Dr. Reddy has made significant contributions to pharmaceutical sciences. His expertise spans nanodrug delivery, diabetic wound healing, and biomaterials, with a particular focus on natural products and innovative formulations. He has published over 100 peer-reviewed research articles, authored several book chapters, and holds multiple patents related to advanced drug delivery systems. Dr. Reddy also serves as a research coordinator, an active Ph.D. guide, and is on the editorial board and review panels of reputed international journals.

Chenmala Karthika is Postdoctoral Fellow at the National Institute of Health in the USA. She completed her Ph.D. at JSS College of Pharmacy, Ooty, and has published more than 35 national and international research articles in journals of high repute. Over the past five years, her research has focused on cancer and its therapeutic approaches, with an additional interest in aging, COVID-19, and neurodegenerative disorders.

Ali T. Zari is Associate Professor in the Department of Biological Sciences at the Faculty of Sciences, King Abdulaziz University in Jeddah, Saudi Arabia. He obtained his Bachelor of Science in Biology in 2007 from King Abdulaziz University and completed his Master of Science in Biology in 2014 at Laurentian University in Canada. He earned his Ph.D. in biomolecular sciences from Laurentian University in Canada in 2019. Dr. Zari's research interests encompass the examination of the effects of medicinal plants and their bioactive components on various cancer cells and other diseases using biomolecular techniques in vitro and in vivo to investigate their potential effectiveness as pharmaceutical sources. Additionally, he investigates the potential benefits and risks of using nanomaterials for therapeutic purposes. Dr. Zari is also involved in in silico prediction of cancer inhibitors and protein modeling for antitumor activity. He has diverse teaching experiences at the graduate and postgraduate levels in the fields of general biology, molecular biology, cell biology, natural products, and animal physiology at King Abdulaziz University.

Gowthamarajan Kuppusamy is Professor and Head, Department of Pharmaceutics (DST-FIST Sponsored), JSS College of Pharmacy, Ooty (JSS Academy of Higher Education & Research, Mysuru). He holds M.Pharm (1996) and Ph.D. (2007) degrees from the Tamil Nadu Dr. MGR Medical University, Chennai. With over two decades of academic and research experience, he has developed industry-relevant academic programs and led collaborative projects in intranasal and nanotechnology-based drug delivery. His current focus is on optimized and personalized drug delivery systems. He has guided over 50 PG and 18 Ph.D. scholars. He has executed several government-funded projects in areas such as Alzheimer's disease, colorectal cancer, cardiovascular diseases, and space medicine, receiving grants from DST-SERB, AICTE, ICMR, DBT, and SPARC. He has played a significant role in technology transfers and serves as a consultant for multiple pharmaceutical industries. Dr. Gowthamarajan has been consistently listed among the top 2% scientists globally by Stanford University and Elsevier for the past three consecutive years (2022–2024). He has over 150 peer-reviewed publications and 23 patents and is a member of national academic bodies, editorial boards, and regulatory committees.

Part I

Neurodegenerative Disorder and Correlation with Natural Ingredients

1 General Introduction of Neurodegenerative Disorders

Nayudu Teja[1], *Sounderrajan Vijayakhiruba*[1], *Vadabam Valasa Reddy*[2], *Neelufar Shaik*[3], *Aparna Baliwada*[4], *and Amol S. Rakte*[5]

[1]Department of Pharmaceutics, JSS College of Pharmacy, JSS Academy of Higher Education and Research, The Nilgiris, Tamil Nadu, India

[2]Assistant Professor, Department of Pharmaceutical Chemistry, Adarsa College of Pharmacy, G. Kothapalli, Andhra Pradesh, India

[3]Department of Pharmaceutical Chemistry, JSS College of Pharmacy, JSS Academy of Higher Education and Research, The Nilgiris, Tamil Nadu, India

[4]Associate Professor, Department of Pharmaceutical chemistry, Malla Reddy Pharmacy College, Hyderabad

[5]Professor, Indrayani Vidya Mandirs, Krishnarao Bhegade Institute of Pharmaceutical Education and Research, Talegaon Dabhade, Tal Maval District, Pune, Maharashtra

1.1 INTRODUCTION

1.1.1 Neurodegenerative Disorders

Neurodegeneration refers to the progressive malfunction and loss of neuronal structure and function, ultimately leading to the death of neuronal cells. Neurodegenerative disorders (NDDs) encompass diseases affecting the central nervous system (CNS) and peripheral nervous system, impacting the brain's structure and functionality. These conditions result in the gradual loss of neural structures essential for proper brain functions. Globally, neurodegenerative illnesses are a leading cause of disability and mortality. Prominent NDDs include Parkinson's disease (PD), Alzheimer's disease (AD), Huntington's disease (HD), multiple sclerosis, amyotrophic lateral sclerosis (ALS), vascular dementia, epilepsy, cerebral ischemia, behavioral disorders, and mental illnesses [1].

The clinical presentation of NDDs, whether acute or chronic, depends on the loss of specific neuronal populations within functional neural networks. Acute neurodegeneration arises from sudden damage to neurons, often resulting in rapid cell death due to events such as head injuries, strokes, traumatic brain injuries, cerebral or subarachnoid hemorrhages, and ischemic brain damage. In contrast, chronic neurodegeneration is a long-term condition characterized by a gradual and progressive deterioration of neurons in the nervous system. This process, often driven by multifactorial causes, leads to the irreversible destruction of specific neuronal populations over time. Disorders like ALS, HD, PD, and AD are classified as chronic NDDs [2, 3].

1.1.2 History

NDDs have been linked, both directly and indirectly, to climate conditions exacerbated by global warming, which are degrading natural habitats and ecosystems. According to Cheshire, climate

DOI: 10.1201/9781003478652-2

change can contribute to mental illnesses by impairing the body's ability to regulate temperature. A climate change report predicts that, between 2030 and 2050, an additional 250,000 deaths per year could occur due to factors associated with climate change, including an estimated 38,000 deaths among the elderly caused by heat exposure and related neurological complications [1].

Neurological disorders currently affect approximately 15% of the global population and are the leading cause of both physical and cognitive disability worldwide. Over the past 30 years, the number of individuals affected by these conditions has risen significantly. Furthermore, it is projected that the burden of chronic NDDs will double in the next 20 years, largely due to the aging global population. This trend will present significant challenges to maintaining universal access to neurological care. Organizations such as the World Health Organization (WHO) and the National Health Service of the United Kingdom have issued warnings, stating that "neurological services are not sustainable in their current form and redesign is needed" and that "available resources for neurological services are insufficient in most countries of the world compared with the global need for neurological care" [4].

In 2019, approximately 50 million people globally were affected by non-communicable dementia, which often leads to cognitive decline. These conditions impact all age groups and geographic regions. In Europe, the prevalence of non-communicable dementia in the 65–69 age group is 1.6% in men and 1% in women, increasing to 11% and 12.6% in the 85–89 age group, respectively. A WHO report emphasizes the urgency of the situation, warning that, without immediate global action, the burden of neurological disorders will pose an even greater and potentially unmanageable threat to public health [5].

1.1.3 Cause of Neurodegeneration

Neurological disorders can result from various factors, including ischemia, oxidative and extracellular stress, and inflammation; abnormal protein deposition in neural tissue; autoimmune-mediated neuronal loss; and infections caused by viruses. Depending on the affected site, neuronal loss or demyelization can lead to motor deficits, behavioral disturbances, and cognitive decline [6]. For many elderly individuals, NDDs such as Parkinson's and Alzheimer's diseases remain significant clinical challenges. Neurodegeneration is associated with dysfunction in neuronal networks and synapses, along with the accumulation of physiologically altered protein variants in the brain. While most neurons are produced by neural stem cells during childhood, their numbers are significantly reduced in adulthood [7].

Stroke is the leading cause of non-traumatic deaths in industrialized nations, accounting for the highest number of NDD-related fatalities. NDDs are second only to headache disorders in prevalence, and, as global life expectancies rise and populations grow, the incidence of NDDs like PD and AD is expected to increase worldwide [6]. The development of NDDs is linked to various biological phenomena commonly observed in patients, including neuron inflammation, oxidative stress (OS), cognitive decline, the accumulation of neurofibrillary tangles (NFTs), aberrant amyloid-beta (Aβ) peptide deposition, impaired or insufficient neurotransmitter synthesis, and abnormal ubiquitination. Aging plays a pivotal role in the progression of NDDs, which are irreversible and impose significant societal and economic burdens, with limited treatment options available [3].

Psychological disorders, such as schizophrenia and other cognitive and mood-related impairments, are neurological conditions associated with intellectual disability. These impairments are linked to alterations in gamma-aminobutyric acid, low dopamine (DA) levels, and serotonin [1]. The blood–brain barrier (BBB), while highly effective at protecting the brain, poses a significant challenge to the management of NDDs. Despite advances in invasive and non-invasive procedures, clinical adoption remains limited due to concerns about potential BBB damage and their long-term efficacy. Nanotherapeutics, capable of crossing the BBB without causing harm, have emerged as a promising alternative for halting or reversing neurodegeneration. Their growing popularity is attributed to their advantages over traditional treatment methods [7].

Neurodegeneration is associated with several biological mechanisms, including OS, neuroinflammation, excitotoxicity, mitochondrial dysfunction, abnormal protein misfolding and aggregation, and apoptosis. Numerous studies have aimed to understand the mechanisms and potential treatment targets for NDDs. Neuroprotection strategies that interact with these pathophysiological processes are the most effective for preventing or slowing neurodegeneration. NDDs are characterized by protein aggregation, OS, and inflammation within the CNS. Molecular processes linked to these conditions include abnormal ubiquitination, OS, and neurotransmitter depletion or insufficient synthesis. Common features of NDDs include inflammation, mitochondrial dysfunction, abnormal cellular transport and protein deposition, excitotoxicity, intracellular calcium (Ca^{2+}) excess, and the unchecked production of reactive oxygen species (ROS).

One of the hallmark features of NDDs is the presence of large protein aggregates within the brain, extracellular spaces, or neurons, often referred to as "amyloidal plaques." Genetic evidence suggests that the transformation of native and soluble proteins into aggregates and their precursor oligomers is a major contributing factor in NDDs [3]. Natural products have been valued since ancient times for their therapeutic properties. In recent years, they and their bioactive compounds have been extensively studied for their biological activities, nutritional benefits, and potential health and therapeutic applications [2].

Genetic factors play major roles in the pathogenesis of neurodegeneration as monogenic causes for Mendelian disorders or susceptibility modifiers for sporadic complex diseases. Additionally, copy number variants caused by duplication, insertion, or deletion of DNA segments usually larger than 50 base pairs, and epigenetics, which builds a bridge between the environmental information and gene expression through epigenetic regulations such as DNA methylation, RNA-mediated mechanisms, and histone protein modifications, also lead to disease phenotypes. The widespread applications of high-throughput technologies in biology and medicine, such as next-generation sequencing and DNA microarray (gene chip) technology, are generating an increasing body of evidence which broadens the understanding of the genetic and epigenetic basis of NDDs [8].

The genetic and epigenetic code interplay has been extensively examined in learning and memory. Pioneering studies have confirmed the critical involvement of epigenetic changes in processes related to learning, memory, and synaptic plasticity. It is also evident that disruptions in epigenetic mechanisms not only hinder normal brain functioning but are also linked to various neurological diseases, particularly AD and PD [9].

1.2 ENVIRONMENTAL TOXINS AND LIFESTYLE FACTORS IN NDDS

NDDs, such as AD, PD, and ALS, are influenced not only by genetic predispositions but also significantly by environmental toxins and lifestyle factors [10].

Environmental toxins – including heavy metals (lead, mercury, arsenic), pesticides (like paraquat and rotenone), industrial chemicals (e.g., polychlorinated biphenyls or PCBs), and air pollution – have been shown to damage neurons. These toxins can trigger OS, increase inflammation in the brain, disrupt mitochondrial function, and cause protein misfolding and aggregation, all of which are key contributors to neurodegeneration [11].

Lifestyle factors also play a major role. Poor dietary habits (high sugar, processed foods), lack of physical activity, chronic stress, smoking, excessive alcohol use, and poor sleep hygiene can negatively affect brain health over time. These behaviors can increase systemic inflammation, impair the brain's ability to clear toxic proteins, and accelerate aging-related cognitive decline [12].

On the other hand, **protective lifestyle factors** such as regular exercise, a nutrient-rich diet (like the Mediterranean or MIND diets), mental stimulation, adequate sleep, and stress management have been shown to support brain function, reduce inflammation, and delay the onset or progression of NDDs (Figure 1.1).

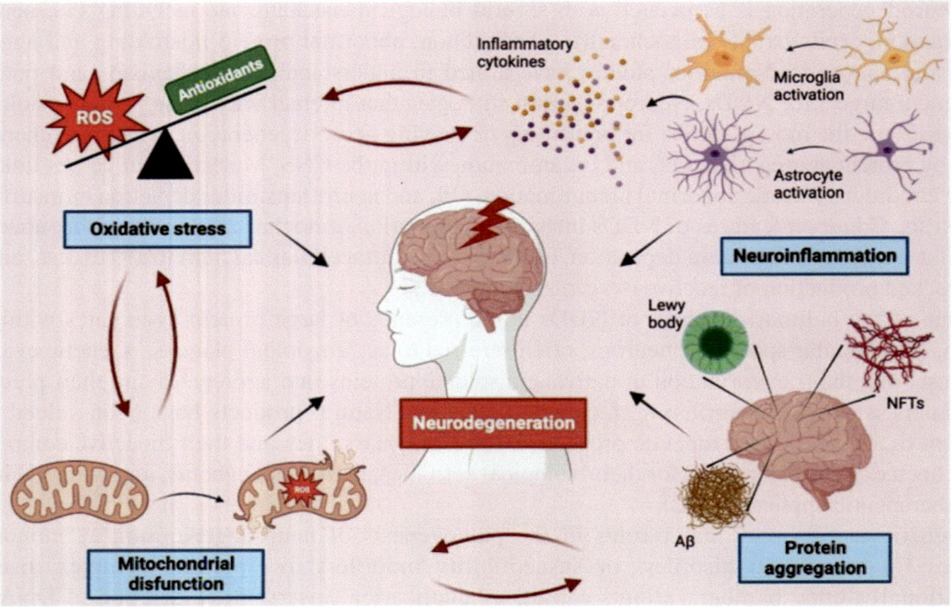

FIGURE 1.1 A schematic representation of the key mechanisms involved in neurodegeneration, including reactive oxygen species (ROS) and neurofibrillary tangles (NFTs).

1.2.1 Pathophysiology of Neurodegeneration

Neurodegeneration, characterized by the progressive loss of neural tissue, including neuronal death, is a complex process with multifactorial causes that predominantly affect human brain neurons. A key feature of neurodegeneration is the disruption of protein homeostasis, leading to the accumulation of inclusion bodies and insoluble deposits in various brain regions. Closely associated with aging, this condition is also linked to numerous other pathogenic pathways, including OS, neuroinflammation, mitochondrial dysfunction, calcium dysregulation, axonal transport deficiencies, and DNA damage. The prolonged persistence of these factors eventually overwhelms the body's defense mechanisms, creating an imbalance between survival and cell death. This imbalance triggers programmed cell death through various pathways, such as necrosis, autophagy, apoptosis, and parthanatos [13].

Although our understanding of the genetics and pathology of NDD has improved rapidly in recent years, the biology of neurodegeneration is complex, and there remains an unsatisfying disconnect between the initiators that can be observed as a component of disease pathology and the subsequent activation of neuronal cell death executioners. Determining which cell death pathways are relevant during disease progression and the mechanisms by which they are activated remains a key question in the field, and unraveling these mechanisms has the potential to drive the development of a new generation of targeted therapeutics that modify NDD progression [14].

1.2.1.1 Oxidative Stress

The biochemical and biomolecular changes observed in various NDDs, such as AD, ALS, HD, and PD, are widely recognized to be driven by OS. Both high and low levels of ROS are crucial, as they play a key role in molecular mechanisms underlying signal transduction, synaptic plasticity, and memory formation. However, excessive production of ROS can damage cellular components such as proteins, lipids, and DNA, leading to OS and neuronal cell death. OS can result from both

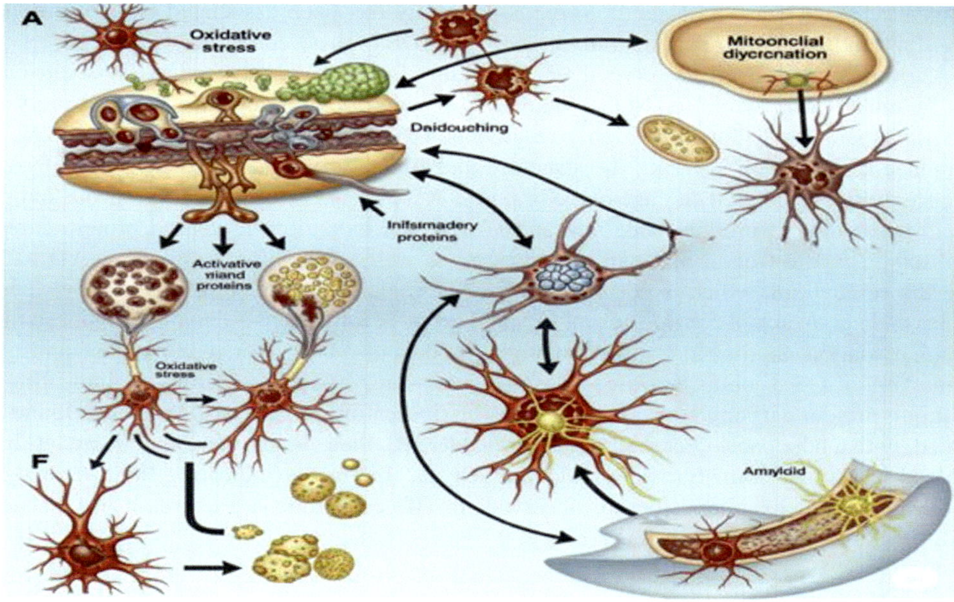

FIGURE 1.2 Neuroinflammation.

the increased generation of ROS and reactive nitrogen species (RNS) and the reduced activity of antioxidant enzymes, both enzymatic and non-enzymatic. To maintain neuronal cell function, it is essential to maintain a delicate balance between oxidant and antioxidant responses [13] (Figure 1.2).

Disruptions in metal homeostasis, leading to increased free radical generation, are believed to directly cause oxidative damage in NDDs. One significant factor influencing ROS-mediated cellular damage, besides the production of O_2^- and H_2O_2, is the availability of redox-active iron. Both the aging brain and several NDDs exhibit elevated levels of redox-active iron, primarily produced by degenerating mitochondria. Iron catalyzes the formation of highly reactive radicals from lipid peroxides and H_2O_2, contributing to OS. Although most iron in the brain is stored in ferritin, "catalytic" iron can be easily released from damaged brain tissue. In animal models of AD, enhanced lipid per oxidation occurs before the formation of amyloidal-beta (Aβ) plaques, and abnormal metal homeostasis may contribute to the generation of ROS and toxic Aβ oligomers. Imbalances in iron homeostasis serve as precursors to neurodegeneration, particularly in AD. Mutations in genes involved in iron metabolism have been linked to disorders like neuroferritinopathy and Friedreich's ataxia, which result in OS and cell death. Iron accumulation is evident in brain regions affected by AD and PD, with iron–DA complexes playing a key role in PD pathophysiology [15].

NADPH oxidase (NOX), a multisubunit enzyme complex, plays a primary role in ROS production, generating hydrogen peroxide and superoxide as the two main ROS. NOX enzymes are expressed in microglia, neurons, astrocytes, and the neurovascular system, with reliable data available for NOX expression in oligodendrocytes. NOX enzymes are essential for the development, biology, and maturation of neurons in the CNS. While other NOX isoforms may also be important, NOX2 appears to be most involved in producing pathological OS in the brain. Evidence suggests that NOX enzymes contribute to a number of NDDs, including Parkinson's, Huntington's, and Alzheimer's diseases. The observation that the inhibition of NOX benefits various NDDs further supports the role of NADPH oxidase in neurodegeneration. ROS production in the CNS is also attributed to xanthine oxidase, an enzyme involved in purine metabolism, which oxidizes hypoxanthine and xanthine to produce uric acid. This process generates ROS such as superoxide and hydrogen peroxide. Increased xanthine oxidase activity has been associated with disorders like Parkinson's and Alzheimer's, potentially exacerbating OS in these conditions [13].

L-Arginine and L-citrulline serve as the precursors for the catalytic production of nitric oxide (NO) by nitric oxide synthase (NOS). While NO plays critical biological roles, it rapidly interacts with superoxide to form the highly reactive peroxynitrite anion under OS. Like ROS, RNS can induce nitrosylation, altering the structure and function of lipids, proteins, and DNA [13]. Evidence of ROS-mediated damage in the brains of AD patients is supported by various indicators of lipid peroxidation, protein oxidation, and DNA/RNA oxidation. Increased protein carbonyls are observed in the frontal and parietal cortices, as well as the hippocampus, of AD patients, while the cerebellum, which is not affected by AD, remains spared. Comparisons between AD samples and age-matched controls reveal elevated levels of 8-hydroxy-2-deoxyguanosine, a marker of DNA oxidation. Studies in transgenic animal models of AD further support these findings. In PD, markers of OS, such as 8-hydroxy-2-deoxyguanosine, 4-hydroxy-2-nonenal (indicating lipid oxidation), protein carbonyls, and 3-nitro-tyrosine (indicating protein oxidation), are found to be elevated, with a noticeable decrease in glutathione (GSH), the primary natural antioxidant. In HD, increased levels of 3-nitro-tyrosine, malondialdehyde, lipofuscin (indicating lipid peroxidation), and 8-hydroxy-2-deoxyguanosine are observed. In HD cell models, a decrease in GSH is also noted, similar to what is seen in PD [13].

1.2.1.2 Neuroinflammation

Inflammation is the immune system's defense mechanism against infections, trauma, toxin buildup, and other pathological conditions. In the CNS, this response is known as neuroinflammation, a complex process involving various cell types, including oligodendrocytes, astrocytes, and microglia, which collaborate in a coordinated manner. Neurotrophic factors, cytokines, ions, and neurotransmitters all contribute to the interaction between these cells. Microglia and astrocytes are the main drivers of neuroinflammation. Similar to macrophages, microglia cells help maintain brain homeostasis under normal conditions and perform several reparative tasks, such as phagocytosing abnormal proteins, organizing synapses, responding to biotoxins, and removing dendritic debris. For example, microglia can prevent neurodegeneration in the early stages of AD.

AD can be mitigated by clearing amyloid-β (Aβ) through phagocytosis mediated by activated microglia and other immune mechanisms. Activated microglia also promote astrocyte growth, leading to neuroprotective effects and aiding in the repair of damaged neuronal tissue. After brain injury, activated astrocytes show increased cell proliferation and mobility, upregulation of intermediate filaments, and hypertrophy. While neuroinflammation is a fundamental protective response, chronic or excessive inflammation can lead to harmful brain damage. Aging, metabolic disorders, and viral infections are among the risk factors that can cause persistent neuroinflammation and contribute to neurodegeneration. Neuroinflammation is often marked by the activation of nuclear factor kappa B (NF-κB), a transcription factor that regulates genes involved in inflammation, apoptosis, cell survival, and neuronal differentiation [13] (Figure 1.3).

Activated microglia can be compared to macrophages, where two distinct phenotypes are observed: the M1 proinflammatory phenotype and the M2 anti-inflammatory phenotype. The M1 phenotype is activated through TLR and IFN-γ-mediated signaling pathways, leading to tissue damage and the production of proinflammatory cytokines and chemokines. This activation also results in the overexpression of inflammatory genes, including iNOS and COX-2, and proinflammatory mediators such as TNF-α, IL-6, IL-1, ROS, and BACE1. Additionally, M1 microglia overproduce enzymes like inducible NOS and NADPH oxidase, which generate superoxide anions. In M1 microglia, phagocytic activity is inhibited, which negatively affects brain progenitor cell differentiation and maturation, ultimately impairing synapse formation and plasticity [13].

Conversely, the M2 phenotype, activated by cytokines like IL-4, IL-10, IL-13, and TGF-β, has neuroprotective properties. M2 microglia produce factors such as FIZZ1, arginase 1, Ym1, CD206, IGF-1, and Fzd1, which are associated with enhanced phagocytic capacity, synaptic transmission,

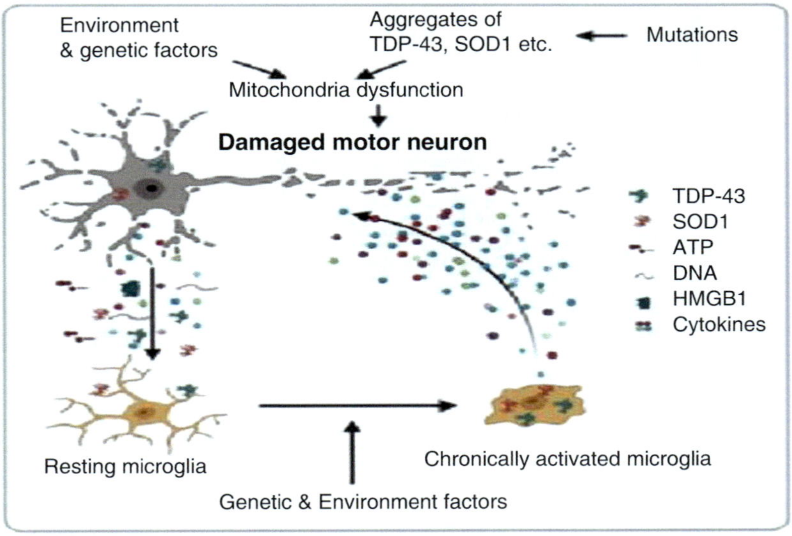

FIGURE 1.3 Mitochondrial dysfunction.

plasticity, and neural repair. IL-4, for example, has been shown to prevent the release of proinflammatory cytokines like TNF-α, IL-6, and NO [13].

In a similar manner, reactive astrocytes are categorized into A1 and A2 phenotypes. M1 microglia can release cytokines that activate A1 astrocytes, which lose typical astrocytic functions such as synapse maintenance. A1 astrocytes may also secrete neurotoxin substances that harm mature oligodendrocytes and some neurons. These proinflammatory A1 astrocytes contribute to neuroinflammation by upregulating genes involved in the complement cascade and producing proinflammatory molecules like IL-1β, TNF-α, and NO. On the other hand, A2 astrocytes, which are activated by anti-inflammatory cytokines like TGF-β, IL-10, IL-13, and IL-4, have neuroprotective functions. They secrete neurotrophic or anti-inflammatory factors that promote neuron survival and aid in tissue repair. While reactive astrocytes and microglia often display mixed or intermediate phenotypes, the M1/M2 and A1/A2 model provides a helpful framework for understanding their roles in CNS disorders. Neuroinflammation, driven by these cellular responses, is a significant factor in NDDs like PD, AD, and ALS, making it a key target for counteracting neurodegeneration [13].

1.2.1.3 Mitochondrial Dysfunction

The brain requires a significant amount of energy, relying heavily on a constant oxygen supply to meet its high metabolic demands. Approximately 20% of the body's baseline oxygen is used by the brain, with the Na^+/K^+ A10TPase alone consuming nearly half of the brain's energy to maintain the resting membrane potential of excitatory cells.

The ATP generated through oxidative phosphorylation fuels the brain's metabolic activity, with cortical neurons consuming around 4.7 billion ATP molecules every second. Beyond energy production, mitochondria are also essential in regulating neurotransmission, the cell cycle, and cell death. Ensuring mitochondrial efficiency is vital due to the critical nature of these functions. Various processes, such as biogenesis, fission, fusion, and mitophagy (the removal of damaged mitochondria to prevent apoptosis), help maintain mitochondrial integrity. Mitochondrial dysfunction can lead to synaptic and neuronal degeneration, as neurons have limited regenerative capacity [13] (Figure 1.4).

Mitochondria are the primary source of ROS due to electron leakage in the electron transport chain, and ROS also directly damages mitochondrial components, perpetuating a harmful cycle.

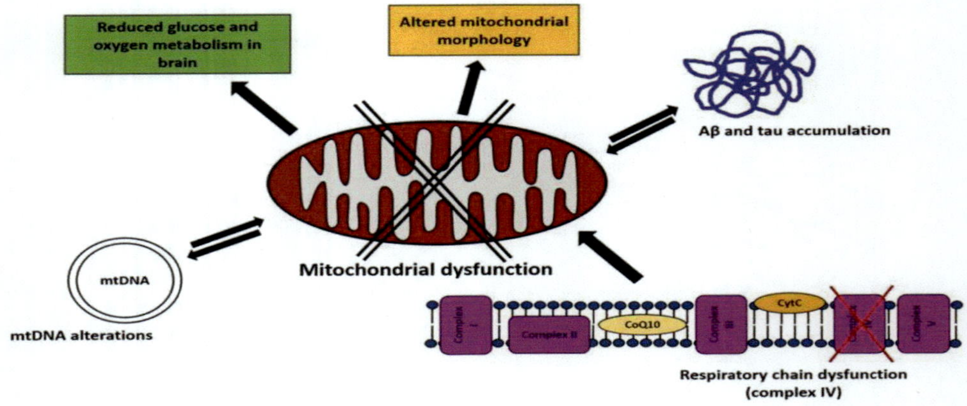

FIGURE 1.4 Structure and functional components of blood–brain barrier.

Acute ROS exposure can deactivate the iron-sulfur centers of electron transport chain complexes, while prolonged exposure damages proteins, DNA, and lipids. The inner mitochondrial membrane, located close to the site of ROS synthesis, is particularly vulnerable. ROS-induced lipid peroxidations can increase membrane permeability, impairing transporters and enzymes and halting mitochondrial energy production. Recent research suggests that damaged mitochondria may also contribute to inflammation by triggering risk-signaling reactions. Extracellular vesicles generated in response to stress can transfer mitochondrial DNA (mtDNA) to the extracellular space, where they function as damage-associated molecular patterns, activating an innate inflammatory immune response [13].

Mitochondrial damage has been linked to several NDDs. In PD, this manifests as changes in mitochondrial shape, mitochondrial DNA alterations, disrupted calcium homeostasis, and impaired electron transport chain function. Post-mortem tissues from AD patients show altered mitochondrial morphology, reduced ATP generation, and inhibited complex IV activity. Similarly, ALS causes mitochondrial damage, affecting dynamics, structure, bioenergetics, and calcium homeostasis, with blood samples from ALS patients revealing abnormalities in electron transport chain proteins, including lower levels of FAD synthase, riboflavin kinase, cytochrome C1, and succinate dehydrogenase complex [13].

Protein aggregates, commonly seen in NDDs, are associated with mitochondrial damage. Studies in AD and PD models have shown that mitochondrial membranes contain both amyloid-beta (Aβ) and α-synuclein. α-Synuclein appears to impair electron transport chain complex I efficiency, leading to mitochondrial damage and dopaminergic neuron death. Additionally, pathogenic mutations in α-synuclein promote mitochondrial fragmentation, suggesting its role in modifying mitochondrial morphology. Tau protein interactions with mitochondrial proteins and alterations in mitochondrial homeostasis contribute to neurodegeneration and the potential onset of tauopathies [13]. In conclusion, given the brain's high energy needs, mitochondria are crucial to its metabolism, and even minor dysfunction can have severe consequences, leading to neurodegeneration [13].

1.2.1.4 Protein Aggregation

Misfolded proteins are believed to be a central factor in the development of various NDDs, as their gradual accumulation and aggregation are hallmark features of these conditions. The aggregation process typically begins with a seed or oligomers, where specific misfolded protein fragments interact to form a non-native conformation. This misfolding can then propagate, converting other proteins into toxic forms. Misfolded proteins generally adopt intermolecular structures rich in β-sheets, which arise when soluble oligomers aggregate into small protofibrils. This aggregation leads to the formation of extracellular or intracellular inclusions, a characteristic feature of NDDs. Different types of protein aggregation are associated with specific NDDs. For example, pathological inclusions of α-synuclein are characteristic of PD and other conditions such as Lewy body dementia, the Lewy body variant of AD, and multisystem atrophy. Similarly, β-amyloid plaques, a hallmark of AD, and tau, a microtubule-associated protein (MAP), are commonly found in diseases like AD, PD, and HD. Additionally, TAR DNA-binding protein 43 (TDP-43) is associated with frontotemporal dementia and ALS [13].

Misfolded proteins are considered a primary cause of various NDDs, characterized by their gradual accumulation and aggregation. The aggregation process typically begins with β- and γ-secretase enzymes synthesizing amyloid-β (Aβ) from the transmembrane amyloid precursor protein. Aβ exists in two main forms: Aβ40 (40 residues) and Aβ42 (42 residues), differing by two additional residues at the C-terminus in Aβ42. Aβ monomers aggregate to form oligomers, protofibrils, and amyloid fibrils. While soluble amyloid oligomers can diffuse throughout the brain, larger, insoluble amyloid fibrils further aggregate into amyloid plaques, primarily consisting of Aβ42 in AD. Emerging research highlights that oligomers, rather than fibrils, are the primary agents of toxicity and disease propagation. Aβ accumulation induces ROS production and activates the Nf-κB inflammatory pathway, creating a detrimental cycle that leads to neurodegeneration, cognitive decline, and mitochondrial dysfunction [13].

Another protein implicated in NDDs is tau, which is associated with microtubules. Humans have six tau isoforms (50–70 kDa), with all six present in the adult brain but only the shortest isoform in the fetal brain. Tau contains multiple phosphorylation sites activated by about 20 different kinases. Under normal conditions, tau stabilizes microtubules, earning it the classification of a MAP. However, abnormal hyperphosphorylation of tau reduces its microtubule affinity and promotes aggregation into NFTs, which destabilize microtubules and disrupt neuronal cytoskeletons [13].

In PD and related disorders, α-synuclein is the primary pathogenic protein. Encoded by the SNCA gene, α-synuclein is a 140-amino acid protein with an intrinsically disordered structure and is predominantly expressed in the CNS. It comprises three main domains: the N-terminal domain (residues 1–60), which can form α-helices and bind to amphipathic membranes; the central domain (residues 61–95), which is highly amyloidogenic and drives aggregation into insoluble protofibrils and fibrils; and the C-terminal domain (residues 96–140), which undergoes phosphorylation and truncation. Aggregated α-Synuclein combines with other molecules such as ubiquitin, neurofilament protein, and lipid membranes to form Lewy bodies. These aggregates are implicated in neurodegenerative conditions like PD, multisystem atrophy, and dementia with Lewy bodies [13].

TDP-43, a protein of approximately 400 amino acids, is primarily nuclear but can form cytoplasmic aggregates in NDDs, often with post-translational modifications. TDP-43 is essential for RNA regulation, including mRNA stability and splicing, critical for brain development. In 2006, two key studies first identified hyperphosphorylated and ubiquitinated cytoplasmic TDP-43 in the anterior horn motor neurons of ALS patients. These aggregates are now recognized as pathological hallmarks of frontotemporal dementia and ALS, forming through progressively disordered intermediates [13].

To prevent protein aggregation, cells maintain proteostasis – a complex network involving molecular chaperones for protein folding and disaggregation, as well as proteasomal and autophagic systems for clearing misfolded proteins. Given the importance of proteostasis in preserving cellular integrity, understanding its regulation in neurodegeneration could pave the way for innovative treatment [13].

1.2.1.5 Epigenetic Aggregation

Epigenetics, an intricate field at the crossroads of genetics and environmental influences, is a pivotal avenue for unraveling the complexity inherent in gene regulation and cellular function. Going beyond the traditional focus on the linear sequence of DNA, epigenetics delves into reversible modifications on a cell's DNA bases or histones, affecting gene expression that transpires without modifying the fundamental genetic code [16].

Epigenetic regulation is crucial for normal development and cellular differentiation. It plays a significant role in diverse biological processes, such as embryonic development, tissue-specific gene expression, and response to environmental stimuli. Epigenetic mechanisms can modify gene expression outcomes, thereby influencing the phenotypic characteristics of the cell [17].

With the increase in research in recent years, it was observed that alterations in gene expression did not always originate from changes in the genetic sequence, which has led to understanding the role of epigenetics in NDDs, including AD and PD.

1.3 TRADITIONAL AND ADVANCED THERAPIES TO TREAT NEURODEGENERATIVE DISEASES

A. **Traditional therapies** for NDDs focus on symptom management and improving quality of life. Common approaches include medications (e.g., levodopa, donepezil), physical and occupational therapy, speech and swallowing therapy, supportive care, assistive devices, and lifestyle modifications [18–21].

However, these treatments have **several limitations** (1–4)

1. They do not cure or reverse the disease.
2. Efficacy in slowing disease progression is limited.
3. Pharmacological treatments often cause side effects.
4. Therapies are not personalized.
5. Many symptoms remain partially managed.

B. **Advanced therapies** for NDDs

Therapy	Description
Gene therapy	Involves delivering therapeutic genes to target cells or tissues to correct genetic disorders, provide therapeutic effects, or modulate cellular functions [22].
Steroids	Neuroactive steroids and sex steroids promote survival, neurogenesis, and memory function by limiting neuronal apoptosis, oxidative stress, and other dysfunctions [23].
Immunotherapies	Use of monoclonal antibodies and specific antigens to target misfolded and aggregated proteins involved in neurodegenerative diseases [24].
Molecular targeted therapies	Focuses on developing therapies that target specific molecules involved in disease progression, based on advanced neurobiology [25].
Pharmacological therapy	Drugs targeting beta-amyloid, nerve growth factor, anti-inflammatory properties, and dopamine uptake to treat diseases like Alzheimer's, Parkinson's, and Huntington's [26].
Current symptomatic therapies	Cholinesterase inhibitors (donepezil, rivastigmine, galantamine) and NMDA receptor antagonists (memantine) to enhance cognitive performance in Alzheimer's disease (AD) [27].

1.4 CONCLUSION

This review summarizes the current understanding of the multiple pathways that contribute to the cascade of events leading to the manifestation of NDDs. It emphasizes that pathways and networks serve as critical frameworks for interpreting diverse findings related to neurodegeneration. Therefore, a deeper understanding of the mechanisms underlying NDDs is essential for the development of more effective therapeutics. The urgent priority is to elucidate the biological pathways involved in disease progression. The challenge ahead lies in identifying effective drugs capable of targeting dysregulated pathways for treating NDDs. With the wealth of information available, particularly on AD and PD, there is optimism that an ideal therapeutic drug will be developed shortly by targeting specific pathways associated with NDDs.

REFERENCES

1. Ayeni EA, Aldossary AM, Ayejoto DA, Gbadegesin LA, Alshehri AA, Alfassam HA, Afewerky HK. Neurodegenerative diseases: Implications of environmental and climatic influences on neurotransmitters and neuronal hormones activities. *Int J Environ Res Public Health.* 2022; 19(19): 12495. https://doi.org/10.3390/ijerph191912495.
2. Mohd Sairazi NS, Sirajudeen KNS. Natural products and their bioactive compounds: Neuroprotective potentials against neurodegenerative diseases. *Evidence-Based Complementary and Alternative Medicine.* 2020 Feb 14; 2020: 6565396. https://doi.org/10.1155/2020/6565396
3. Agnati LF, Baldelli E, Andreoli N, Woods AS, Vellani V, Marcellino D, Guidolin D, Fuxe KL. On the key role played by altered protein conformation in Parkinson's disease. *J Neural Transm.* 2008; 115(9): 1285–1299.
4. Goyal R, Papadakis M, Ashraf GM. Natural products in the management of neurodegenerative diseases. *Nutrition & Metabolism.* 2024 Oct 21;21(1):1–15.
5. Van Schependom J, D'haeseleer M. Advances in neurodegenerative diseases. *J Clin Med.* 2023 Feb 21; 12(5):1709. https://doi.org/10.3390/jcm12051709
6. Sharifi RM, Lankatillake C, Dias DA, Docea AO, Mahomoodally MF, Lobine D, Chazot PL, Kurt B, Boyunegmez Tumer T, Catarina Moreira A, et al. Impact of natural compounds on neurodegenerative disorders: From preclinical to pharmacotherapeutics. *Journal of Clinical Medicine.* 2020; 9(4): 1061. https://doi.org/10.3390/jcm9041061.
7. Lamptey RNL, Chaulagain B, Trivedi R, Gothwal A, Layek B, Singh J. A review of the common neurodegenerative disorders: Current therapeutic approaches and the potential role of nanotherapeutics. *International Journal of Molecular Sciences.* 2022; 23(3): 1851. https://doi.org/10.3390/ijms23031851.
8. Yuan L, Guo Y, Wen S, Deng H. Editorial: Genetic and epigenetic basis of neurodegenerative diseases. *Front Aging Neurosci.* 2022; 14: 958847.
9. Coda DM, Graff J. Neurogenetic and neuroepigenetic mechanisms in cognitive health and disease. *Front Mol Neurosci.* 2020; 13: 205.
10. Oulton ME, Lee HG. Environmental factors and neurodegenerative diseases: A review. *Curr Environ Health Rep.* 2018; 5(4): 342–349.
11. Coppedè F, Migliore L. Genetic and environmental factors in neurodegenerative diseases: The example of Parkinson's disease. *Toxicol Appl Pharmacol.* 2009; 241(2): 243–252.
12. Monti P, Biganzoli E, Bollati V. Impact of air pollution and occupational inhalation exposures on neurodegenerative disorders: An epigenetic perspective. iScience. 2025 Jun 4.
13. Angeloni C, Malaguti M, Prata C, Freschi M, Barbalace MC, Hrelia S. Mechanisms underlying neurodegenerative disorders and potential neuroprotective activity of agrifood by-products. *Antioxidants (Basel).* 2022 Dec 30; 12(1): 94. https://doi.org/10.3390/antiox12010094
14. Andreone BJ, Larhammar M, Lewcock JW. Cell death and neurodegeneration. *Cold Spring Harb Perspect Biol.* 2020; 12(2): a036434.
15. Jellinger KA. Basic mechanisms of neurodegeneration: A critical update. *J Cell Mol Med.* 2010 Mar; 14(3): 45787. https://doi.org/10.1111/j.15824934.2010.01010.x
16. Ghosh P, Saadat A. Neurodegeneration and epigenetics: A review. *Neurologia.* 2021; 38: e62–e68.
17. Paul S. Impact of epigenetics on human health and possible tool for remediation. *Nucleus.* 2021; 64: 255–258.
18. Lee S, Bang SM, Lee JW, Cho KS. Evaluation of traditional medicines for neurodegenerative diseases using Drosophila models. *Evid Based Complementary Altern Med.* 2014; 2014.
19. Ramanathan S, Archunan G, Sivakumar M, Tamil Selvan S, Fred AL, Kumar S, Gulyás B, Padmanabhan P. Theranostic applications of nanoparticles in neurodegenerative disorders. *Int J Nanomedicine.* 2018; 13:5561–5576.
20. Vissers MF, Heuberger JA, Groeneveld GJ. Targeting for success: Demonstrating proof-of-concept with mechanistic early phase clinical pharmacology studies for disease-modification in neurodegenerative disorders. *Int J Mol Sci.* 2021; 22: 1615.
21. Speed LJ, van Dam WO, Hirath P, Vigliocco G, Desai RH. Impaired comprehension of speed verbs in Parkinson's disease. *J Int Neuropsychol Soc.* 2017; 23: 412–420.
22. Sudhakar V, Richardson RM. Gene therapy for neurodegenerative diseases. *Neurotherapeutics.* 2019; 16: 166–175.

23. Akwa Y. Steroids and Alzheimer's disease: Changes associated with pathology and therapeutic potential. *Int J Mol Sci*. 2020; 21: 4812.
24. Mortada I, Farah R, Nabha S, Ojcius DM, Fares Y, Almawi WY, Sadier NS. Immunotherapies for neurodegenerative diseases. *Front Neurol*. 2021; 12: 654739.
25. Manjula R, Anuja K, Alcain FJ. SIRT1 and SIRT2 activity control in neurodegenerative diseases. *Front Pharmacol*. 2021; 11: 585821.
26. Krzysztoforska K, Mirowska-Guzel D, Widy-Tyszkiewicz E. Pharmacological effects of protocatechuic acid and its therapeutic potential in neurodegenerative diseases: Review on the basis of in vitro and in vivo studies in rodents and humans. *Nutr Neurosci*. 2019; 22: 72–82.
27. Marucci G, Buccioni M, Dal Ben D, Lambertucci C, Volpini R, Amenta F. Efficacy of acetylcholinesterase inhibitors in Alzheimer's disease. *Neuropharmacology*. 2021; 190: 108352.

2 Role of Natural Herbs in Alzheimer's Disease and Their Future Prospects

Gifty M. Jojo,[1] Mr. Jobin Johnson,[1] Gowthamarajan Kuppusamy,[2] and Karri V. V. S. Narayana Reddy[2]

[1]Nirmala College of Pharmacy, Muvattupuzha, Kerala, India

[2]JSS College of Pharmacy, Ooty, Tamil Nadu, India

2.1 INTRODUCTION

Alzheimer's disease (AD) is a progressive neurodegenerative disorder characterized by cognitive decline, memory loss, and behavioral changes. As the most common cause of dementia, AD affects millions of individuals worldwide, posing significant challenges to patients, caregivers, and healthcare systems. Despite extensive research, the exact cause of AD remains elusive, and current treatments primarily offer symptomatic relief without addressing the underlying pathology. The pathophysiology of AD involves the accumulation of amyloid-beta plaques and neurofibrillary tangles composed of hyperphosphorylated tau protein in the brain. These abnormal protein aggregations disrupt neuronal communication and trigger neuroinflammation, oxidative stress, and synaptic dysfunction, ultimately leading to neuronal death. Key brain regions affected include the hippocampus, essential for memory formation, and the cerebral cortex, responsible for higher-order cognitive functions (Alzheimer's Association, 2023; Sperling et al., 2016; Schultz and Del Tredici, 2004; Jack et al., 2018; Isaacson et al., 2018; Isaacson et al., 2019).

Several hypotheses have been proposed to explain the underlying mechanisms of AD. The amyloid cascade hypothesis suggests that the accumulation of amyloid-beta peptides initiates a cascade of events leading to neurodegeneration (Hardy and Higgins, 1992; Hardy and Allsop, 1991). The tau hypothesis posits that tau protein abnormalities are the primary drivers of neuronal damage (Arnsten et al., 2021; Gallardo and Holtzman, 2019). Additionally, the cholinergic hypothesis highlights the role of acetylcholine deficiency in cognitive decline (Wilcock et al., 1982; Nitsch, 1996; Yan and Feng, 2004). Modern terminology termed AD as type III diabetes mellitus (type 3 DM), as recent studies have identified a chronic resistance and or deficiency of insulin or insulin-like-growth factor as the underlying cause of AD (Watson and Craft, 2003; Candeias et al., 2012; Liu et al., 2011).

Current treatments for AD primarily focus on alleviating symptoms rather than curing or halting disease progression. Pharmacological options include cholinesterase inhibitors (such as donepezil, rivastigmine, and galantamine) and the NMDA receptor antagonist memantine, which aim to improve cognitive function and daily activities. While these medications can offer some benefits, their effects are often modest and temporary, and they come with side effects like nausea, dizziness, and insomnia. Non-pharmacological interventions, such as cognitive therapy, behavioral management, and physical exercise, can also provide supportive benefits but require consistent engagement and significant caregiver involvement. The major limitation of current treatments is their inability to address the underlying causes of AD, offering only symptomatic relief without altering the

DOI: 10.1201/9781003478652-3

disease's progression. The quest for effective, disease-modifying therapies continues, highlighting the urgent need for more advanced and comprehensive treatment approaches (Yiannopoulou and Papageorgiou, 2020; Weller and Budson, 2018; Tiwari et al., 2019).

Natural ingredients, particularly medicinal herbs, have shown promise in modulating various aspects of AD pathology. These herbs can exert their beneficial effects through multiple mechanisms, making them valuable candidates for complementary therapies. The following outlines the general mechanisms through which herbs may impact AD.

2.1.1 Antioxidant Activity

Oxidative stress plays a significant role in the progression of AD. The excessive production of reactive oxygen species (ROS) can damage cellular components, leading to neuronal death. Many medicinal herbs possess potent antioxidant properties, which help neutralize ROS and protect neurons from oxidative damage. For instance, *Curcuma longa* (turmeric) contains curcumin, a compound known for its strong antioxidant effects. By reducing oxidative stress, these herbs can mitigate one of the key pathological processes in AD (Firdaus and Singh, 2021; Albarracin et al., 2012; Hasan et al., 2023).

2.1.2 Anti-Inflammatory Effects

Chronic inflammation is a hallmark of AD, contributing to neuronal injury and cognitive decline. Medicinal herbs often contain bioactive compounds with anti-inflammatory properties that can modulate the brain's inflammatory response. Herbs like *Ginkgo biloba* have been shown to reduce the production of pro-inflammatory cytokines and inhibit the activation of microglia, the brain's immune cells. By dampening the inflammatory response, these herbs can help protect neurons and slow disease progression (Gregory et al., 2021; Cooper and Ma, 2017; Mehla et al., 2020).

2.1.3 Inhibition of Amyloid-Beta Aggregation

The accumulation of amyloid-beta plaques is a central feature of AD. Some herbs contain compounds that can prevent the formation or aggregation of amyloid-beta peptides. Herbs like *Bacopa monnieri* have been found to inhibit amyloid-beta aggregation and promote its clearance from the brain (Gregory et al., 2021; Malishev et al., 2017; Chen et al., 2021). By targeting amyloid-beta, these herbs can address one of the primary pathological processes in AD.

2.1.4 Modulation of Neurotransmitter Levels

Neurotransmitter imbalances, particularly the deficiency of acetylcholine, are linked to cognitive decline in AD. Certain herbs can enhance neurotransmitter levels or modulate their activity. *Huperzia serrata* contains huperzine A, an alkaloid that inhibits acetylcholinesterase, the enzyme responsible for breaking down acetylcholine. By preserving acetylcholine levels, huperzine A can improve cognitive function and memory in AD patients (Friedli and Inestrosa, 2021).

2.1.5 Enhancement of Cerebral Blood Flow

Impaired cerebral blood flow and reduced oxygen supply to the brain are associated with AD. Some medicinal herbs can enhance cerebral circulation, ensuring that neurons receive adequate nutrients and oxygen. *Ginkgo biloba*, for instance, has vasodilatory properties that improve blood flow to the brain. Enhanced cerebral circulation can support neuronal health and function, potentially alleviating some symptoms of AD (Mashayekh et al., 2011; Wang and Zhang, 2019).

2.1.6 Neuroprotection and Neurogenesis

Several herbs have neuroprotective effects, promoting the survival of neurons and supporting the growth of new neurons (neurogenesis). For example, *Withania somnifera* (Ashwagandha) has been shown to promote neurogenesis and protect neurons from various forms of stress. Neuroprotective herbs can help maintain neuronal integrity and function, potentially slowing the progression of AD (Mikulska et al., 2023).

2.1.7 Metal Chelation

Metal ions such as copper and iron can catalyze the formation of ROS, contributing to oxidative stress and amyloid-beta aggregation. Some herbs possess metal-chelating properties, which can reduce the availability of these ions and mitigate their harmful effects. For instance, green tea polyphenols can chelate metal ions and reduce oxidative damage in the brain (Chen et al., 2021).

By targeting multiple pathways, natural ingredients offer a multifaceted approach to managing the disease. As research progresses, the integration of these herbs into comprehensive treatment strategies holds promise for improving outcomes for AD patients and offering new avenues for therapeutic development. The most investigated plant in AD is described in the following session.

2.2 ASHWAGANDHA

Ashwagandha, also known as *Withania somnifera*, is a small shrub that belongs to the family Solanaceae. The plant is native to India, the Middle East, and parts of Africa. The plant's roots and berries are a prominent herb in traditional Ayurvedic medicine, renowned for its adaptogenic properties, which help the body manage stress. Recent studies suggest that it may offer benefits for memory and cognitive function, making it relevant for conditions like AD (Kushwah et al., 2023). Ashwagandha contains several active ingredients such as alkaloids (isopelletierine, anaferine, cuseohygrine, anahygrine, etc.), steroidal lactones (withanolides, withaferins), and saponins (Saleem et al., 2020).

Withanolide A present in the herbal extract was found to reverse Aβ-peptide-induced memory deficits and inhibited the degeneration of axons, dendrites, and synapses in Aβ-peptide-induced AD of mice (Kuboyama, Tohda and Komatsu, 2005). Ashwagandha aqueous extracts were observed to elevate ACh levels and enhance choline acetyl transferase activity in rats, potentially contributing to the observed cognitive enhancement and memory improvement (Schliebs et al., 1997). Administration of the root extract led to an upregulation of the low-density lipoprotein receptor-related protein, facilitating Aβ clearance and ameliorating AD pathology in middle-aged and elderly APP/PS1 mice (Sehgal et al., 2012). Also, it was reported that the oral administration of semi-purified ashwagandha extract reverses the behavioral deficits and inhibits Aβ peptides accumulation in an APP/PS1 mouse model of AD (Sehgal et al., 2012). Despite the extensive literature on the therapeutic effects of ashwagandha, there is limited data on its clinical application for cognitive impairment.

In a prospective, randomized, double-blind, placebo-controlled pilot study involving 50 participants with mild cognitive impairment, individuals were administered either ashwagandha root extract (300 mg twice daily) or a placebo for a duration of eight weeks. Following the eight-week period, the group receiving ashwagandha treatment exhibited notable enhancements in both immediate and general memory assessments in comparison to the placebo group. Additionally, the treatment group displayed marked progress in executive function, sustained attention, and information-processing speed (Choudhary, Bhattacharyya and Bose, 2017). However, further clinical trials are needed to confirm the potential benefits of the plant in AD management (Gregory et al., 2021).

2.3 BRAHMI

B. monnieri, commonly known as brahmi (Bm), is a perennial medicinal plant that thrives in the damp and marshy wetlands of Southern and Eastern India, Africa, Europe, Australia, Asia, as well as North and South America. The plant contains bioactive phytochemicals such as saponins; bacopasides III, IV, and V; bacosides A and B; bacosaponins A, B, C, D, E, and F; alkaloids; sterols; betulic acid; polyphenols; and sulfhydryl compounds, which are believed to underlie its neuroprotective properties (Fatima et al., 2022).

Various studies conducted in vitro and in vivo models have indicated that the phytochemicals present in the plants are potent antioxidants, free radical scavengers, and anti-inflammatory agents (Gregory et al., 2021). Additionally, the plant impacts the cholinergic system, a vital component for memory and cognitive functions, by boosting the synthesis and release of acetylcholine while inhibiting acetylcholinesterase, an enzyme responsible for acetylcholine breakdown. This dual action may enhance synaptic transmission and cognitive processes (Uabundit et al., 2010; Saini, Singh and Sandhir, 2012). Studies propose that brahmi also could potentially aid in diminishing the development of beta-amyloid plaques by inhibiting the aggregation of beta-amyloid proteins and facilitating their elimination from the brain (Gregory et al., 2021).

Joshi et al. (2006) evaluate the effects of brahmi on cognitive deficits induced by scopolamine. The study demonstrated the overall improvement in memory with marked reduction in oxidative stress biomarkers in the brain upon treatment with 100 and 200 mg/kg kg^{-1} for eight successive days (Joshi and Parle, 2012). Nongnut in 2010 reported cognitive enhancement and neuroprotection upon two weeks of treatment with alcoholic extract of *B. monnieri* in the animal model of AD (Uabundit et al., 2010). Another research with standardized herbal extract of brahmi reported the reversal of colchicine-induced cognitive deficits, along with the reinstatement of antioxidant enzyme activities in an animal model (Saini, Singh and Sandhir, 2012).

Stough et al. (Cohen, 2014) documented a notable enhancement in verbal learning, memory consolidation, and the pace of early information processing subsequent to the administration of *Bacopa* (comprising 55% combined bacosides) over a 12-week period, with a daily dosage of 300 mg, in a double-blind placebo-controlled trial involving healthy volunteers aged between 18 and 60 years (Stough et al., 2001). In a separate randomized, double-blind, placebo-controlled study conducted with 54 elderly individuals lacking clinical indications of dementia (mean age 73.5 years), a similar *Bacopa* regimen led to an improvement in an auditory verbal learning assessment, delayed word recall memory performance, and a Stroop test in comparison to the placebo (Calabrese et al., 2008). For individuals over the age of 55 experiencing memory deficiencies, a standardized *Bacopa* extract of 125 mg was administered twice daily over a 12-week period in a double-blind, placebo-controlled fashion, resulting in a marked enhancement in cognitive functions such as mental control, logical memory, and paired-associate learning. Moreover, the consumption of *Bacopa* extract at a dosage of 300 mg/kg per day for 12 weeks exhibited enhancements in memory acquisition and retention among the elderly Australian population (Raghav et al., 2006).

While *B. monnieri* has shown cognitive-enhancing effects in animal studies and small clinical trials, larger and more rigorous studies are needed to confirm its benefits in AD.

2.4 GOTU KOLA

Gotu Kola, scientifically known as *Centella asiatica*, is a small, herbaceous, perennial plant from the family Apiaceae. *C. asiatica*, also known as Indian pennywort, Asiatic pennywort, or gotu kola, is an herbaceous, perennial plant belonging to the family Apiaceae. The plant originates in tropical areas of Africa, Asia, Australia, and islands within the western Pacific Ocean. The plant contains compounds such as triterpenoids and flavonoids that exhibit strong antioxidant and inflammatory properties (Gray et al., 2018; Kumar and Gupta, 2003; Xu et al., 2008).

In addition to a potent antioxidant, the herbal extract was found to modulate various neurotransmitters such as dopamine, 5-hydroxytryptamine, and noradrenaline in the brain and thereby enhance learning and memory in rats (Nalini et al., 1995; Soumyanath et al., 2005). There has been evidence for the improvement in Aβ burden and related neurodegeneration at the preclinical level (Dhanasekaran et al., 2009). A randomized, placebo-controlled, double-blind clinical study was conducted with the herbal extract in 28 healthy volunteers. The results indicated that a high dose of the plant extract improved working memory. Additionally, self-rated mood improvements were observed following the treatment, suggesting that the herb may help mitigate the age-related cognitive decline and mood swings in elderly individuals (Soumyanath et al., 2012). However, the safety profile of the plant is not well validated. Also, there is limited clinical evidence for supporting the bioavailability of active components.

2.5 SAFFRON

Crocus sativus, commonly known as saffron crocus or autumn crocus, is a species of flowering plant belonging to the family Iridaceae. Safranal, a carboxaldehyde, is the primary component of saffron. Both in vitro and in vivo studies demonstrate that the phytochemicals in saffron exhibit antioxidant, anti-inflammatory, and anti-amyloidogenic effects (El Midaoui et al., 2022).

The efficiency of the herb in treating mild to moderate AD was investigated in 46 patients who were randomly assigned to receive either 30 mg/day of saffron or a placebo. After 16 weeks of treatment, saffron significantly outperformed the placebo in improving cognitive performance (Akhondzadeh et al., 2010a). In a follow-up study, researchers compared saffron extract with the cholinesterase inhibitor donepezil in patients with mild to moderate AD. In a 22-week double-blind, randomized, controlled trial, 54 participants received either 30 mg/day of saffron or 10 mg/day of donepezil. By the end of the study, saffron showed similar improvements in cognitive function to donepezil but with fewer side effects. The efficiency of the herb was further compared with memantine, another FDA-approved drug for the management of AD. In addition to demonstrating a low rate of adverse effects, saffron extract was comparable to memantine in reducing cognitive decline in patients with moderate to severe AD (Akhondzadeh et al., 2010b).

2.6 TURMERIC

Turmeric, or *C. longa*, is a flowering plant in the ginger family Zingiberaceae. The active components of turmeric are turmerone oil and water-soluble curcuminoids such as curcumin, demethoxycurcumin (DMC), bisdemethoxycurcumin (BDMC), and cyclocurcumin (Aggarwal et al., 2007). Curcumin is known for its antioxidant, anti-inflammatory properties, which are linked to a reduced risk of AD. Curcumin administration is found to be beneficial in improving cognitive performance along with amelioration of oxidative stress and Aβ pathology in an animal model. Additionally, curcumin has been found to have the ability to chelate with metal ions such as copper, zinc, and iron, which are found to be concentrated in the AD brain and trigger amyloid aggregation and neurotoxicity (Begum et al., 2008; Lim et al., 2001; Yang et al., 2005; Wang et al., 2011; Yanagisawa et al., 2015). Due to the diminished bioavailability of curcumin and inadequate blood–brain barrier (BBB) permeability, numerous analogs of curcumin underwent evaluation in terms of their bioavailability and impact on animal models of AD. However, only a few clinical studies have investigated the impact of curcumin on human cognitive function, and their results have been inconclusive. Researchers largely agree that combining curcumin with other dietary supplements enhances its neuroprotective effects. Therefore, further research is needed to improve curcumin's systemic bioavailability and its permeability across the BBB (Gregory et al., 2021).

2.7 LIQUORICE

Liquorice (*Glycyrrhiza glabra*) has been studied for its potential benefits in AD. The active compounds in liquorice, such as glycyrrhizin and glabridin, exhibit anti-inflammatory, antioxidant, and neuroprotective properties. Various studies indicate that liquorice extract can also ameliorate the formation of beta-amyloid plaques, a hallmark of AD, and enhance cognitive function (Pastorino et al., 2018; Han et al., 2014; Bisht et al., 2022; Furusawa et al., 2009; Guo et al., 2022). However, small-scale clinical trials with the herb exhibited no significant improvement in cognitive function. Even though the herb is safe in healthy individuals in small doses, it should be avoided by anyone at risk for hypertension or cardiovascular disease due to its mineralocorticoid effects. Hence, extended clinical trials are needed to establish the efficacy and safety profile of liquorice in the AD (Alzheimer's Drug Discovery Foundation, 2024).

2.8 SHANKHPUSHPI

Four botanicals, such as *Cascara decussata* Schult. (CD), *Clitorea ternatea* Linn. (CT), *Convolvulus pluricaulis* Choisy. (CP), and *Evolvulus alsinoides* Linn. (EA), are considered as sources of shankhpushpi by Indian practitioners on the basis of their morphological descriptions given in ancient texts (Sethiya et al., 2019). The plant has been used as a nervine tonic for improving cognitive and memory function. However, the secondary metabolites such as flavanol glycosides, triterpenoids, steroids, and anthocyanins present in the plant were found to be beneficial in improving various biomarkers of AD. Treatment with plant extract was reported with improvement in oxidative stress, Aβ toxicity, and tau protein-induced neurodegeneration (Bihaqi et al., 2009; Nahata, Patil and Dixit, 2008; Nahata, Patil and Dixit, 2010). Activation of the cholinergic neurotransmitter system upon treatment with the herbal extract in the AD model may be beneficial in improving cognitive function (Rai et al., 2002; Taranalli and Cheeramkuzhy, 2000; Rai et al., 2001). However, no clinical evidence has confirmed these preclinical findings yet.

2.9 OTHER HERBS

Giloy (*Tinospora cordifolia*), a climbing shrub that grows on other trees, from the botanical family Menispermaceae has historically been used as an adaptogen or rejuvenator. Various preclinical studies have demonstrated the beneficial effects of the plant in improving oxidative stress, neuroinflammation, tau pathology, memory, and cognition. However, extensive clinical trials are required to confirm this evidence (Prakash et al., 2017; Mishra et al., 2016; Birla et al., 2021).

Ginger (*Zingiber officinale*) has been utilized in India as a spice as well as for the management of indigestion and inflammation. Preclinical model of AD has shown significant reduction in inflammatory biomarkers and Aβ along with the downregulation of acetyl choline esterase enzyme (Altman and Marcussen, 2001; Kiuchi, Shibuya and Sankawa, 1982; Tjendraputra et al., 2001; Cuya, Baptista and França, 2018).

Similarly, Thulsi (*Ocimum sanctum* belongs to the family Lamiaceae) is a well-reported antioxidant and anti-inflammatory herb. The plant has shown a notable decrease in transfer latency and an increase in step-down latency upon administration of the *O. sanctum* extract in elderly mice (Joshi and Parle, 2006). The ethanolic extract of the plant stimulated the production of choline acetyltransferase expression in human cerebral microvascular endothelial cells (Joshi and Parle, 2006; Hening et al., 2018).

Azadirachta indica, commonly known as neem, is a perennial tree that can be found extensively in India and its adjacent regions. A research endeavor conducted by Maiti et al., involving rats, demonstrated potential antidepressant properties associated herbal extract (Maiti et al., 2013). Limonoids extracted from neem fruits exhibited a reduction in the extensive aggregation of tau in vitro, consequently fostering the development of delicate, short, and fragile aggregates (Gorantla et al., 2019).

Allium sativum, belonging to the Amaryllidaceae family, has been utilized as an herbal medicine since ancient times. Garlic contains various bioactive compounds, such as allicin, which have been shown to have antioxidant, anti-inflammatory, and neuroprotective properties. Garlic treatment demonstrated amelioration of Aβ pathology, neurodegeneration, and cognitive impairment in preclinical trials (Gupta, Indi and Rao, 2009; Sripanidkulchai, 2020; Thorajak et al., 2017).

2.10 FUTURE PROSPECTS OF HERBS IN AD

While various studies show promising results in animal models and small-scale clinical trials, the cognitive benefits in humans are often modest and inconsistent. This is because the efficacy of herbal medicines can vary widely among individuals due to factors such as genetic variability, disease stage, and overall health. The lack of standardization of herbal extracts can also be a reason for variable results in various clinical trials.

Current research must be further extended to focus on the identification, isolation, and standardization of the active compounds in herbal extracts. Advanced technologies like molecular docking and high-throughput screening can be employed to rapidly identify potential therapeutic compounds from vast libraries of herbal extracts. For instance, molecular docking studies have been utilized to predict the binding affinity of herbal compounds like curcumin, resveratrol, and ginsenosides to amyloid-beta and tau proteins, which are hallmarks of AD pathology. These computational approaches help in narrowing down the list of promising candidates for further in vitro and in vivo testing.

Initiatives must be taken for the standardization of these herbal extracts to ensure consistency in the quality and concentration of the active ingredients, which is crucial for achieving reliable and reproducible therapeutic effects. Standardization involves quantifying specific bioactive markers within the extracts, which has been demonstrated in studies involving Ginkgo biloba, where the active components such as ginkgolides and bilobalide are standardized to specific concentrations to ensure efficacy.

The therapeutic outcome of many active compounds is further limited by their poor BBB permeability. The BBB is a significant obstacle in delivering therapeutic agents to the brain due to its selective permeability. Some herbal compounds, like quercetin and berberine, have shown poor BBB penetration, which limits their efficacy in treating neurological conditions. To overcome this, investigators should focus on assessing the BBB permeability of active ingredients and the application of advanced drug delivery systems, especially nanotechnology, to enhance the delivery and efficacy of these compounds. Nanotechnology-based delivery systems, such as liposomes, nanoparticles, and nanoemulsions, have been shown to improve the bioavailability and BBB penetration of herbal compounds. For instance, nanoformulations of curcumin and resveratrol have demonstrated improved brain uptake and enhanced therapeutic effects in preclinical models of AD.

Despite promising preclinical and early clinical studies, robust large-scale clinical trials with standardized herbal extracts in appropriate dosage forms are required to establish the efficacy and safety of herbal medicines for AD management. For example, recent meta-analyses of clinical trials on Ginkgo biloba for AD have shown mixed results, highlighting the need for more rigorous, standardized studies to confirm its benefits. Additionally, studies on other promising herbal extracts, such as *B. monnieri* and *Salvia officinalis*, need to be conducted on a larger scale with standardized formulations to draw definitive conclusions about their efficacy.

Furthermore, integrating pharmacogenomics into the research can help tailor herbal treatments to individual genetic profiles, potentially improving outcomes. By identifying genetic markers that influence the metabolism and efficacy of herbal compounds, personalized treatment strategies can be developed. This approach has already been explored in conventional medicine and holds promise for herbal therapeutics as well.

In conclusion, while the potential of herbal medicines for AD is evident from preclinical and early clinical studies, significant challenges remain. The focus must be on the standardization of extracts, enhancing BBB permeability, and conducting large-scale clinical trials to establish definitive evidence of efficacy and safety. The integration of advanced technologies and personalized medicine approaches will be crucial in unlocking the full therapeutic potential of herbal medicines for AD (Gregory et al., 2021; Mehla et al., 2020; Nishteswar, Joshi and Karra, 2014).

2.11 CONCLUSION

AD continues to challenge patients, caregivers, and healthcare systems globally due to its complex and multifactorial nature. Despite advances in understanding AD pathology, current treatments remain largely palliative, emphasizing the urgent need for novel therapeutic approaches. This review has highlighted the potential of natural ingredients, particularly medicinal herbs, in addressing various aspects of AD pathology through diverse mechanisms, including antioxidant activity, anti-inflammatory effects, inhibition of amyloid-beta aggregation, modulation of neurotransmitter levels, enhancement of cerebral blood flow, neuroprotection, neurogenesis, and metal chelation.

Key herbs such as ashwagandha, brahmi, gotu kola, saffron, turmeric, liquorice, shankhpushpi, giloy, ginger, tulsi, neem, and garlic have shown promising results in preclinical and early clinical studies. However, significant challenges remain, including the need for standardization of herbal extracts, identification and isolation of active compounds, and development of advanced drug delivery systems to enhance their bioavailability and efficacy.

Future research should focus on large-scale clinical trials to establish the safety and effectiveness of these herbal medicines in AD. Additionally, employing advanced technologies for the identification of therapeutic compounds and enhancing their delivery through innovative drug delivery systems could pave the way for more effective treatments. By integrating these natural ingredients into comprehensive treatment strategies, we may unlock new therapeutic avenues for managing and potentially mitigating the progression of AD, offering hope to millions affected by this debilitating condition.

REFERENCES

Aggarwal, B.B., Sundaram, C., Malani, N. and Ichikawa, H. (2007) 'Curcumin: The Indian solid gold', *Advances in Experimental Medicine and Biology*, 595, pp. 1–75.

Akhondzadeh, S., Sabet, M.S., Harirchian, M.H., Togha, M., Cheraghmakani, H., Razeghi, S., Hejazi, S.Sh., Yousefi, M.H., Alimardani, R., Jamshidi, A., Zare, F. and Moradi, A. (2010b) 'Saffron in the treatment of patients with mild to moderate Alzheimer's disease: A 16-week, randomized and placebo-controlled trial', *Journal of Clinical Pharmacy and Therapeutics*, 35, pp. 581–588.

Akhondzadeh, S., Shafiee, S.M., Harirchian, M.H., Togha, M., Cheraghmakani, H., Razeghi, S., Hejazi, S.S., Yousefi, M.H., Alimardani, R., Jamshidi, A., Rezazadeh, S.A., Yousefi, A., Zare, F., Moradi, A. and Vossoughi, A. (2010a) 'A 22-week, multicenter, randomized, double-blind controlled trial of *Crocus sativus* in the treatment of mild-to-moderate Alzheimer's disease', *Psychopharmacology*, 207, pp. 637–643.

Albarracin, S.L., Stab, B., Casas, Z., Sutachan, J.J., Samudio, I., Gonzalez, J., Gonzalo, L., Capani, F., Morales, L. and Barreto, G.E. (2012) 'Effects of natural antioxidants in neurodegenerative disease', *Nutritional Neuroscience*, 15(1), pp. 1–9. https://doi.org/10.1179/1476830511Y.0000000028.

Altman, R.D. and Marcussen, K.C. (2001) 'Effects of a ginger extract on knee pain in patients with osteoarthritis', *Arthritis and Rheumatism*, 44(11), pp. 2531–2538. https://doi.org/10.1002/1529-0131(200111)44:11<2531::AID-ART433>3.0.CO;2-J.

Alzheimer's Association (2023) '2023 Alzheimer's disease facts and figures', *Alzheimer's & Dementia*, 19(4). https://doi.org/10.1002/alz.13016.

Alzheimer's Drug Discovery Foundation (2024) *Licorice root cognitive vitality for researchers*. Available at: https://www.alzdiscovery.org/uploads/cognitive_vitality_media/Licorice-Root-Cognitive-Vitality-For-Researchers.pdf (Accessed: 11 June 2024).

Arnsten, A.F.T., Datta, D., Del Tredici, K. and Braak, H. (2021) 'Hypothesis: Tau pathology is an initiating factor in sporadic Alzheimer's disease', *Alzheimer's & Dementia*, 17(1), pp. 115–124. https://doi.org/10.1002/alz.12192.

Begum, A.N., Jones. M.R., Lim, G.P., Morihara, T., Kim, P., Heath, D.D., Rock, C.L., Pruitt, M.A., Yang, F., Hudspeth, B., Hu, S., Faull, K.F., Teter, B., Cole, G.M. and Frautschy, S.A. (2008) 'Curcumin

structure-function, bioavailability, and efficacy in models of neuroinflammation and Alzheimer's disease', *Journal of Pharmacology and Experimental Therapeutics*, 326, pp. 196–208.
Bihaqi, S.W., Sharma, M., Singh, A.P. and Tiwari, M. (2009) 'Neuroprotective role of *Convolvulus pluricaulis* on aluminium induced neurotoxicity in rat brain', *Journal of Ethnopharmacology*, 124, pp. 409–415.
Birla, H., Keswani, C., Singh, S.S., Zahra, W., Dilnashin, H., Rathore, A.S., Singh, R., Rajput, M., Keshri, P. and Singh, S.P. (2021) 'Unraveling the neuroprotective effect of *Tinospora cordifolia* in a parkinsonian mouse model through the proteomics approach', *ACS Chemical Neuroscience*, 12(22), pp. 4319–4335.
Bisht, D., Rashid, M., Arya, R.K.K., Kumar, D., Chaudhary, S.K., Rana, V.S. and Sethiya, N.K. (2022) 'Revisiting liquorice (*Glycyrrhiza glabra* L.) as anti-inflammatory, antivirals and immunomodulators: Potential pharmacological applications with mechanistic insight', *Phytomedicine Plus*, 2, p. 100206. https://doi.org/10.1016/j.phyplu.2021.100206.
Calabrese, C., Gregory, W.L., Leo, M., Kraemer, D., Bone, K. and Oken, B. (2008) 'Effects of a standardized *Bacopa monnieri* extract on cognitive performance, anxiety, and depression in the elderly: A randomized, double-blind, placebo-controlled trial', *Journal of Alternative and Complementary Medicine*, 14, pp. 707–713.
Candeias, E., Duarte, A.I., Carvalho, C., Correia, S.C., Cardoso, S., Santos, R.X., Plácido, A.I., Perry, G. and Moreira, P.I. (2012) 'The impairment of insulin signaling in Alzheimer's disease', *IUBMB Life*, 64(12), pp. 951–957.
Chen, X., Drew, J., Berney, W. and Lei, W. (2021) 'Neuroprotective natural products for Alzheimer's disease', *Cells*, 10(6), p. 1309. https://doi.org/10.3390/cells10061309.
Choudhary, D., Bhattacharyya, S. and Bose, S. (2017) 'Efficacy and safety of ashwagandha (*Withania somnifera* (L.) Dunal) root extract in improving memory and cognitive functions', *Journal of Dietary Supplements*, 14, pp. 599–612.
Cohen, M. (2014) 'Tulsi – *Ocimum sanctum*: A herb for all reasons', *Journal of Ayurveda and Integrative Medicine*, 5(4), pp. 251–259.
Cooper, E.L. and Ma, M.J. (2017) 'Alzheimer disease: Clues from traditional and complementary medicine', *Journal of Traditional and Complementary Medicine*, 7, pp. 380–385.
Cuya, T., Baptista, L. and Celmar Costa França, T. (2018) 'A molecular dynamics study of components of the ginger (*Zingiber officinale*) extract inside human acetylcholinesterase: Implications for Alzheimer disease', *Journal of Biomolecular Structure and Dynamics*, 36(14), pp. 3843–3855. https://doi.org/10.1080/07391102.2017.1401004.
Dhanasekaran, M., Holcomb, L.A., Hitt, A.R., Tharakan, B., Porter, J.W., Young, K.A. and Manyam, B.V. (2009) '*Centella asiatica* extract selectively decreases amyloid β levels in hippocampus of Alzheimer's disease animal model', *Phytotherapy Research*, 23(1), pp. 14–19. https://doi.org/10.1002/ptr.2405.
El Midaoui, A., Ghzaiel, I., Vervandier-Fasseur, D., Ksila, M., Zarrouk, A., Nury, T., Khallouki, F., El Hessni, A., Ibrahimi, S.O., Latruffe, N., Couture, R., Kharoubi, O., Brahmi, F., Hammami, S., Masmoudi-Kouki, O., Hammami, M., Ghrairi, T., Vejux, A. and Lizard, G. (2022) 'Saffron (*Crocus sativus* L.): A source of nutrients for health and for the treatment of neuropsychiatric and age-related diseases', *Nutrients*, 14(3), p. 597. https://doi.org/10.3390/nu14030597.
Fatima, U., Roy, S., Ahmad, S., Ali, S., Elkady, W.M., Khan, I., Alsaffar, R.M., Adnan, M., Islam, A. and Hassan, M.I. (2022) 'Pharmacological attributes of *Bacopa monnieri* extract: Current updates and clinical manifestation', *Frontiers in Nutrition*, 9, p. 972379. https://doi.org/10.3389/fnut.2022.972379.
Firdaus, Z. and Singh, T.D. (2021) 'An insight in pathophysiological mechanism of Alzheimer's disease and its management using plant natural products', *Mini Reviews in Medicinal Chemistry*, 21(1), pp. 35–57. https://doi.org/10.2174/1389557520666200730155928.
Friedli, M.J. and Inestrosa, N.C. (2021) 'Huperzine A and its neuroprotective molecular signaling in Alzheimer's disease', *Molecules*, 26(21), p. 6531. https://doi.org/10.3390/molecules26216531.
Furusawa, J.-I., Funakoshi-Tago, M., Tago, K., Mashino, T., Inoue, H., Sonoda, Y. and Kasahara, T. (2009) 'Licochalcone A significantly suppresses LPS signaling pathway through the inhibition of NF-κB p65 phosphorylation at serine 276', *Cellular Signalling*, 21, pp. 778–785. https://doi.org/10.1016/j.cellsig.2009.01.021.
Gallardo, G. and Holtzman, D.M. (2019) 'Amyloid-β and tau at the crossroads of Alzheimer's disease', *Advances in Experimental Medicine and Biology*, 1184, pp. 187–203. https://doi.org/10.1007/978-981-32-9358-8_16.
Gorantla, N.V., Das, R., Mulani, F.A., Thulasiram, H.V. and Chinnathambi, S. (2019) 'Neem derivatives inhibit tau aggregation', *Journal of Alzheimer's Disease Reports*, 3(1), pp. 169–178.
Gray, N.E., Alcazar Magana, A., Lak, P., Wright, K.M., Quinn, J., Stevens, J.F., Maier, C.S. and Soumyanath, A. (2018) 'Centella asiatica – phytochemistry and mechanisms of neuroprotection and cognitive enhancement', *Phytochemistry Reviews*, 17(1), pp. 161–194. https://doi.org/10.1007/s11101-017-9528-y.

Gregory, J., Vengalasetti, Y.V., Bredesen, D.E. and Rao, R.V. (2021) 'Neuroprotective herbs for the management of Alzheimer's disease', *Biomolecules*, 11(4), p. 543. https://doi.org/10.3390/biom11040543.

Guo, J., Xue, J., Ding, Z., Li, X., Wang, X. and Xue, H. (2022) 'Activated phosphoinositide 3-kinase/Akt/mammalian target of rapamycin signal and suppressed autophagy participate in protection offered by licochalcone A against amyloid-β peptide fragment 25–35–induced injury in SH-SY5Y cells', *World Neurosurgery*, 157, pp. e390–e400. https://doi.org/10.1016/j.wneu.2021.10.098.

Gupta, V.B., Indi, S.S. and Rao, K.S.J. (2009) 'Garlic extract exhibits antiamyloidogenic activity on amyloid-beta fibrillogenesis: Relevance to Alzheimer's disease', *Phytotherapy Research*, 23(1), pp. 111–115. https://doi.org/10.1002/ptr.2574.

Han, J.Y., Park, S.H., Yang, J.H., Kim, M.G., Cho, S.S., Yoon, G., Cheon, S.H. and Ki, S.H. (2014) 'Licochalcone suppresses LXRα-induced hepatic lipogenic gene expression through AMPK/Sirt1 pathway activation', *Toxicological Research*, 30, pp. 19–25. https://doi.org/10.5487/TR.2014.30.1.019.

Hardy, J. and Allsop, D. (1991) 'Amyloid deposition as the central event in the aetiology of Alzheimer's disease', *Trends in Pharmacological Sciences*, 12, pp. 383–388.

Hardy, J. and Higgins, G. (1992) 'Alzheimer's disease: The amyloid cascade hypothesis', *Science*, 256(5054), pp. 184–185.

Hasan, S., Khatri, N., Rahman, Z.N., Menezes, A.A., Martini, J., Shehjar, F., Mujeeb, N. and Shah, Z.A. (2023) 'Neuroprotective potential of flavonoids in brain disorders', *Brain Sciences*, 13(9), p. 1258. https://doi.org/10.3390/brainsci13091258.

Hening, P., Mataram Auriva, M.B., Wijayanti, N., Kusindarta, D.L. and Wihadmadyatami, H. (2018) 'The neuroprotective effect of *Ocimum sanctum* Linn. ethanolic extract on human embryonic kidney-293 cells as in vitro model of neurodegenerative disease', *Veterinary World*, 11(9), pp. 1237–1243. https://doi.org/10.14202/vetworld.2018.1237–1243.

Isaacson, R.S., Ganzer, C.A., Hristov, H., Hackett, K., Caesar, E., Cohen, R., Kachko, R., Meléndez-Cabrero, J., Rahman, A., Scheyer, O., Hwang, M.J., Berkowitz, C., Hendrix, S., Mureb, M., Schelke, M.W., Mosconi, L., Seifan, A. and Krikorian, R. (2018) 'The clinical practice of risk reduction for Alzheimer's disease: A precision medicine approach', *Alzheimer's & Dementia*, 14, pp. 1663–1673.

Isaacson, R.S., Hristov, H., Saif, N., Hackett, K., Hendrix, S., Melendez, J., Safdieh, J., Fink, M., Thambisetty, M., Sadek, G., Bellara, S., Lee, P., Berkowitz, C., Rahman, A., Meléndez-Cabrero, J., Caesar, E., Cohen, R., Lu, P.L., Dickson, S.P., Hwang, M.J., Scheyer, O., Mureb, M., Schelke, M.W., Niotis, K., Greer, C.E., Attia, P., Mosconi, L. and Krikorian, R. (2019) 'Individualized clinical management of patients at risk for Alzheimer's dementia', *Alzheimer's & Dementia*, 15, pp. 1588–1602.

Jack, C.R. Jr, Bennett, D.A., Blennow, K., Carrillo, M.C., Dunn, B., Haeberlein, S.B., Holtzman, D.M., Jagust, W., Jessen, F., Karlawish, J., Liu, E., Molinuevo, J.L., Montine, T., Phelps, C., Rankin, K.P., Rowe, C.C., Scheltens, P., Siemers, E., Snyder, H.M. and Sperling, R. (2018) 'NIA-AA Research framework: Toward a biological definition of Alzheimer's disease', *Alzheimer's & Dementia*, 14, pp. 535–562.

Joshi, H. and Parle, M. (2006) 'Evaluation of nootropic potential of *Ocimum sanctum* Linn. in mice', *Indian Journal of Experimental Biology*, 44(2), pp. 133–136.

Joshi, H. and Parle, M. (2012) 'Effects of memorhis, a poly herbal formulation in the management memory dysfunctions', *Research Journal of Pharmacognosy and Phytochemistry*, 4(2), pp. 97–103.

Kiuchi, F., Shibuya, M. and Sankawa, U. (1982) 'Inhibitors of prostaglandin biosynthesis from ginger', *Chemical and Pharmaceutical Bulletin (Tokyo)*, 30(2), pp. 754–757. https://doi.org/10.1248/cpb.30.754.

Kuboyama, T., Tohda, C. and Komatsu, K. (2005) 'Neuritic regeneration and synaptic reconstruction induced by withanolide A', *British Journal of Pharmacology*, 144, pp. 961–971.

Kumar, V. and Gupta, Y.K. (2003) 'Effect of *Centella asiatica* on cognition and oxidative stress in an intracerebroventricular streptozotocin model of Alzheimer's disease in rats', *Clinical and Experimental Pharmacology and Physiology*, 30, pp. 336–342.

Kushwah, S., Maurya, N.S., Kushwaha, S., Scotti, L., Chawade, A. and Mani, A. (2023) 'Herbal therapeutics for Alzheimer's disease: Ancient Indian medicine system from the modern viewpoint', *Current Neuropharmacology*, 21(4), pp. 764–776. https://doi.org/10.2174/1570159X21666230216094353.

Lim, G.P., Chu, T., Yang, F., Beech, W., Frautschy, S.A. and Cole, G.M. (2001) 'The curry spice curcumin reduces oxidative damage and amyloid pathology in an Alzheimer transgenic mouse', *Journal of Neuroscience*, 21, pp. 8370–8377.

Liu, Y., Liu, F., Grundke-Iqbal, I., Iqbal, K. and Gong, C.X. (2011) 'Deficient brain insulin signalling pathway in Alzheimer's disease and diabetes', *Journal of Pathology*, 225(1), pp. 54–62.

Maiti, R., Raghavendra, M., Kumar, S. and Acharya, S.B. (2013) 'Role of aqueous extract of *Azadirachta indica* leaves in an experimental model of Alzheimer's disease in rats', *International Journal of Applied and Basic Medical Research*, 3(1), pp. 37–47. https://doi.org/10.4103/2229-516X.112239.

Malishev, R., Shaham-Niv, S., Nandi, S., Kolusheva, S., Gazit, E. and Jelinek, R. (2017) 'Bacoside-A, an Indian traditional-medicine substance, inhibits β-amyloid cytotoxicity, fibrillation, and membrane interactions', *ACS Chemical Neuroscience*, 8, pp. 884–891.

Mashayekh, A., Pham, D.L., Yousem, D.M., Dizon, M., Barker, P.B. and Lin, D.D. (2011) 'Effects of *Ginkgo biloba* on cerebral blood flow assessed by quantitative MR perfusion imaging: A pilot study', *Neuroradiology*, 53(3), pp. 185–191. https://doi.org/10.1007/s00234-010-0790-6.

Mehla, J., Gupta, P., Pahuja, M., Diwan, D. and Diksha, D. (2020) 'Indian medicinal herbs and formulations for Alzheimer's disease, from traditional knowledge to scientific assessment', *Brain Sciences*, 10(12), p. 964. https://doi.org/10.3390/brainsci10120964.

Mikulska, P., Malinowska, M., Ignacyk, M., Szustowski, P., Nowak, J., Pesta, K., Szeląg, M., Szklanny, D., Judasz, E., Kaczmarek, G., Ejiohuo, O.P., Paczkowska-Walendowska, M., Gościniak, A. and Cielecka-Piontek, J. (2023) 'Ashwagandha (*Withania somnifera*) - current research on the health-promoting activities: A narrative review', *Pharmaceutics*, 15(4), p. 1057. https://doi.org/10.3390/pharmaceutics15041057.

Mishra, R., Manchanda, S., Gupta, M., Kaur, T., Saini, V., Sharma, A. and Kaur, G. (2016) '*Tinospora cordifolia* ameliorates anxiety-like behavior and improves cognitive functions in acute sleep deprived rats', *Scientific Reports*, 6(1), p. 25564.

Nahata, A., Patil, U.K. and Dixit, V.K. (2008) 'Effect of *Convulvulus pluricaulis* Choisy. on learning behaviour and memory enhancement activity in rodents', *Natural Product Research*, 22, pp. 1472–1482.

Nahata, A., Patil, U.K. and Dixit, V.K. (2010) 'Effect of *Evolvulus alsinoides* Linn. on learning behavior and memory enhancement activity in rodents', *Phytotherapy Research*, 24, pp. 486–493.

Nalini, K., Karanth, K.S., Rao, A. and Aroor, A.R. (1995) 'Effects of *Celastrus paniculatus* on passive avoidance performance and biogenic amine turnover in albino rats', *Journal of Ethnopharmacology*, 47, pp. 101–108.

Nishteswar, K., Joshi, H. and Karra, R.D. (2014) 'Role of indigenous herbs in the management of Alzheimer's disease', *Ancient Science of Life*, 34(1), pp. 3–7. https://doi.org/10.4103/0257-7941.150763.

Nitsch, R.M. (1996) 'From acetylcholine to amyloid: Neurotransmitters and the pathology of Alzheimer's disease', *Neurodegeneration*, 5, pp. 477–482.

Pastorino, G., Cornara, L., Soares, S., Rodrigues, F. and Oliveira, M.B.P.P. (2018) 'Liquorice (*Glycyrrhiza glabra*): A phytochemical and pharmacological review', *Phytotherapy Research*, 32(12), pp. 2323–2339. https://doi.org/10.1002/ptr.6178.

Prakash, R., Ramya, N., Dhivya, R. and Priyadarshini, M. (2017) 'Neuroprotective activity of ethanolic extract of *Tinospora cordifolia* on LPS induced neuroinflammation', *Translational Biomedicine*, 8(4).

Raghav, S., Singh, H., Dalal, P.K., Srivastava, J.S. and Asthana, O.P. (2006) 'Randomized controlled trial of standardized Bacopa monniera extract in age associated memory impairment', *Indian Journal of Psychiatry*, 48, pp. 238–242.

Rai, K.S., Murthy, K.D., Karanth, K.S. and Nalini, K. (2002) '*Clitoria ternatea* root extract enhances acetylcholine content in rat hippocampus', *Fitoterapia*, 73, pp. 685–689.

Rai, K.S., Murthy, K.D., Karanth, K.S. and Rao, M.S. (2001) '*Clitoria ternatea* (Linn) root extract treatment during growth spurt period enhances learning and memory in rats', *Indian Journal of Physiology and Pharmacology*, 45, pp. 305–313.

Saini, N., Singh, D. and Sandhir, R. (2012) 'Neuroprotective effects of *Bacopa monnieri* in experimental model of dementia', *Neurochemical Research*, 37(9), pp. 1928–1937. https://doi.org/10.1007/s11064-012-0811-4.

Saleem, S., Muhammad, G., Hussain, M.A., Altaf, M. and Bukhari, S.N.A. (2020) '*Withania somnifera* L.: Insights into the phytochemical profile, therapeutic potential, clinical trials, and future prospective', *Iranian Journal of Basic Medical Sciences*, 23, pp. 1501–1526. https://doi.org/10.22038/IJBMS.2020.44254.10378.

Schliebs, R., Liebmann, A., Bhattacharya, S.K. and Kumar, A. (1997) 'Systemic administration of defined extracts from *Withania somnifera* (Indian Ginseng) and Shilajit differentially affects cholinergic but not glutamatergic and GABAergic markers in rat brain', *Neurochemistry International*, 30, pp. 181–190.

Schultz, C. and Del Tredici, K.H.B. (2004) 'Neuropathology of Alzheimer's disease', in R.R. and B.R. (eds) *Alzheimer's Disease Current Clinical Neurology*. Totowa, NJ: Humana Press.

Sehgal, N., Gupta, A., Valli, R.K., Joshi, S.D., Mills, J.T., Hamel, E., Khanna, P., Jain, S.C., Thakur, S.S. and Ravindranath, V. (2012) '*Withania somnifera* reverses Alzheimer's disease pathology by enhancing low-density lipoprotein receptor-related protein in liver', *Proceedings of the National Academy of Sciences of the United States of America*, 109, pp. 3510–3515.

Sethiya, N.K., Nahata, A., Singh, P.K. and Mishra, S.H. (2019) 'Neuropharmacological evaluation on four traditional herbs used as nervine tonic and commonly available as Shankhpushpi in India', *Journal of Ayurveda and Integrative Medicine*, 10(1), pp. 25–31. https://doi.org/10.1016/j.jaim.2017.08.012.

Soumyanath, A., Zhong, Y.P., Henson, E., Wadsworth, T., Bishop, J., Gold, B.G., Quinn, J.F. (2012) '*Centella asiatica* extract improves behavioral deficits in a mouse model of Alzheimer's disease: Investigation of a possible mechanism of action', *International Journal of Alzheimer's Disease*, 2012, p. 381974.

Soumyanath, A., Zhong, Y.P., Yu, X., Bourdette, D., Koop, D.R., Gold, S.A. and Gold, B.G. (2005) '*Centella asiatica* accelerates nerve regeneration upon oral administration and contains multiple active fractions increasing neurite elongation in-vitro', *Journal of Pharmacy and Pharmacology*, 57, pp. 1221–1229.

Sperling, R.A., Aisen, P.S., Beckett, L.A., Bennett, D.A., Craft, S., Fagan, A.M., Iwatsubo, T., Jack, C.R., Kaye, J., Montine, T.J., Park, D.C., Reiman, E.M., Rowe, C.C., Siemers, E., Stern, Y., Yaffe, K., Carrillo, M.C., Thies, B., Morrison-Bogorad, M., Wagster, M.V. and Phelps, C.H. (2016) 'Preclinical Alzheimer's disease: Definition, natural history, and diagnostic criteria', *Alzheimer's & Dementia*, 12(3), pp. 292–323. https://doi.org/10.1016/j.jalz.2016.02.002.

Sripanidkulchai, B. (2020) 'Benefits of aged garlic extract on Alzheimer's disease: Possible mechanisms of action', *Experimental and Therapeutic Medicine*, 19(2), pp. 1560–1564.

Stough, C., Lloyd, J., Clarke, J., Downey, L.A., Hutchison, C.W., Rodgers, T. and Nathan, P.J. (2001) 'The chronic effects of an extract of *Bacopa monniera* (Brahmi) on cognitive function in healthy human subjects', *Psychopharmacology*, 156, pp. 481–484.

Taranalli, A.D. and Cheeramkuzhy, T.C. (2000) 'Influence of *Clitoria ternatea* extracts on memory and central cholinergic activity in rats', *Pharmaceutical Biology*, 38, pp. 51–56.

Thorajak, P., Pannangrong, W., Welbat, J.U., Chaijaroonkhanarak, W., Sripanidkulchai, K. and Sripanidkulchai, B. (2017) 'Effects of aged garlic extract on cholinergic, glutamatergic and GABAergic systems with regard to cognitive impairment in Aβ-induced rats', *Nutrients*, 9(7), p. 686.

Tiwari, S., Atluri, V., Kaushik, A., Yndart, A. and Nair, M. (2019) 'Alzheimer's disease: Pathogenesis, diagnostics, and therapeutics', *International Journal of Nanomedicine*, 14, pp. 5541–5554. https://doi.org/10.2147/IJN.S200490.

Tjendraputra, E., Tran, V.H., Liu-Brennan, D., Roufogalis, B.D. and Duke, C.C. (2001) 'Effect of ginger constituents and synthetic analogues on cyclooxygenase-2 enzyme in intact cells', *Bioorganic Chemistry*, 29(3), pp. 156–163.

Uabundit, N., Wattanathorn, J., Mucimapura, S. and Ingkaninan, K. (2010) 'Cognitive enhancement and neuroprotective effects of *Bacopa monnieri* in Alzheimer's disease model', *Journal of Ethnopharmacology*, 127(1), pp. 26–31. https://doi.org/10.1016/j.jep.2009.09.056.

Wang, H.Y. and Zhang, Y.Q. (2019) 'The main active constituents and detoxification process of *Ginkgo biloba* seeds and their potential use in functional health foods', *Journal of Food Composition and Analysis*, 83. https://doi.org/10.1016/j.jfca.2019.103247.

Watson, G.S. and Craft, S. (2003) 'The role of insulin resistance in the pathogenesis of Alzheimer's disease: Implications for treatment', *CNS Drugs*, 17(1), pp. 27–45.

Weller, J. and Budson, A. (2018) 'Current understanding of Alzheimer's disease diagnosis and treatment', *F1000Research*, 7, p. F1000 Faculty Rev-1161. https://doi.org/10.12688/f1000research.14506.1.

Wilcock, G.K., Esiri, M.M., Bowen, D.M. and Smith, C.C. (1982) 'Alzheimer's disease. Correlation of cortical choline acetyltransferase activity with the severity of dementia and histological abnormalities', *Journal of Neurological Sciences*, 57(2–3), pp. 407–417.

Xu, Y., Cao, Z., Khan, I. and Luo, Y. (2008) 'Gotu Kola (*Centella asiatica*) extract enhances phosphorylation of cyclic AMP response element binding protein in neuroblastoma cells expressing amyloid beta peptide', *Journal of Alzheimer's Disease*, 13, pp. 341–349.

Yan, Z. and Feng, J. (2004) 'Alzheimer's disease: Interactions between cholinergic functions and A-amyloid', *Current Alzheimer's Research*, 1, pp. 241–248.

Yanagisawa, D., Ibrahim, N.F., Taguchi, H., Morikawa, S., Hirao, K., Shirai, N., Sogabe, T. and Tooyama, I. (2015) 'Curcumin derivative with the substitution at C-4 position, but not curcumin, is effective against amyloid pathology in APP/PS1 mice', *Neurobiology of Aging*, 36, pp. 201–210.

Yang, F., Lim, G.P., Begum, A.N., Ubeda, O.J., Simmons, M.R., Ambegaokar, S.S., Chen, P.P., Kayed, R., Glabe, C.G., Frautschy, S.A. and Cole, G.M. (2005) 'Curcumin inhibits formation of amyloid beta oligomers and fibrils, binds plaques, and reduces amyloid in vivo', *Journal of Biological Chemistry*, 280, pp. 5892–5901.

Yiannopoulou, K.G. and Papageorgiou, S.G. (2020) 'Current and future treatments in Alzheimer disease: An update', *Journal of Central Nervous System Disease*, 12, p. 1179573520907397. https://doi.org/10.1177/1179573520907397.

Yin, H.L., Wang, Y.L., Li, J.F., Han, B., Zhang, X.X., Wang, Y.T. and Geng, S. (2011) 'Effects of curcumin on hippocampal Bax and Bcl-2 expression and cognitive function of a rat model of Alzheimer's disease', *Neural Regeneration Research*, 6, pp. 1845–1849.

3 Polyphenols and Other Agents with Their Uses for Amyotrophic Lateral Sclerosis

Subhasri Bogadi, Gowthamarajan Kuppusamy, and Karri V. V. S. Narayana Reddy
Department of Pharmaceutics, JSS College of Pharmacy, JSS Academy of Higher Education & Research, Ooty, Tamil Nadu, India

3.1 INTRODUCTION

Amyotrophic lateral sclerosis (ALS), often known as Lou Gehrig's disease, is a neurological condition that worsens over time and mainly affects the motor neurons that control voluntary muscles. The brain's capacity to start and regulate muscular movement is impacted by this debilitating disorder, which causes these motor neurons to deteriorate gradually (1). Lower and upper motor neuron degeneration, which affects the spinal cord and the motor cortex of the brain, is the defining feature of ALS. People with ALS eventually experience muscle atrophy as well as stiffness and weakness in their muscles as the disease advances. Affected areas of the body, like the hands or legs, are usually the first to experience symptoms before they spread to other sites. Respiratory failure may result from ALS's advanced phases, which impact the muscles involved in breathing (2). Although a tiny fraction of ALS cases are associated with genetic variables, the specific origin of the disease is still largely unclear in the majority of cases. Any age, gender, or ethnicity can be affected by the condition, though people between the ages of 40 and 70 are the ones who are diagnosed with it most frequently (3). The present treatments for ALS attempt to enhance the quality of life for those afflicted, reduce symptoms, and delay the disease's development. Sadly, there is currently no known cure for ALS (4). To help patients and their families deal with the practical, emotional, and physical difficulties brought on by the disease, ALS is a problematic condition that calls for all-encompassing care. Treatment for ALS is fraught with problems, chief among which is the lack of a cure and the scarcity of available therapies. The variety of conditions makes treatment plans more complex because different people experience different symptoms and rates of advancement.

In many cases, delayed diagnosis impedes early actions. The lack of clarity surrounding the complex mechanisms behind motor neuron degeneration hampered the development of tailored therapeutics (5). There is a shortage of biomarkers for treatment monitoring and early identification and restricted access to specialized care. The high care expenses and the financial strain on people worsen the difficulties. The disease's rarity makes conducting clinical studies difficult, and partial symptomatic alleviation is nevertheless present. Research and advocacy initiatives are underway to overcome these challenges and improve ALS treatment strategies (6).

The need for creative and varied ways to overcome the shortcomings of existing treatment techniques is the foundation for investigating polyphenols and other compounds as possible therapeutics for ALS. Richly distributed in many plant-based meals, polyphenols have antioxidant, neuroprotective, and anti-inflammatory qualities that may mitigate the intricate pathophysiology of ALS. Furthermore, developing novel medicines outside established therapies provides a more comprehensive therapeutic toolset, acknowledging the complex nature of ALS progression (7). Since the advantages of current treatments

DOI: 10.1201/9781003478652-4

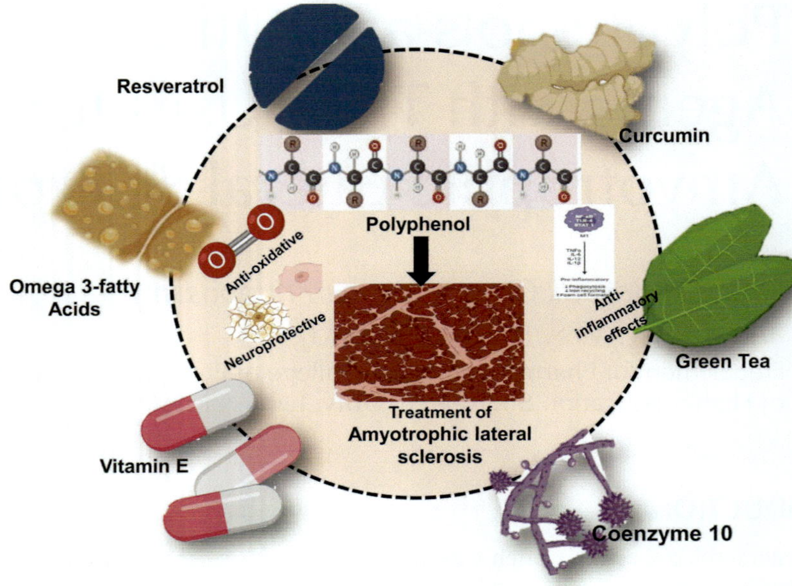

FIGURE 3.1 Graphical abstract illustrating the therapeutic potential of various polyphenolic and antioxidant agents—including resveratrol, curcumin, green tea polyphenols, coenzyme Q10, vitamin E, and omega-3 fatty acids—in the treatment of amyotrophic lateral sclerosis through antioxidant, neuroprotective, and anti-inflammatory mechanisms.

are somewhat limited, the goal is to improve outcomes and the quality of life for ALS patients by exploring alternative options that may offer more focused and effective interventions (8–12). We highlight the potential of polyphenols and new compounds as prospective therapeutic pathways in this chapter, emphasizing the need for a thorough investigation to address the ongoing problems with the treatment of ALS. By carefully analyzing their anti-inflammatory and neuroprotective qualities, we hope to add to the current discussion on creative approaches and provide an overview of how ALS treatments develop. The polyphenol-based therapeutic strategies for ALS are summarized in Figure 3.1.

3.2 PATHOPHYSIOLOGY

The neurodegenerative disease known as ALS is typified by the gradual death of motor neurons in the brain and spinal cord (13). The pathogenic events that culminate in the unrelenting degeneration of these vital neurons are part of the intricate molecular and cellular pathways that underlie ALS. The abnormal aggregation of proteins within damaged neurons is a significant characteristic of the pathophysiology of ALS. Essential for the processing and transporting of RNA, TDP-43, and FUS mislocalize and create insoluble aggregates that impair cellular processes. A characteristic of ALS, this proteinopathy can occur in familial and sporadic instances (14). The pathophysiology of ALS also heavily depends on mitochondrial dysfunction. The vulnerability of motor neurons is caused by reduced energy generation and elevated oxidative stress. Neuronal damage is exacerbated by excitotoxicity caused by an imbalance in glutamate levels. This increased excitability plays a role in the motor neurons' selective susceptibility in ALS. The participation of glial cells, such as microglia and astrocytes, highlights the importance of neuroinflammation in ALS. Dysregulated glial activity can create a toxic milieu for motor neurons, producing pro-inflammatory cytokines and neurotoxic substances (15). Gene mutations have been connected to cases of familial ALS, suggesting that genetic factors may play a significant role. The disruption of cellular processes caused by gene mutations like SOD1, C9orf72, and others makes motor neurons more susceptible to degeneration. It isn't easy to comprehend the pathophysiology of ALS because of the interaction of these several

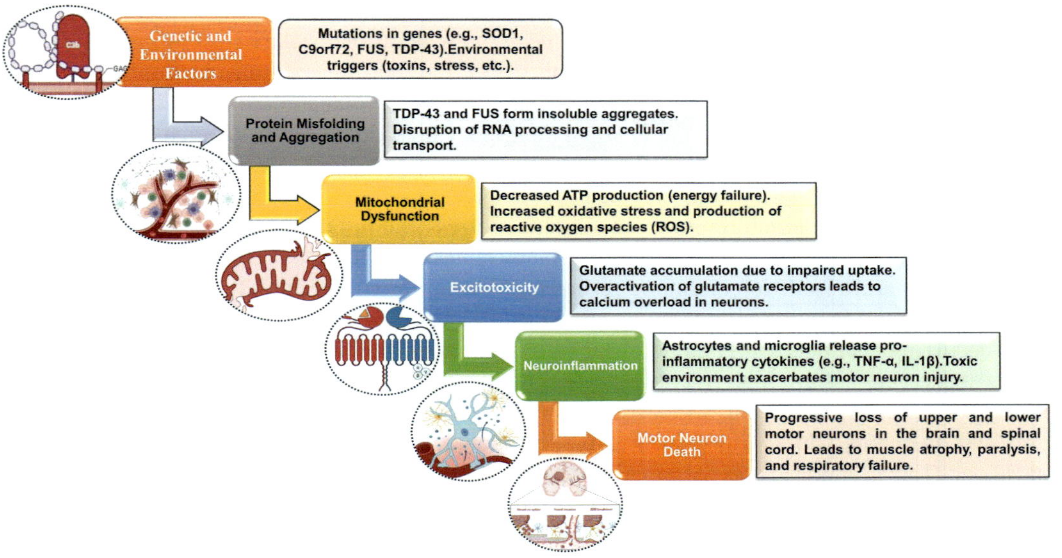

FIGURE 3.2 Pathophysiology of ALS.

molecular and cellular pathways seen in Figure 3.2. Developing treatment options offers a complex task when focusing on these systems (16). The development of focused therapies aiming at delaying or stopping the course of ALS is made possible by advances in our understanding of the complex interplay between inflammatory, proteinopathic, and genetic processes.

The intricacies of these pathways are still being uncovered by research, which gives rise to optimism for novel therapies that can deal with the underlying causes of this debilitating neurodegenerative illness (17). A comprehensive understanding of ALS's molecular and cellular mechanisms is necessary to identify critical targets for therapeutic intervention. Important targets include protein aggregation, mitochondrial malfunction, excitotoxicity, and neuroinflammation. Therapeutic development may benefit from approaches that reduce abnormal protein accumulation, repair mitochondrial function, adjust glutamate levels, and minimize neuroinflammation (18). The shortcomings of present methods are shown through a critical analysis of therapeutic modalities. The FDA-approved treatments for ALS, riluzole, and edaravone, have limited advantages and concentrate more on treating symptoms than underlying causes. These therapies demonstrate the critical need for more potent interventions that focus on the recognized critical pathways to mitigate symptoms and impede or reverse the course of the disease (19). Precision medicine methods that consider the variability of ALS and tailored treatment plans based on genetic and molecular profiles are beginning to emerge as research progresses (20, 21). There is promise for more significant interventions with the combination of novel medicines that target critical pathways found in the pathophysiology of ALS. This could change the therapy landscape for this challenging neurodegenerative illness.

3.3 POLYPHENOLS

Polyphenols, distinguished by numerous phenolic rings, are a large and diversified group of naturally occurring chemicals in plant-based foods. Polyphenols are classified based on their chemical structures, resulting in diverse subclasses with various functions (22). Flavonoids stand out among the primary classes, which include flavonols, flavones, flavanones, and flavan-3-ols, often known as catechins. Examples of structural variety within this class include quercetin in onions, apigenin in parsley, and epicatechin in green tea. Another important group is phenolic acids, which comprise hydroxybenzoic and hydroxycinnamic acids, which contain chemicals such as gallic acid and caffeine (23). Polyphenolic amides, such as capsaicinoids found in chili peppers, and other compounds, such as stilbenes (e.g., resveratrol) and lignans (e.g., secoisolariciresinol, matairesinol), add to the complex classification of polyphenols, reflecting their diverse roles in plant biology and potential health benefits. This classification not only helps comprehend the structural diversity of polyphenols but also offers information about their bioavailability and physiological consequences.

The subclasses have distinct action modes, including antioxidant capabilities and anti-inflammatory actions. As a result, polyphenols have received a lot of interest due to their possible health benefits (24). Their ability to reduce oxidative stress and inflammation has implications for a variety of chronic diseases, including neurodegenerative disorders such as ALS. Recognizing polyphenols' diverse nature leads to continuing research efforts to maximize their therapeutic potential in improving overall health and treating specific medical diseases. Polyphenols, a broad set of naturally occurring chemicals with numerous health advantages, are found in many plant-based diets. Fruits and vegetables are high in polyphenols, giving people access to these beneficial substances. Blueberries, strawberries, and raspberries are exceptionally high in flavonoids such as anthocyanins and quercetin. Citrus fruits, such as oranges, lemons, and grapefruits, provide the diet with flavanones like hesperidin and naringenin. Additional fruit sources include apples and grapes with flavonols like quercetin and catechins like epicatechin.

Tea is a high-polyphenol-content beverage. Both green and black teas include catechins, such as epigallocatechin gallate (EGCG), known for their antioxidant qualities. Coffee made from coffee beans contains chlorogenic acids, another type of polyphenol with possible health advantages. Red wine from grapes is known for its high concentration of resveratrol, a stilbene with antioxidant qualities. Whole grains, nuts, and seeds all add to polyphenol consumption. Buckwheat and oats are grains rich in rutin and avenanthramides, respectively (25). Lignans and other polyphenolic components can be found in nuts like almonds, walnuts, flaxseeds, and chia seeds.

Additionally, herbs and spices such as cinnamon, turmeric, and oregano are high in polyphenols, improving the flavor and nutritional profile of numerous cuisines. Incorporating various plant-based foods into the diet offers a balanced intake of polyphenols, contributing to the potential health advantages of these bioactive substances. Polyphenol bioavailability and metabolism are complex processes with important implications for therapeutic potential and health benefits. Bioavailability is the extent and rate at which the body absorbs and utilizes polyphenols following ingestion. Several factors influence polyphenol bioavailability, including chemical structure, food matrix, and interactions with other dietary components. Polyphenols undergo various changes after intake (26). The structural variety of polyphenols regulates their absorption in the digestive tract. Small, water-soluble polyphenols like flavonoids are commonly absorbed in the small intestine via passive diffusion. Larger, more complex polyphenols may enter the colon, where gut microbes can break them down into smaller, more absorbable molecules. The resultant metabolites can subsequently enter the bloodstream and travel throughout the body (27). Polyphenols are extensively metabolized in the liver and conjugated with glucuronic acid, sulfate, or methyl groups. These alterations increase water solubility and promote elimination via urine or bile. Polyphenols' metabolic destiny differs between individuals due to genetics, gut microbiota composition, and overall health (28). The timing of polyphenol consumption relative to meals can also influence bioavailability. Polyphenol absorption may be enhanced or inhibited when specific nutrients or dietary components are consumed together. For example, the availability of nutritional fibers can change polyphenol transit time in the gut, affecting absorption and metabolism.

3.4 MECHANISM OF ACTION

3.4.1 Anti-Inflammatory Effects

Polyphenols have potent anti-inflammatory effects due to their subtle interactions with molecular pathways involved in the inflammatory response (28). Their action method inhibits pro-inflammatory enzymes, specifically cyclooxygenase (COX) and lipoxygenase (LOX), essential in manufacturing inflammatory mediators such as prostaglandins and leukotrienes. Furthermore, polyphenols influence the expression and activity of inflammatory mediators such as cytokines, chemokines, and adhesion molecules, which regulate immune cell recruitment and activation (29). Their intense antioxidant activity helps to reduce inflammation by scavenging reactive oxygen species (ROS) and reducing oxidative stress-induced damage to cells and tissues.

Polyphenols block the activation and transcription of pro-inflammatory genes via the nuclear factor-kappa B (NF-κB) pathway, a key regulator of inflammation. Furthermore, these chemicals stimulate anti-inflammatory pathways, including the nuclear factor erythroid 2-related factor 2 (Nrf2)-antioxidant response element (ARE) pathway, which improves cellular defense systems against oxidative stress and inflammation (30). Polyphenols fine-tune the generation of inflammatory mediators by altering immune cell signaling pathways such as mitogen-activated protein kinases (MAPKs) and Janus kinase/signal transducer and activator of transcription (JAK/STAT). Polyphenols' complex nature enables them to target multiple stages in the inflammatory cascade, highlighting their potential as therapeutic agents for chronic inflammation-related ailments such as neurodegenerative diseases, cardiovascular disorders, and metabolic syndromes (31, 32). Understanding the molecular processes behind polyphenols' anti-inflammatory properties is critical for realizing their full therapeutic potential in preventive and medicinal therapies.

3.4.2 Antioxidants

Polyphenols have strong antioxidant properties, which are essential for their biological function. The principal mechanism involves neutralizing ROS and other free radicals produced by regular cellular activities and environmental stresses (33). Polyphenols' antioxidant activities help protect cells from oxidative stress, regulating various physiological processes. Polyphenols work as direct antioxidants, donating electrons to free radicals and effectively neutralizing their reactivity. Furthermore, they increase the activity of natural antioxidant enzymes such as superoxide dismutase, catalase, and glutathione peroxidase, thus strengthening the cellular defense against oxidative damage (34). This comprehensive strategy contributes to the delicate balance between ROS generation and cellular antioxidant capability.

In addition to direct scavenging of free radicals, polyphenols have metal-chelating capabilities, which prevent metal-catalyzed processes that produce damaging radicals. They can also influence signaling pathways involved with oxidative stress, such as the Nrf2 pathway, whose activation increases the expression of antioxidant and detoxifying enzymes, which strengthens the cell's defense against oxidative damage. Polyphenols' antioxidant activities are relevant to various health issues, including cardiovascular disease, neurological disorders, and cancer. Polyphenols help to prevent cellular damage and inflammation, both of which play essential roles in the etiology of many chronic diseases by reducing oxidative stress (35). Understanding the complex mechanisms by which polyphenols provide antioxidant protection is critical for unlocking their therapeutic potential and incorporating them into methods for boosting overall health and well-being.

3.4.3 Neuroprotective Mechanisms

Polyphenols have necessary neuroprotective actions, which maintain brain health and may reduce the risk of neurodegenerative illnesses (36). A critical component of their neuroprotective impact is their capacity to combat oxidative stress in the central nervous system. Polyphenols' powerful antioxidant activities neutralize ROS and free radicals linked to neurodegeneration-related oxidative damage. Furthermore, polyphenols have anti-inflammatory properties within the brain (37). They contribute to a less conducive environment for chronic inflammation, which has been linked to advancing numerous neurodegenerative illnesses by modifying inflammatory pathways and lowering the generation of pro-inflammatory mediators. Polyphenols also interact with cellular signaling pathways involved in neuronal survival and growth. They can regulate kinases and transcription factors involved in neuronal function and plasticity, which helps to maintain synaptic connections and general brain integrity (38).

Furthermore, polyphenols may increase neurogenesis or the formation of new neurons, especially in brain regions important for learning and memory. Moreover, these chemicals have metal-chelating capabilities, which may prevent the accumulation of metal ions (such as iron and copper) that might contribute to oxidative stress and protein aggregation in the brain, which are hallmarks of neurodegenerative illnesses. Certain polyphenols, such as flavonoids, have been shown to pass the blood–brain barrier and interact directly with neural tissues. Once within the brain, polyphenols may alter cellular processes, improve neuronal resilience, and contribute to neuroprotection (39). Polyphenols, with their diverse neuroprotective actions, are exciting candidates for therapies to prevent or reduce the progression of neurodegenerative diseases. While more research is needed to fully understand the specific effects of different polyphenols on neurological health, current evidence suggests that including a variety of these compounds in the diet may provide a more holistic approach to supporting brain function and lowering the risk of cognitive decline.

3.4.3.1 Potential Impact on ALS Pathophysiology

Polyphenols show promise in affecting the pathophysiology of ALS, providing possible advantages through various pathways that may help neuroprotection (40). While research in this arena is continuing and evolving, numerous elements of polyphenol activity point to their potential impact on ALS.

The oxidative stress linked with ALS causes neuronal degeneration. Polyphenols, with their strong antioxidant qualities, may assist in combating oxidative stress by neutralizing free radicals and lowering cellular damage, potentially alleviating one of the causes of ALS progression. In anti-inflammatory effects, chronic inflammation is linked to the pathogenesis of ALS. Polyphenols, renowned for their anti-inflammatory characteristics, may modify inflammatory pathways, inhibit the release of pro-inflammatory molecules, and produce an environment less conducive to neuroinflammation (41). Polyphenols have neuroprotective properties by affecting cellular signaling pathways, improving neuronal survival, and possibly increasing synaptic plasticity (42). These strategies may help to maintain the structural and functional integrity of motor neurons in ALS. In mitochondrial function, ALS is connected with mitochondrial dysfunction, which reduces neuron energy production. By altering mitochondrial biogenesis and function, some polyphenols may help improve cellular energy metabolism and reduce the effects of mitochondrial dysfunction in ALS. Protein misfolding and aggregation, as seen by the buildup of misfolded proteins such as TDP-43, are defining hallmarks of ALS. Certain polyphenols have anti-aggregation capabilities that may affect the abnormal protein aggregates linked with ALS pathogenesis. Metal ion dysregulation, notably in copper and iron, has been linked to ALS. With their metal-chelating properties, polyphenols may aid in metal ion homeostasis, potentially lowering metal-induced oxidative stress in the central nervous system.

While the precise impact of polyphenols on ALS pathophysiology requires further investigation in pre-clinical and clinical studies, the cumulative evidence suggests that their multifaceted properties may provide a comprehensive approach to addressing various aspects of the complex neurodegenerative processes associated with ALS (43). Integrating polyphenol-rich foods or supplements into the diet may be an adjuvant treatment for people with ALS to support overall neuroprotection and may influence the disease's progression.

3.5 POLYPHENOLS IN ALS RESEARCH

A comprehensive review of studies on the effect of polyphenols in ALS models indicates a growing body of evidence pointing to possible therapeutic benefits. Pre-clinical studies employing cellular and animal models of ALS investigated the impact of different polyphenols on disease development and underlying processes (40). Flavonoids, an essential class of polyphenols, have been the subject of investigation, with substances like quercetin and resveratrol showing potential in treating neurodegenerative processes related to ALS. Research utilizing ALS models has regularly demonstrated the antioxidant characteristics of polyphenols and their capacity to mitigate oxidative stress in neural tissues (44). Furthermore, polyphenols have shown anti-inflammatory properties by regulating inflammatory pathways and reducing neuroinflammation, which is a significant factor in the pathophysiology of ALS. These results imply that polyphenols may affect several aspects of ALS pathophysiology, addressing the inflammatory and oxidative aspects of the illness.

Furthermore, studies have explored the possible neuroprotective effects of polyphenols in models of ALS. There is evidence that they can improve pathways for cellular survival, control protein aggregation, and support mitochondrial activity, all of which help to preserve motor neurons (45). Though cautious interpretation is necessary due to variances in study designs and the specific polyphenols studied, the overall trend toward positive outcomes emphasizes the therapeutic potential of polyphenols in models of ALS. Notwithstanding the encouraging results, issues like bioavailability, ideal dosage, and adaptability to human trials still need to be resolved (46). However, the combined knowledge from research on polyphenols in ALS models highlights their potential as subjects for additional study in clinical settings, providing a fresh and comprehensive method of addressing the intricate pathophysiology of ALS (40). An increasing interest in evaluating the possible therapeutic benefits of polyphenols in human subjects has led to a progressive expansion of the landscape of observational studies and clinical trials examining the role of these substances in ALS. Clinical trials, which frequently involve dietary adjustments or supplementation techniques, have attempted to evaluate polyphenol treatments' safety, tolerability, and possible efficacy.

Conversely, observational studies aim to clarify correlations between the intake of polyphenols and the occurrence, development, or treatment of ALS symptoms. Numerous scientific studies have examined the effects of certain polyphenols, like curcumin and resveratrol, on the course of ALS. In these trials, polyphenols' potential neuroprotective effects are evaluated by neurological function, inflammation, and oxidative stress measurements (47). The outcomes of these studies provide important new information about the viability and early effectiveness of polyphenol therapies for individuals with ALS. Observational studies seek to identify trends and correlations between dietary polyphenol consumption and outcomes related to ALS by utilizing sizable cohorts of patients. These studies consider variables including general nutritional practices, the consumption of foods high in polyphenols, and their possible relationship to the course of the disease or survival rates (48). Although observational, these studies offer insightful, real-world information that could guide future interventional trials.

In polyphenols and ALS, the challenges associated with performing observational research and clinical trials include the disease's variability, different study designs, and the requirement for long-term follow-up to capture the disease's gradual development (49). Furthermore, when planning and analyzing this research, factors like bioavailability, ideal doses, and choosing suitable polyphenols should be considered carefully. The amalgamation of results from observational studies and clinical trials will advance the field by thoroughly comprehending the possible advantages and difficulties linked to polyphenol treatments in ALS. The body of knowledge gathered from these studies will direct future investigations and influence the field of ALS patients' treatment approaches. Research on polyphenols for ALS faces several obstacles that highlight how difficult it is to convert encouraging pre-clinical results into practical clinical treatments. A significant barrier is the bioavailability of polyphenols; many of these chemicals have low absorption rates, which restricts their ability to reach target tissues, especially the central nervous system. This bioavailability problem calls for creative approaches to improve efficacy and delivery (9). Another level of complexity is added by the wide variety of polyphenols, each with a distinct chemical structure, bioavailability profile, and mode of action. It is difficult to extrapolate results across the polyphenol class since determining the precise effects of different polyphenols necessitates a detailed methodology. The intrinsic variability of ALS presents an additional obstacle to the study of polyphenols. Different clinical subgroups of the disease exist, each with unique hereditary variables and rates of disease progression. Optimizing treatment strategies requires customizing polyphenol therapies to account for this variation (50).

Further complicating the process of synthesizing results is the diversity in study designs, which include variations in polyphenol administration techniques, dosage, and duration. Standardizing research procedures and approaches may improve comparability among various studies and lead to a more coherent knowledge of the effects of polyphenols on ALS. Large-scale, carefully planned clinical trials are necessary to confirm the minimal clinical evidence supporting polyphenols' usefulness in treating ALS despite encouraging pre-clinical results (43, 51). Unlocking the full therapeutic potential of polyphenols in ALS will require teamwork, careful study design, and creative thinking to overcome these obstacles.

3.6 OTHER AGENTS WITH POTENTIAL BENEFITS

3.6.1 Curcumin

Because of its varied pharmacological characteristics, curcumin, a polyphenolic molecule derived from turmeric, has attracted attention for its possible advantages in ALS. Its powerful anti-inflammatory properties are its most notable quality, and they might be essential for ALS patients, as neuroinflammation plays a role in the disease's progression (40). Curcumin may reduce the inflammatory cascade in the central nervous system by modifying critical inflammatory pathways, providing a possible treatment route. Another essential factor consistent with the pathophysiology of ALS is curcumin's antioxidant action. It is critical to reduce the oxidative damage linked to motor neuron degeneration because of its ability to neutralize free radicals and lower oxidative stress (41). Curcumin may lessen the effects of the compromised mitochondrial function seen in ALS patients by increasing cellular energy output. The adaptability of curcumin also includes its capacity to regulate protein aggregation, which is a prevalent characteristic of neurodegenerative illnesses (52). Curcumin has the potential to support the preservation of protein homeostasis in neurons by disrupting the misfolding and aggregation of proteins like TDP-43. Curcumin's neuroprotective actions further establish it as a substance with potential advantages for ALS by altering many cellular signaling pathways linked to neuronal survival and synaptic function (53). Furthermore, in some formulations, its capacity to pass the blood–brain barrier increases its potential for targeting the central nervous system. Even though there are still issues with curcumin's bioavailability, research efforts are being made to improve dosing schedules and formulations to gain a more profound knowledge of curcumin's potential as a treatment for neurodegenerative diseases like ALS (40, 54).

3.6.2 Resveratrol

The polyphenolic chemical resveratrol, widely distributed in plant sources, has gained attention as a potential treatment for neurological diseases such as ALS. Resveratrol is a compound with many uses (55). It is well-known for its potent antioxidant and anti-inflammatory capabilities, which target essential players in the pathogenesis of ALS. It may play a role in improving cellular resilience and health because of its capacity to activate sirtuins, essential proteins involved in cellular processes like stress response and DNA repair. According to studies, resveratrol may have neuroprotective effects by promoting the production of heat shock proteins, helping cells deal with stress, and avoiding protein misfolding, an essential factor in neurodegenerative diseases (56).

Additionally, the effect of resveratrol on mitochondrial activity is consistent with identifying mitochondrial malfunction in ALS. Resveratrol can influence the course of disease by improving energy metabolism in neurons by promoting mitochondrial biogenesis and function (57). Its anti-aggregation characteristics further enhance its potential therapeutic effects, which prevent the creation of protein aggregates linked to ALS (58). Resveratrol's capacity to penetrate the blood–brain barrier increases its applicability to neurological disorders by guaranteeing it reaches the central nervous system. Although encouraging, issues with absorption and the best dosages highlight the need for more investigation to fully understand the precise mechanisms and possible therapeutic applications of resveratrol in ALS and other neurodegenerative illnesses.

3.6.3 Green Tea

Green tea extract, made from *Camellia sinensis* leaves, has drawn interest because of its high concentration of polyphenolic chemicals, especially catechins, and because it may have therapeutic advantages, including the ability to treat neurodegenerative diseases like ALS (40). Catechins have anti-inflammatory and antioxidant qualities, particularly EGCG. The potential of green tea extract to reduce oxidative stress and neuroinflammation is particularly intriguing in the context of ALS, as these processes are known to accelerate the disease's course (59). According to studies, the polyphenols in green tea may have neuroprotective effects by decreasing inflammation, scavenging free radicals, and altering cellular signaling pathways linked to neuronal survival.

Additionally, green tea extract has shown promise in affecting protein misfolding and aggregation, pathological features of ALS. Green tea extract may help preserve protein homeostasis in neurons by preventing the aggregation of proteins like TDP-43. Its capacity to penetrate the blood–brain barrier further increases its applicability to neurological disorders by enabling it to act directly on the central nervous system (41). Although studies on the precise effects of green tea extract in ALS are still in progress, its exemplary safety record and possible range of advantages make it a promising candidate for new treatment approaches to neurodegenerative diseases (60). However, dosage standardization and bioavailability concerns must be addressed to maximize its potential as a therapeutic adjunct.

3.6.4 Omega-3 Fatty Acids

Omega-3 fatty acids have been extensively studied for their potential health benefits, including their implications for neurodegenerative diseases like ALS (61). Two notable omega-3 fatty acids are eicosapentaenoic acid (EPA) and docosahexaenoic acid (DHA), which are abundant in fatty fish, flaxseeds, and walnuts. These vital fatty acids are present in all cell membranes and are critical for preserving membrane fluidity and regulating several biological functions (62). Because of their anti-inflammatory and neuroprotective qualities, omega-3 fatty acids are essential in ALS, where motor neurons gradually degenerate. According to studies, omega-3 fatty acids may influence the synthesis of inflammatory mediators and encourage the resolution of inflammation, which may assist in regulating inflammation (63). The pathophysiology of ALS is partially attributed to chronic inflammation, and the anti-inflammatory properties of omega-3 fatty acids may attenuate the neuroinflammatory processes seen in ALS patients.

Additionally, these fatty acids aid in synthesizing specific pro-resolving mediators (SPMs), aiding tissue repair and resolving inflammation. Another feature of ALS that is impacted by omega-3 fatty acids is oxidative stress. They assist in reducing oxidative damage to cells and neutralizing ROS by functioning as antioxidants (64). These fatty acids may also affect brain function by regulating the release of neurotransmitters and synaptic plasticity. Although investigations into the precise role of omega-3 fatty acids in ALS are still underway, these fatty acids are of interest in the larger context of neurodegenerative illnesses due to their overall neuroprotective and anti-inflammatory characteristics (65). A prospective dietary approach to promote general brain health is to include foods or supplements high in omega-3 fatty acids; current research endeavors to clarify their precise function in preventing and treating ALS.

3.6.5 Vitamin E

As a fat-soluble antioxidant, vitamin E comprises several substances, the most physiologically active of which is alpha-tocopherol (66). Vitamin E, which may be found in various foods like nuts, seeds, and vegetable oils, has been researched for possible health advantages, including its connection to neurodegenerative conditions like ALS. As an antioxidant, its primary function is to counteract free radicals, shielding cells from oxidative damage. Vitamin E's antioxidant qualities make it a promising candidate for neuroprotective benefits in ALS, where oxidative stress is linked to motor neuron degeneration (41).

Research on the effects of vitamin E on inflammatory processes has shown that it can decrease the release of pro-inflammatory chemicals and regulate immunological responses. Since neuroinflammation plays a role in the pathogenesis of ALS, its anti-inflammatory potential is essential. Because of its capacity to reduce inflammation, vitamin E may contribute to developing a neuroprotective environment, which could affect how ALS develops (52). The impact of vitamin E on protecting mitochondrial activity has also been studied. The pathophysiology of ALS is linked to mitochondrial dysfunction, and vitamin E's role in preserving mitochondrial integrity may support cellular energy metabolism and brain health in general.

Furthermore, vitamin E may have other routes to protect neurons in ALS by influencing gene expression and signaling pathways linked to cell survival and death (67). Although pre-clinical research and specific observational data have indicated the potential benefits of vitamin E, outcomes from clinical trials examining its effectiveness in ALS have been conflicting. Difficulties include the optimal dosage, possible drug interactions, and the intricate interactions between various types of vitamin E. Further research enhances our comprehension of the distinct processes by which vitamin E functions in ALS and its potential as a supplemental therapy strategy (44). This underscores the necessity of customized interventions to tackle the complex features of ALS pathogenesis.

3.6.6 Coenzyme 10

Coenzyme Q10 (CoQ10) is an antioxidant and an essential part of the electron transport chain. It has attracted attention because it may be able to treat several illnesses, including neurological diseases like ALS. CoQ10 is a naturally occurring substance in the body and diet, essential for synthesizing cellular energy (68). It is also a potent antioxidant that shields cells from oxidative damage. CoQ10 is a prospective option for neuroprotective therapies in ALS because of its dual roles as an energy facilitator and antioxidant, contributing to the degeneration of motor neurons caused by oxidative stress and mitochondrial malfunction (69). According to pre-clinical research, coQ10 supplementation may slow the course of ALS by improving cellular energy metabolism, enhancing mitochondrial activity, and reducing oxidative stress. Its capacity to combat free radicals and promote mitochondrial health is consistent with the complex pathophysiology of ALS, in which heightened oxidative stress and disturbances in cellular energy production are important factors (52). Clinical trials investigating the effectiveness of CoQ10 in ALS have produced conflicting results despite favorable pre-clinical data. This highlights the difficulty in extrapolating pre-clinical insights to human outcomes. Choosing the proper dosages, dealing with different CoQ10 formulations, and taking possible drug interactions into account are some of the challenges (70). Current studies are aimed at improving our comprehension of the precise role that CoQ10 plays in ALS, highlighting the necessity of conducting thorough clinical trials to clarify its potential for treatment (43, 67). CoQ10 continues to be a fascinating topic of study as the field of ALS research develops, providing insights into the complex mechanisms driving neurodegeneration and possible therapeutic options.

3.7 COMBINING THERAPIES

3.7.1 Synergistic Effects of Polyphenols and Other Agents

Combining medicines is a potential strategy in the search for efficient treatments for complicated neurodegenerative diseases like ALS, especially given the synergistic effects of polyphenols with other medications (71). The distinct and diverse modes of action demonstrated by polyphenols, such as their anti-inflammatory, antioxidant, and neuroprotective characteristics, enhance the benefits of other medicinal agents. Their capacity to concurrently address several facets of ALS pathology, resulting in a thorough and all-encompassing approach to illness care, represents their potential synergy. Studies investigating the combination of polyphenols with other agents, including vitamins, omega-3 fatty acids, and coenzyme Q10, seek to capitalize on the many benefits each component provides (72). For example, combining polyphenols and omega-3 fatty acids may benefit from their respective anti-inflammatory capabilities, addressing neuroinflammation, a hallmark of ALS.

Furthermore, when combined, the antioxidant actions of polyphenols and vitamin E may provide a strong defense against oxidative stress, improving protection for susceptible motor neurons. The synergistic strategy includes addressing bioavailability issues and optimizing therapeutic doses. Combining polyphenols with drugs that increase absorption or prolong their stay in the body may overcome the limits of individual substances (73). This collaborative method acknowledges the complexities of ALS pathogenesis, in which various factors contribute to disease progression, and highlights the potential benefits of a synergistic therapy arsenal. While combining medicines is appealing, it requires careful evaluation of possible interactions, ideal dosage regimes, and individual patient differences (74). Clinical trials to determine the safety and efficacy of these combination techniques are critical for turning pre-clinical findings into relevant treatment solutions. As research in this field progresses, discovering synergistic effects between polyphenols and other medicines opens the door to creative and personalized therapies, promising better results in ALS patients. Identifying possible combinations for treating ALS requires deliberately integrating therapeutic drugs to target several aspects of the disease's complicated pathogenesis.

One intriguing combination involves the synergistic effects of polyphenols and other neuroprotective drugs (75). For example, combining polyphenols, renowned for their anti-inflammatory and antioxidant effects, with omega-3 fatty acids could result in a more holistic strategy for addressing neuroinflammation and oxidative stress, both common in ALS. The anti-aggregation characteristics of polyphenols and the possible mitochondrial support offered by omega-3 fatty acids may provide a complex defense against the many degenerative processes associated with ALS. Another possible combination is the addition of polyphenols to coenzyme Q10, a molecule essential for mitochondrial activity and energy production. Polyphenols' antioxidant properties may enhance CoQ10's role in cellular energy metabolism and oxidative stress reduction (76). The combined action of these medicines can potentially improve mitochondrial function and give comprehensive assistance to motor neurons in ALS.

Furthermore, investigating combinations with vitamins like E may boost antioxidant defense, resulting in more robust protection against oxidative damage. Integrating potential combinations for ALS treatment includes medicinal techniques that control inflammation. Combining anti-inflammatory drugs like curcumin with neuronal function-supporting chemicals like omega-3 fatty acids may produce a more balanced approach to modulating the inflammatory response while boosting neuroprotection (77). Furthermore, the combined usage of polyphenols and vitamins, such as vitamin E, may have synergistic effects on inflammation and oxidative stress reduction. While the concept of prospective combinations has promise, it is critical to approach cautiously when considering appropriate doses, bioavailability, and potential interactions (78). Rigorous pre-clinical and clinical studies are required to confirm the safety and efficacy of these combinations. Tailoring therapy to the distinct needs of ALS subtypes and individual patient profiles will be critical for improving therapeutic outcomes. Investigating prospective combinations is a dynamic and growing component of ALS research, providing hope for more successful and personalized therapeutic techniques.

3.8 TRANSLATIONAL CHALLENGES

Translational problems in ALS research frequently concentrate on bridging the gap between pre-clinical and clinical investigations, an essential step in improving prospective treatment strategies. Pre-clinical investigations, commonly undertaken in cellular or animal models, provide crucial insights into the mechanisms of action and possible efficacy of new therapeutics (79). However, translating these discoveries into significant therapeutic applications presents numerous challenges. One major problem is the intricacy of ALS as a human disease. While pre-clinical models can imitate some elements of ALS pathology, they may not convey the disease's variability and complexities as seen in actual patients (80). Bridging this gap requires a thorough grasp of pre-clinical models' limitations and a critical assessment of their applicability to human ALS.

Another translational problem is determining optimal dosages and formulations for human usage. Effective doses reported in pre-clinical investigations may not immediately convert into safe and effective ones in humans. Bioavailability, metabolism, and potential adverse effects must all be carefully studied and evaluated in clinical trials to guarantee that the therapeutic benefits seen in pre-clinical models are effectively realized in patients. Furthermore, identifying robust biomarkers presents a critical translational barrier (81). Biomarkers that reliably identify disease development or responsiveness to treatment in pre-clinical models may not always correlate with human patient outcomes. Validating biomarkers that appropriately reflect ALS pathophysiology and treatment response is critical for determining the effectiveness of innovative therapies in clinical settings. Overcoming these translational hurdles requires concerted efforts from researchers, doctors, and pharmaceutical producers (82). Improved pre-clinical models that more accurately mimic human ALS, novel trial designs, and a thorough grasp of the disease's biochemical foundations are all necessary for effective translational research. Addressing these issues is critical for realizing the potential benefits of promising therapeutic techniques revealed in pre-clinical investigations and transforming them into actual improvements in the lives of ALS patients through effective clinical interventions (83). Identifying accurate biomarkers for assessing the efficiency of polyphenol treatments in ALS is critical to furthering translational research and aiding the transition from pre-clinical to clinical trials. Biomarkers are quantitative signs of disease progression, treatment response, or underlying molecular alterations, and they provide essential insights into polyphenols' effects on ALS pathology (84).

The varied nature of ALS makes it challenging to establish indicators for polyphenol effectiveness. Various molecular and cellular changes characterize ALS, including oxidative stress, inflammation, protein misfolding, and mitochondrial dysfunction (85). Identifying biomarkers that adequately reflect the pathways targeted by polyphenols is critical for determining their efficacy. Oxidative stress biomarkers, such as lipid peroxidation or protein oxidation, could indicate polyphenol efficacy. Polyphenols are well-known antioxidants; therefore, changes in oxidative stress markers may tell their role in decreasing free radicals and mitigating oxidative damage, which are critical processes in ALS development (86). Inflammation-related biomarkers, such as cytokines and inflammatory mediators, may provide insight into polyphenols' anti-inflammatory properties. Modulating neuroinflammation is an essential therapeutic approach in ALS, and discovering biomarkers that mimic the effect of polyphenols on inflammatory pathways can be a helpful tool for assessing therapy response. Biomarkers for protein aggregation and misfolding, both frequent in ALS, are also important (87). Polyphenols have shown anti-aggregation characteristics in pre-clinical trials, and biomarkers indicating changes in protein aggregation dynamics may provide insight into their usefulness in reducing this element of ALS pathology. Validating these indicators necessitates thorough investigation in both pre-clinical models and clinical trials. Establishing a link between changes in biomarker levels and therapy results will help determine the efficacy of polyphenol therapies. Furthermore, given the variability of ALS, developing biomarkers that can distinguish distinct subtypes or stages of the illness will improve the precision of treatment assessment. While hurdles remain, ongoing research attempts to establish biomarkers for polyphenol efficacy show promise for improving our understanding of their influence on ALS and aiding the development of personalized and targeted therapy methods.

3.9 FUTURE DIRECTIONS AND CONCLUSION

Future directions in ALS research are defined by ongoing examination of novel compounds and agents, demonstrating a dedication to enhancing therapeutic techniques and improving our understanding of neurodegenerative processes (88). The discovery and analysis of new polyphenols can potentially increase the arsenal of neuroprotective chemicals. As technology and research approaches advance, researchers aggressively investigate polyphenols' varied chemical structures and properties to identify those that may provide unique benefits in ALS, potentially outperforming currently investigated chemicals. In addition to exploring novel polyphenols, the attention shifts to a more extensive range of medicines that may have promise in ALS treatment. This contains substances with anti-inflammatory, antioxidant, and neuroprotective effects that go beyond polyphenols (89). Investigating these compounds provides a complete strategy for tackling the diverse character of ALS disease, and they may have synergistic benefits when paired with existing therapy techniques. Future efforts include improving understanding of the interaction of genetics and environmental variables in ALS. Identifying genetic factors that determine individual responses to polyphenols and other neuroprotective drugs may pave the way for more personalized treatment options. Integrating genetic data into clinical trials may improve patient classification, leading to more effective and customized therapies.

Furthermore, advances in medication delivery technologies and formulations are expected to play a critical role in the future of ALS treatments. Innovative techniques to increase the bioavailability and targeted delivery of polyphenols and other medicines to the central nervous system may overcome existing absorption and distribution limitations, thereby improving therapeutic outcomes. The future of ALS treatment is increasingly defined by a trend toward personalized treatments that recognize the disease's inherent heterogeneity and each patient's unique genetic and physiological characteristics (90). One critical component of personalized ALS treatment is the identification of particular genetic markers and variations that influence disease susceptibility and development. As genetic research in ALS progresses, recognizing these markers permits the creation of targeted medicines tailored to the underlying genetic factors influencing an individual's disease history. Personalized genetic information can help guide treatment decisions, providing a more detailed picture of prospective therapeutic outcomes (91).

Similarly, using biomarkers in personalized treatments is critical. Biomarkers, which indicate illness progression or treatment response, provide objective measures for monitoring and adjusting customized treatment strategies. Identifying accurate biomarkers specific to distinct subtypes of ALS allows clinicians to personalize therapies to each patient's unique biological traits, resulting in a more precise and effective treatment plan. The development of new imaging technology also helps to personalize ALS treatment (92). Functional magnetic resonance imaging (fMRI) and positron emission tomography (PET) enable thorough assessments of neuronal activity and anatomical changes in the brain and spinal cord. Incorporating imaging data into personalized treatment regimens provides valuable insights into neurodegeneration's geographical and temporal patterns, allowing for the optimization of therapies depending on the precise regions affected in each patient.

Furthermore, personalized techniques in ALS treatment consider lifestyle factors, comorbidities, and individual preferences (93). Tailoring treatment programs to a patient's circumstances results in a more comprehensive and patient-centered approach. This could include personalized food recommendations, exercise routines, and psychosocial assistance, taking into account the various requirements and preferences of individuals living with ALS. As technological breakthroughs continue to transform medical practice, combining artificial intelligence and machine learning holds the potential for improving personalized ALS treatment (94). Predictive algorithms, which analyze large datasets containing genetic, clinical, and imaging information, can help foresee disease trajectories and optimize treatment interventions individually. Finally, the pursuit of personalized treatments in ALS treatment reflects a shift toward precision medicine, which recognizes the intrinsic diversity in disease presentation and development among individuals. The combination of genetic discoveries, biomarkers, imaging technology, and lifestyle considerations ushers in a more personalized and tailored approach to ALS care, giving patients confronting this complex neurodegenerative condition fresh hope for better outcomes and quality of life (95).

To summarize, studying polyphenols and other compounds as prospective therapeutic approaches for ALS has provided valuable insights into their various modes of action and neuroprotective qualities. An overview of significant data emphasizes the multidimensional character of ALS pathophysiology, with polyphenols exhibiting anti-inflammatory, antioxidant, and possibly disease-modifying properties. Identifying certain polyphenols, like those present in green tea, curcumin, and resveratrol, suggests the possibility of targeted therapies that address different components of the disease (96). The implications for future ALS therapies are enormous, highlighting the importance of a comprehensive and personalized approach. The excellent pre-clinical studies imply that combining polyphenols with additional neuroprotective medications could provide a synergistic strategy for combating the complex and heterogeneous character of ALS.

Furthermore, personalized treatment regimens driven by genetic markers, biomarkers, and imaging technologies can improve therapy outcomes and customize interventions to unique patient profiles. As we progress, a resounding call to action for additional study emerges to unravel ALS's complexity and hone polyphenols' therapeutic potential. Large-scale clinical trials are required to validate pre-clinical results and determine the safety and efficacy of polyphenol treatments in humans. Continued research into emerging polyphenols, agents, and creative drug delivery technologies is critical for developing new treatments and a better understanding of ALS pathophysiology.

ABBREVIATIONS

ALS amyotrophic lateral sclerosis
ROS reactive oxygen species

REFERENCES

1. Morris J. ALS and related motor neuron diseases: An overview. *The Neurodiagnostic Journal*. 2015 Sep 1;55(3):180–94.
2. Masrori P, Van Damme P. ALS: A clinical review. *European Journal of Neurology*. 2020 Oct;27(10):1918–29.
3. Saudagar R, Garge L. ALS: An overview. *Journal of Drug Delivery and Therapeutics*. 2019 May 15;9(3):613–6.
4. Oskarsson B, Gendron TF, Staff NP. ALS: An update for 2018. In *Mayo Clinic Proceedings* 2018 Nov 1 (Vol. 93, No. 11, pp. 1617–28). Elsevier.
5. Je G, Keyhanian K, Ghasemi M. Overview of stem cells therapy in ALS. *Neurological Research*. 2021 Aug 3;43(8):616–32.
6. Strong M, Rosenfeld J. ALS: A review of current concepts. *ALS and Other Motor Neuron Disorders*. 2003 Sep 1;4(3):136–43.
7. Khamaysa M, Pradat PF. Status of ALS treatment, insights into therapeutic challenges and dilemmas. *Journal of Personalized Medicine*. 2022 Sep 28;12(10):1601.
8. Salmon K, Genge A. Clinical trials in ALS–current challenges and strategies for future directions. In *Spectrums of ALS: Heterogeneity, Pathogenesis and Therapeutic Directions* 2021 Apr 27 (pp. 161–80). John Wiley & Sons.
9. Soares P, Silva C, Chavarria D, Silva FS, Oliveira PJ, Borges F. Drug discovery and ALS: Emerging challenges and therapeutic opportunities. *Ageing Research Reviews*. 2023 Jan 1;83:101790.
10. Thomasset SC, Berry DP, Garcea G, Marczylo T, Steward WP, Gescher AJ. Dietary polyphenolic phytochemicals – promising cancer chemopreventive agents in humans? A review of their clinical properties. *International Journal of Cancer*. 2007 Feb 1;120(3):451–8.
11. Elmets CA, Singh D, Tubesing K, Matsui M, Katiyar S, Mukhtar H. Cutaneous photoprotection from ultraviolet injury by green tea polyphenols. *Journal of the American Academy of Dermatology*. 2001 Mar 1;44(3):425–32.

12. Nihal M, Roelke CT, Wood GS. Anti-melanoma effects of vorinostat in combination with polyphenolic antioxidant (−)-epigallocatechin-3-gallate (EGCG). *Pharmaceutical Research*. 2010 Jun;27:1103–14.
13. Le Gall L, Anakor E, Connolly O, Vijayakumar UG, Duddy WJ, Duguez S. Molecular and cellular mechanisms affected in ALS. *Journal of Personalized Medicine*. 2020 Aug 25;10(3):101.
14. Peters OM, Ghasemi M, Brown RH. Emerging mechanisms of molecular pathology in ALS. *The Journal of Clinical Investigation*. 2015 May 1;125(5):1767–79.
15. Wood A, Gurfinkel Y, Polain N, Lamont W, Lyn Rea S. Molecular mechanisms underlying TDP-43 pathology in cellular and animal models of ALS and FTLD. *International Journal of Molecular Sciences*. 2021 Apr 29;22(9):4705.
16. Hellman L. Regulation of IgE homeostasis, and the identification of potential targets for therapeutic intervention. *Biomedicine & Pharmacotherapy*. 2007 Jan 1;61(1):34–49.
17. Kwon BK, Borisoff JF, Tetzlaff W. Molecular targets for therapeutic intervention after spinal cord injury. *Molecular Interventions*. 2002 Jul 1;2(4):244.
18. Slevin M, Krupinski J, Badimon L. Controlling the angiogenic switch in developing atherosclerotic plaques: Possible targets for therapeutic intervention. *Journal of Angiogenesis Research*. 2009 Dec;1(1):4.
19. Steinmetz SK. Wife beating: A critique and reformulation of existing theory. *Bulletin of the American Academy of Psychiatry and the Law*. 1978 Jan 1;6(3):322–34.
20. Maiuro RD, Hagar TS, Lin HH, Olson N. Are current state standards for domestic violence perpetrator treatment adequately informed by research? A question of questions. In *Domestic Violence Offenders* 2014 Jan 2 (pp. 21–44). Routledge.
21. Snell-Johns J, Mendez JL, Smith BH. Evidence-based solutions for overcoming access barriers, decreasing attrition, and promoting change with underserved families. *Journal of Family Psychology*. 2004 Mar;18(1):19.
22. Singla RK, Dubey AK, Garg A, Sharma RK, Fiorino M, Ameen SM, Haddad MA, Al-Hiary M. Natural polyphenols: Chemical classification, definition of classes, subcategories, and structures. *Journal of AOAC International*. 2019 Sep 1;102(5):1397–400.
23. Prabhu S, Molath A, Choksi H, Kumar S, Mehra R. Classifications of polyphenols and their potential application in human health and diseases. *International Journal of Physiology, Nutrition and Physical Education*. 2021;6(1):293–301.
24. Truzzi F, Tibaldi C, Zhang Y, Dinelli G, D′ Amen E. An overview on dietary polyphenols and their biopharmaceutical classification system (BCS). *International Journal of Molecular Sciences*. 2021 May 24;22(11):5514.
25. Bertelli A, Biagi M, Corsini M, Baini G, Cappellucci G, Miraldi E. Polyphenols: From theory to practice. *Foods*. 2021 Oct 27;10(11):2595.
26. D'Archivio M, Filesi C, Di Benedetto R, Gargiulo R, Giovannini C, Masella R. Polyphenols, dietary sources and bioavailability. *Annali-Istituto Superiore di Sanita*. 2007 Oct;43(4):348.
27. Ovaskainen ML, Torronen R, Koponen JM, Sinkko H, Hellstrom J, Reinivuo H, Mattila P. Dietary intake and major food sources of polyphenols in Finnish adults. *The Journal of nutrition*. 2008 Mar 1;138(3):562–6.
28. Teng H, Chen L. Polyphenols and bioavailability: An update. *Critical Reviews in Food Science and Nutrition*. 2019 Jul 20;59(13):2040–51.
29. Yoon JH, Baek SJ. Molecular targets of dietary polyphenols with anti-inflammatory properties. *Yonsei Medical Journal*. 2005 Oct 31;46(5):585–96.
30. Yahfoufi N, Alsadi N, Jambi M, Matar C. The immunomodulatory and anti-inflammatory role of polyphenols. *Nutrients*. 2018 Nov 2;10(11):1618.
31. Oliviero F, Scanu A, Zamudio-Cuevas Y, Punzi L, Spinella P. Anti-inflammatory effects of polyphenols in arthritis. *Journal of the Science of Food and Agriculture*. 2018 Mar;98(5):1653–9.
32. Kim H, Castellon-Chicas MJ, Arbizu S, Talcott ST, Drury NL, Smith S, Mertens-Talcott SU. Mango (*Mangifera indica* L.) polyphenols: Anti-inflammatory intestinal microbial health benefits, and associated mechanisms of actions. *Molecules*. 2021 May 6;26(9):2732.
33. Lv QZ, Long JT, Gong ZF, Nong KY, Liang XM, Qin T, Huang W, Yang L. Current state of knowledge on the antioxidant effects and mechanisms of action of polyphenolic compounds. *Natural Product Communications*. 2021 Aug;16(7):1934578X211027745.
34. Belščak-Cvitanović A, Durgo K, Huđek A, Bačun-Družina V, Komes D. Overview of polyphenols and their properties. In Belščak-Cvitanović A, Durgo K, Huđek A, Bačun-Družina V, Komes D (eds.), *Polyphenols: Properties, Recovery, and Applications* 2018 Jan 1 (pp. 3–44). Woodhead Publishing.

35. Rice-Evans C. Implications of the mechanisms of action of tea polyphenols as antioxidants in vitro for chemoprevention in humans. *Proceedings of the Society for experimental Biology and Medicine*. 1999 Apr;220(4):262–6.
36. Szwajgier D, Borowiec K, Pustelniak K. The neuroprotective effects of phenolic acids: Molecular mechanism of action. *Nutrients*. 2017 May 10;9(5):477.
37. del Rosario Campos-Esparza M, Adriana Torres-Ramos M. Neuroprotection by natural polyphenols: Molecular mechanisms. *Central Nervous System Agents in Medicinal Chemistry (Formerly Current Medicinal Chemistry-Central Nervous System Agents)*. 2010 Dec 1;10(4):269–77.
38. Frade JG, Ferreira NR, Barbosa RM, Laranjinha J. Mechanisms of neuroprotection by polyphenols. *Current Medicinal Chemistry-Central Nervous System Agents*. 2005 Dec 1;5(4):307–18.
39. Kovacsova M, Barta A, Parohova J, Vrankova S, Pechanova O. Neuroprotective mechanisms of natural polyphenolic compounds. *Activitas Nervosa Superior Rediviva*. 2010;52(3):181–6.
40. Novak V, Rogelj B, Župunski V. Therapeutic potential of polyphenols in ALS and frontotemporal dementia. *Antioxidants*. 2021 Aug 23;10(8):1328.
41. Casani-Cubel J, Benlloch M, Sanchis-Sanchis CE, Marin R, Lajara-Romance JM, de la Rubia Orti JE. The impact of microbiota on the pathogenesis of ALS and the possible benefits of polyphenols. An overview. *Metabolites*. 2021 Feb 20;11(2):120.
42. Cunha-Oliveira T, Montezinho L, Mendes C, Firuzi O, Saso L, Oliveira PJ, Silva FS. Oxidative stress in ALS: Pathophysiology and opportunities for pharmacological intervention. *Oxidative Medicine and Cellular Longevity*. 2020 Nov 15;2020:1–23. https://doi.org/10.1155/2020/5020340
43. Akter R, Rahman H, Behl T, Chowdhury MA, Manirujjaman M, Bulbul IJ, Elshenaw SE, Tit DM, Bungau S. Prospective role of polyphenolic compounds in the treatment of neurodegenerative diseases. *CNS & Neurological Disorders-Drug Targets (Formerly Current Drug Targets-CNS & Neurological Disorders)*. 2021 Jun 1;20(5):430–50.
44. Esposito E, Rotilio D, Di Matteo V, Di Giulio C, Cacchio M, Algeri S. A review of specific dietary antioxidants and the effects on biochemical mechanisms related to neurodegenerative processes. *Neurobiology of Aging*. 2002 Sep 1;23(5):719–35.
45. Rosado-Ramos R, Godinho-Pereira J, Figueira I, Jardim C, Garcia G, Menezes R. Exploring the benefits of cellular models to uncover bioactive polyphenols for neurodegeneration. *Current Pharmaceutical Design*. 2018 May 1;24(19):2076–106.
46. Zhang X, Hong YL, Xu DS, Feng Y, Zhao LJ, Ruan KF, Yang XJ. A review of experimental research on herbal compounds in ALS. *Phytotherapy Research*. 2014 Jan;28(1):9–21.
47. Pohl F, Kong Thoo Lin P. The potential use of plant natural products and plant extracts with antioxidant properties for the prevention/treatment of neurodegenerative diseases: In vitro, in vivo and clinical trials. *Molecules*. 2018 Dec 11;23(12):3283.
48. Cristina Polidori M, Nelles G. Antioxidant clinical trials in mild cognitive impairment and Alzheimer's disease–challenges and perspectives. *Current Pharmaceutical Design*. 2014 Jun 1;20(18):3083–92.
49. Park HR, Yang EJ. Oxidative stress as a therapeutic target in ALS: Opportunities and limitations. *Diagnostics*. 2021 Aug 26;11(9):1546.
50. Yan L, Guo MS, Zhang Y, Yu L, Wu JM, Tang Y, Ai W, Zhu FD, Law BY, Chen Q, Yu CL. Dietary plant polyphenols as the potential drugs in neurodegenerative diseases: Current evidence, advances, and opportunities. *Oxidative Medicine and Cellular Longevity*. 2022 Feb 21;2022:1–18. https://doi.org/10.1155/2022/123456
51. Moradi SZ, Jalili F, Farhadian N, Joshi T, Wang M, Zou L, Cao H, Farzaei MH, Xiao J. Polyphenols and neurodegenerative diseases: Focus on neuronal regeneration. *Critical Reviews in Food Science and Nutrition*. 2022 May 3;62(13):3421–36.
52. Bhullar KS, Rupasinghe HP. Polyphenols: Multipotent therapeutic agents in neurodegenerative diseases. *Oxidative Medicine and Cellular Longevity*. 2013 Oct;2013:1–15. https://doi.org/10.1155/2013/891748
53. Mohseni M, Sahebkar A, Askari G, Johnston TP, Alikiaii B, Bagherniya M. The clinical use of curcumin on neurological disorders: An updated systematic review of clinical trials. *Phytotherapy Research*. 2021 Dec;35(12):6862–82.
54. Maqbool M, Zehravi M. Neuroprotective role of polyphenols in treatment of neurological disorders: A review. *Interventional Pain Medicine and Neuromodulation*. 2021 Dec 31;1(1).
55. Sun AY, Wang Q, Simonyi A, Sun GY. Resveratrol as a therapeutic agent for neurodegenerative diseases. *Molecular Neurobiology*. 2010 Jun;41:375–83.
56. Rocha-González HI, Ambriz-Tututi M, Granados-Soto V. Resveratrol: A natural compound with pharmacological potential in neurodegenerative diseases. *CNS Neuroscience & Therapeutics*. 2008 Sep;14(3):234–47.

57. Yadav E, Yadav P, Khan MM, Singh H, Verma A. Resveratrol: A potential therapeutic natural polyphenol for neurodegenerative diseases associated with mitochondrial dysfunction. *Frontiers in Pharmacology*. 2022 Sep 16;13:922232. https://doi.org/10.3389/fphar.2022.922232
58. Petrella C, Di Certo MG, Gabanella F, Barbato C, Ceci FM, Greco A, Ralli M, Polimeni A, Angeloni A, Severini C, Vitali M. Mediterranean diet, brain and muscle: Olive polyphenols and resveratrol protection in neurodegenerative and neuromuscular disorders. *Current Medicinal Chemistry*. 2021 Nov 1;28(37):7595–613.
59. Mandel SA, Amit T, Weinreb O, Youdim MB. Understanding the broad-spectrum neuroprotective action profile of green tea polyphenols in aging and neurodegenerative diseases. *Journal of Alzheimer's Disease*. 2011 Jan 1;25(2):187–208.
60. Mandel S, Weinreb O, Amit T, Youdim MB. Cell signaling pathways in the neuroprotective actions of the green tea polyphenol (-)-epigallocatechin-3-gallate: Implications for neurodegenerative diseases. *Journal of Neurochemistry*. 2004 Mar;88(6):1555–69.
61. Barros MP, Poppe SC, Bondan EF. Neuroprotective properties of the marine carotenoid astaxanthin and omega-3 fatty acids, and perspectives for the natural combination of both in krill oil. *Nutrients*. 2014 Mar 24;6(3):1293–317.
62. Shirooie S, Nabavi SF, Dehpour AR, Belwal T, Habtemariam S, Argüelles S, Sureda A, Daglia M, Tomczyk M, Sobarzo-Sanchez E, Xu S. Targeting mTORs by omega-3 fatty acids: A possible novel therapeutic strategy for neurodegeneration? *Pharmacological Research*. 2018 Sep 1;135:37–48.
63. Mazza M, Pomponi M, Janiri L, Bria P, Mazza S. Omega-3 fatty acids and antioxidants in neurological and psychiatric diseases: An overview. *Progress in Neuro-Psychopharmacology and Biological Psychiatry*. 2007 Jan 30;31(1):12–26.
64. Darvesh AS, Carroll RT, Bishayee A, Geldenhuys WJ, Van der Schyf CJ. Oxidative stress and Alzheimer's disease: Dietary polyphenols as potential therapeutic agents. *Expert Review of Neurotherapeutics*. 2010 May 1;10(5):729–45.
65. Morén C, deSouza RM, Giraldo DM, Uff C. Antioxidant therapeutic strategies in neurodegenerative diseases. *International Journal of Molecular Sciences*. 2022 Aug 19;23(16):9328.
66. İçer ME, Arslan N, Karadağ G. Effects of vitamin E on neurodegenerative diseases: An update. *Acta Neurobiologiae Experimentalis*. 2021;81.
67. Bhullar KS, Rupasinghe HP. Polyphenols: Multipotent therapeutic agents in neurodegenerative diseases. *Oxidative Medicine and Cellular Longevity*. 2013 Oct;2013.
68. Patel BP, Hamadeh MJ. Nutritional and exercise-based interventions in the treatment of ALS. *Clinical Nutrition*. 2009 Dec 1;28(6):604–17.
69. Pogačnik L, Ota A, Poklar Ulrih N. An overview of crucial dietary substances and their modes of action for prevention of neurodegenerative diseases. *Cells*. 2020 Feb 28;9(3):576.
70. Carrera-Juliá S, Moreno ML, Barrios C, de la Rubia Ortí JE, Drehmer E. Antioxidant alternatives in the treatment of ALS: A comprehensive review. *Frontiers in Physiology*. 2020 Feb 6;11:63.
71. Bhullar KS, Rupasinghe HP. Polyphenols: Multipotent therapeutic agents in neurodegenerative diseases. *Oxidative Medicine and Cellular Longevity*. 2013 Oct;2013.
72. Oppedisano F, Maiuolo J, Gliozzi M, Musolino V, Carresi C, Nucera S, Scicchitano M, Scarano F, Bosco F, Macrì R, Ruga S. The potential for natural antioxidant supplementation in the early stages of neurodegenerative disorders. *International Journal of Molecular Sciences*. 2020 Apr 9;21(7):2618.
73. Mitra S, Tareq AM, Das R, Emran TB, Nainu F, Chakraborty AJ, Ahmad I, Tallei TE, Idris AM, Simal-Gandara J. Polyphenols: A first evidence in the synergism and bioactivities. *Food Reviews International*. 2023 Aug 25;39(7):4419–41.
74. Lewandowska U, Gorlach S, Owczarek K, Hrabec E, Szewczyk K. Synergistic interactions between anticancer chemotherapeutics and phenolic compounds and anticancer synergy between polyphenols. *Advances in Hygiene and Experimental Medicine*. 2014 Jan 2;68:528–40.
75. Daglia M. Polyphenols as antimicrobial agents. *Current Opinion in Biotechnology*. 2012 Apr 1;23(2):174–81.
76. Patra S, Pradhan B, Nayak R, Behera C, Das S, Patra SK, Efferth T, Jena M, Bhutia SK. Dietary polyphenols in chemoprevention and synergistic effect in cancer: Clinical evidences and molecular mechanisms of action. *Phytomedicine*. 2021 Sep 1;90:153554.
77. Mitsumoto H, Olney RK. Drug combination treatment in patients with ALS: Current status and future directions. *Neurology*. 1996 Oct 1;47(4 Suppl 2):103S–7S.
78. Gordon PH, Cheung YK, Levin B, Andrews H, Doorish C, Macarthur RB, Montes J, Bednarz K, Florence J, Rowin J, Boylan K. A novel, efficient, randomized selection trial comparing combinations of drug therapy for ALS. *ALS*. 2008 Jan 1;9(4):212–22.

79. Sun Y, Li X, Bedlack R. An evaluation of the combination of sodium phenylbutyrate and taurursodiol for the treatment of ALS. *Expert Review of Neurotherapeutics*. 2023 Jan 2;23(1):1–7.
80. Wang J, Hu WW, Jiang Z, Feng MJ. Advances in treatment of neurodegenerative diseases: Perspectives for combination of stem cells with neurotrophic factors. *World Journal of Stem Cells*. 2020 May 5;12(5):323.
81. Lambden S, Summers C. Bridging the translational gap: The challenges of novel drug development in critical care. *Annual Update in Intensive Care and Emergency Medicine* 2017. 2017:375–88.
82. Seyhan AA. *Lost in Translation-Bridging the pre-clinical and clinical worlds concepts, examples, successes, and failures in translational medicine* 2019. Elsevier.
83. Seyhan AA. Lost in translation: The valley of death across pre-clinical and clinical divide–identification of problems and overcoming obstacles. *Translational Medicine Communications*. 2019 Dec;4(1):1–9.
84. Banik A, Brown RE, Bamburg J, Lahiri DK, Khurana D, Friedland RP, Chen W, Ding Y, Mudher A, Padjen AL, Mukaetova-Ladinska E. Translation of pre-clinical studies into successful clinical trials for Alzheimer's disease: What are the roadblocks and how can they be overcome? *Journal of Alzheimer's Disease*. 2015 Jan 1;47(4):815–43.
85. Hinder M, Hartl D. Translational medicine–the bridging discipline. Role and tools in the drug development process. In *Principles of Biomedical Sciences and Industry: Translating Ideas into Treatments* 2022 Oct 3 (pp. 119–38). Springer Nature.
86. Kashi DS, Shabir A, Da Boit M, Bailey SJ, Higgins MF. The efficacy of administering fruit-derived polyphenols to improve health biomarkers, exercise performance and related physiological responses. *Nutrients*. 2019 Oct 7;11(10):2389.
87. Gupta RC, Srivastava A, Sinha A, Lall R. Biomarkers of foods and Nutraceuticals: Applications in efficacy, safety, and toxicity. *Nutraceuticals in Veterinary Medicine*. 2019:693–710.
88. Luo H, Cox SB, Gao W, Yu J, Tang L, Wang JS. Metabolic profiling in validation of plasma biomarkers for green tea polyphenols. *Metabolomics*. 2006 Dec;2:235–41.
89. Zhao D, Simon JE, Wu Q. A critical review on grape polyphenols for neuroprotection: Strategies to enhance bioefficacy. *Critical Reviews in Food Science and Nutrition*. 2020 Feb 21;60(4):597–625.
90. Kuppusamy S, Thavamani P, Megharaj M, Naidu R. Bioremediation potential of natural polyphenol rich green wastes: A review of current research and recommendations for future directions. *Environmental Technology & Innovation*. 2015 Oct 1;4:17–28.
91. Pradhan B, Nayak R, Bhuyan PP, Patra S, Behera C, Sahoo S, Ki JS, Quarta A, Ragusa A, Jena M. Algal phlorotannins as novel antibacterial agents with reference to the antioxidant modulation: Current advances and future directions. *Marine Drugs*. 2022 Jun 18;20(6):403.
92. Issa AY, Volate SR, Wargovich MJ. The role of phytochemicals in inhibition of cancer and inflammation: New directions and perspectives. *Journal of Food Composition and Analysis*. 2006 Aug 1;19(5):405–19.
93. Benkler C, Offen D, Melamed E, Kupershmidt L, Amit T, Mandel S, Youdim MB, Weinreb O. Recent advances in ALS research: Perspectives for personalized clinical application. *EPMA Journal*. 2010 Jun;1:343–61.
94. DeLoach A, Cozart M, Kiaei A, Kiaei M. A retrospective review of the progress in ALS drug discovery over the last decade and a look at the latest strategies. *Expert Opinion on Drug Discovery*. 2015 Oct 3;10(10):1099–118.
95. Benkler C, Offen D, Melamed E, Amit T, Mandel S, Youdim MB, Weinreb O. Recent advances in ALS research: Perspectives for personalized clinical application. In *Neurodegenerative Diseases: Integrative PPPM Approach as the Medicine of the Future* 2013 (pp. 235–74). Taylor & Francis.
96. D'Amico E, Patti F, Zanghì A, Zappia M. A personalized approach in progressive multiple sclerosis: The current status of disease modifying therapies (DMTs) and future perspectives. *International Journal of Molecular Sciences*. 2016 Oct 17;17(10):1725.

4 Natural Ingredients for Friedreich's Ataxia

Madhu Tanya Singh,[1] *Sai Varshini Magham,*[1,2]
Raja Shekhar Nunavath,[3] *Sai Kiran S. S. Pindiprolu,*[4]
and Praveen Thaggikuppe Krishnamurthy[1]
[1]Department of Pharmacology, JSS College of Pharmacy, JSS Academy of Higher Education & Research, Ooty, The Nilgiris, Tamil Nadu, India
[2]Department of Pharmacology, Vignan Pharmacy College, Vadlamudi, Guntur, Andhra Pradesh, India
[3]Department of Pharmaceutical Analysis, JSS College of Pharmacy, JSS Academy of Higher Education & Research, Ooty, The Nilgiris, Tamil Nadu, India
[4]Aditya Pharmacy College, Surampalem, Andhra Pradesh, India

4.1 INTRODUCTION

Friedreich's ataxia (FRDA) is one of the most challenging inherited neurodegenerative disorders, characterized by progressive deterioration of the nervous system, cardiac complications, and impaired muscle coordination (1). As an autosomal recessive disease, FRDA is primarily seen in childhood or adolescence, affecting approximately 1 in 50,000 individuals (2). The molecular basis of FRDA lies in the reduced expression of frataxin, a mitochondrial protein crucial for iron homeostasis and cellular energy production, resulting from expanded GAA triplet repeats in the FXN gene (3).

The pathophysiological cascade initiated by frataxin deficiency encompasses multiple cellular perturbations, including mitochondrial dysfunction, oxidative stress accumulation, impaired iron-sulfur cluster biosynthesis, and disrupted cellular energy metabolism (4). These molecular alterations culminate in progressive neurodegeneration, particularly affecting the dorsal root ganglia, spinocerebellar tracts, and heart tissue, leading to the characteristic clinical manifestations of ataxia, loss of proprioception, and hypertrophic cardiomyopathy (5).

In recent years, there has been a paradigm shift toward exploring natural ingredients as potential therapeutic agents for FRDA, driven by their diverse biochemical properties, multiple mechanisms of action, and generally favourable safety profiles (6). Natural compounds offer several advantages, including their ability to target multiple pathological pathways simultaneously, their potential for synergistic interactions, and their long history of traditional use in various medical systems (7). The scientific rationale for investigating natural ingredients in FRDA management is further strengthened by emerging evidence of their antioxidant, anti-inflammatory, and neuroprotective properties, which directly address key pathological features of the disease (8).

Research has identified several promising natural compounds that may influence FRDA pathogenesis through various mechanisms. These include potent antioxidants such as coenzyme Q_{10}, vitamin E, alpha-lipolic acid, and resveratrol, capable of neutralizing reactive oxygen species, compounds that enhance mitochondrial function and energy production, agents that modulate iron homeostasis, and molecules that potentially increase frataxin expression (9). Additionally, natural ingredients often possess pleiotropic effects, potentially addressing multiple aspects of

DOI: 10.1201/9781003478652-5

FRDA pathology simultaneously, which may provide advantages over single-target therapeutic approaches (10).

The complexity of FRDA pathophysiology, coupled with the multifaceted nature of natural compounds, necessitates a comprehensive understanding of their therapeutic potential and mechanisms of action. This review aims to critically evaluate the current evidence supporting the use of natural ingredients in FRDA.

4.2 CLINICAL FEATURES OF FRDA

Nikolaus Friedreich first described FRDA in 1863 as a condition characterized by degenerative atrophy of the posterior columns of the spinal cord. He identified key clinical features of the disease, including ataxia, sensory neuropathy, scoliosis, foot deformities, and cardiomyopathy (11). The genetic basis of FRDA was discovered in 1996, which significantly advanced the understanding of its clinical spectrum (12).

Symptoms of FRDA typically appear between the ages of 10 and 15. Early signs often include gait instability and clumsiness, which gradually progress to trunk and limb ataxia, severely affecting daily functioning. Most patients require wheelchair assistance approximately 11 to 15 years after the onset of symptoms (13). The rate of disease progression varies, with earlier onset usually linked to a faster decline. Atypical forms of FRDA, such as late-onset (symptoms beginning after age 25) and very late-onset (symptoms starting after age 40), tend to progress more slowly (14). The average age of death is 36.5 years (ranging from 12 to 87 years), with cardiac dysfunction being the leading cause of mortality, accounting for 59% of deaths (15).

The severity of FRDA has been assessed using various clinical rating scales. Initially, the International Cooperative Ataxia Rating Scale (ICARS) was employed (16). However, a 2005 evaluation recommended using only the total ICARS score while discontinuing its subscales due to limited reliability (17). Subsequently, the Scale for the Assessment and Rating of Ataxia and Friedreich's Ataxia Rating Scale (FARS) were introduced (18, 19). A modified version of FARS (mFARS) was developed recently, focusing on patients' functional abilities while eliminating certain items. This version has been validated in a study involving over a thousand participants (20, 21).

This progression in diagnostic tools highlights the ongoing efforts to enhance the accuracy and utility of clinical assessments for FRDA.

4.3 NEUROLOGICAL FEATURES

The neurological manifestations of FRDA stem from changes in both the central and peripheral nervous systems. Ataxia, the defining symptom, arises from peripheral sensory neuropathy, spinocerebellar degeneration, and dysfunction of the cerebellum and vestibular system (2). A prominent feature of FRDA is the degeneration of the dorsal root ganglia, marked by smaller neurons and an increase in satellite cells. The atrophy of large myelinated fibers, combined with the preservation of thin myelinated axons in the dorsal roots, contributes to sensory neuropathy and loss of posterior column function, resulting in reduced or absent sensory action potentials (1, 22). Clinically, this manifests as impaired proprioception, diminished reflexes (especially in the lower limbs), and loss of sensory modalities such as vibration and light touch (23, 24).

Degeneration of the pyramidal tracts leads to muscle weakness and hypotonia, indicating motor nerve involvement (25). Signs like the Babinski reflex and occasional spasticity further highlight motor pathway degeneration, which can cause pain, contractures, and spasms (1). In the central nervous system, cerebellar dentate nucleus atrophy is a primary feature, disrupting GABAergic and glutamatergic motor control. Atrophy of the dorsal nuclei of Clarke also impairs spinocerebellar transmission (26). Magnetic resonance imaging studies show reduced gray and white matter in the deep cerebellar nuclei and brainstem (27). Dysarthria, associated with cerebellar damage, worsens as the disease progresses, often making speech unintelligible in advanced stages (28).

The visual and auditory systems are also affected in FRDA. Optic nerve involvement, abnormal visual evoked potentials, and thinning of the retinal nerve fiber layer are notable findings (29). Oculomotor abnormalities, such as fixation instability (e.g., square-wave jerks) and occasional nystagmus, are common. Hearing impairments result from neural conduction deficits in central auditory pathways (30). Dysphagia, a late-stage complication, may require gastrostomy feeding, with aspiration pneumonia being a significant cause of death (31).

4.4 NON-NEUROLOGICAL FEATURES

Although FRDA is classified as a neurological disorder, cardiomyopathy is the leading cause of mortality and affects the majority of patients. Studies involving cardiac imaging in 205 individuals with FRDA revealed that neurological scales do not reliably predict the severity of cardiac dysfunction, and an earlier onset of symptoms is strongly associated with worse cardiac outcomes (32). Notably, patients who succumb to cardiac dysfunction often exhibit hypertrophic traits early in the disease course (18).

A hallmark of FRDA-associated cardiomyopathy is chronic myocarditis, characterized by inflammatory cell infiltration, monocyte adhesion to cardiomyocytes, and myocardial fiber necrosis (33). Left ventricular hypertrophy is a common structural abnormality, though ejection fraction typically remains preserved (34). Electrocardiographic abnormalities, such as T-wave inversion, ST-segment changes, and supraventricular arrhythmias (most notably atrial fibrillation), are frequent and carry negative prognostic implications in advanced stages (35, 36).

FRDA patients are also at increased risk for diabetes mellitus and glucose intolerance, typically developing years after the onset of neurological symptoms. This is attributed to both insulin resistance and β-cell dysfunction caused by pancreatic β-cell apoptosis (37, 38). Abdominal adiposity, associated with impaired β-cell function and increased lipogenesis, may exacerbate insulin resistance, further compounded by reduced physical activity (39, 40).

Orthopedic complications, such as scoliosis and foot deformities like pes cavus and talipes equinovarus, are common in FRDA and negatively impact stability and mobility (41).

4.5 NATURAL ANTIOXIDANTS FOR FRDA

Numerous natural compounds widely found in fruits, vegetables, and plant-based beverages have gained significant interest as potential therapeutic agents for FRDA. These bioactive molecules are recognized for their antioxidant, anti-inflammatory, and neuroprotective properties. By scavenging free radicals, modulating redox pathways, and influencing mitochondrial function, polyphenols may address the underlying oxidative stress and mitochondrial dysfunction central to the pathology of FRDA.

Recent studies have explored the potential of various antioxidants such as coenzyme Q10 and vitamin E, EGb-761, polyunsaturated fatty acids (PUFAs), and polyphenols, such as resveratrol and catechin, for the management of FRDA. These compounds have demonstrated the ability to upregulate mitochondrial biogenesis, enhance frataxin expression, and mitigate oxidative damage. Despite promising results, challenges such as limited bioavailability and systemic delivery remain significant hurdles in translating these findings into clinical applications.

4.6 COENZYME Q10 AND VITAMIN E

Coenzyme Q10 (ubiquinone), a small, lipophilic molecule located in the inner mitochondrial membrane, is known for its antioxidant properties. It scavenges free radicals, preventing oxidative damage to cellular structures and supporting the regeneration of other antioxidants, including vitamins C and E (52, 53). Studies have demonstrated that ubiquinone and its derivatives have significant effects when applied to fibroblasts derived from FRDA patients (54).

S. No.	Natural Compound	Clinical Trial Phase	Study Design	Dose Used	Current Status	Study Outcome	Reference
1.	CoQ_{10} + Vitamin E			300–600 mg/day	No further investigation	Improvement in neurological function and bioenergetics	(42)
2.	Idebenone			2.5–7.5 mg/kg/day 360–2250 mg/day	No further investigation	Variable outcomes in neurological and cardiac functions	(42)
3.	A0001	Phase II	Double-blind, placebo-controlled four weeks	1–1.5 g/day	No further investigation	Significant improvements in FARS scores	(43)
4.	EGb-761	Phase II	Randomized, placebo-controlled, three-month double-blind	240 mg/day	No further investigation	No significant improvement	(44)
5.	Epicatechins	Phase II	Open-label 24-weeks	75–150 mg/day	No further investigation	No significant improvements in neurological functions	(45)
6.	Resveratrol	Phase I/II	Open-label 12 weeks Pilot study	1–5 g/day	**NCT03933163** Enhanced formulation of resveratrol is being studied **Phase:** Phase II **Study design:** double-blind, placebo-controlled two-period crossover trial	No significant difference in frataxin levels. Significant improvement in FARS and ICARS levels. Gastrointestinal side effects were observed	(46, 47)
7.	ALCAR	Phase II	Placebo control Four months triple-phase crossover	2–3 g/day	**NCT01921868** Open-label trial evaluating the cardiovascular outcomes of ALCAR (acetyl-L-carnitine) over a 24-month period is currently underway, with results pending.	No significant improvement in ICARS scores	(48, 49)
8.	RT001	Phase I/II	Double-blind, placebo-controlled trial	1.8–8.64 g/day	**NCT04102501** **Phase:** Phase III **Study design:** Randomized, placebo-controlled, double-blinded **Status:** Ongoing	Significant improvement in neurological function	(50, 51)

Nearly two decades ago, an open-label study was conducted involving 10 FRDA patients to evaluate the effects of a combined antioxidant therapy using coenzyme Q10 and vitamin E. The study assessed outcomes such as ICARS scores and echocardiographic parameters. After six months, no consistent clinical benefits were observed (55). However, the authors noted bioenergetic improvements and attributed the lack of significant effects to the short study duration. Consequently, the same cohort of patients was monitored for 47 months in a longitudinal follow-up study. While posture and gait symptoms continued to deteriorate, ICARS scores remained stable (56).

A two-year randomized double-blind trial was conducted by the same research team, involving 50 patients with FRDA. Participants were divided into two groups receiving either high or low doses of a combined coenzyme Q10 and vitamin E therapy. Notably, 49% of the individuals who completed the study (21 out of 43) demonstrated a reduction in disease progression and showed clinical improvements, as measured by higher ICARS scores. These benefits were observed irrespective of the dosage administered. Additional analysis highlighted a significant relationship between the treatment's effectiveness and the patients' initial levels of coenzyme Q10 and vitamin E (57).

Idebenone, a synthetic derivative of coenzyme Q10, offers advantages such as reduced molecular weight and enhanced water solubility, which contribute to its superior bioavailability. This compound acts as an antioxidant, neutralizing reactive oxygen species (ROS) and inhibiting lipid peroxidation. Additionally, it supports mitochondrial function by promoting electron transport in the mitochondrial respiratory chain, ultimately boosting ATP production (58). Studies have demonstrated its potential to improve longevity, locomotion, and aconitase activity in FRDA fly models and to delay cardiac symptoms in FRDA mouse models (59, 60). The first clinical study on idebenone, conducted two decades ago, reported promising cardiac improvements in three FRDA patients (61). Over the following decade, additional studies were performed to assess its effects on larger FRDA cohorts. While many of these studies reported cardiac improvements, the outcomes related to neurological parameters were inconsistent. Furthermore, four randomized, placebo-controlled phase III clinical trials (MICONOS, IONIA, NICOSIA, and PROTI) failed to provide conclusive evidence of idebenone's efficacy in improving neurological or cardiac function in different FRDA patient groups (62–66).

Despite these mixed results, coenzyme Q10 and idebenone continue to be widely used by FRDA patients as nutraceutical supplements due to their good safety profiles and minimal side effects.

4.7 EGB-761

EGb-761, a standardized extract of *Ginkgo biloba*, has been extensively studied for its potential benefits in Alzheimer's disease. Although preclinical studies using in vitro and in vivo Alzheimer's models have highlighted its neuroprotective properties, clinical trials have failed to demonstrate significant benefits in patients, sparking considerable debate (67).

A phase II randomized, placebo-controlled, double-blind trial was conducted to investigate the efficacy of EGb-761 in 22 participants with FRDA. However, the study did not reveal any significant differences between the treatment and placebo groups. The authors attributed the inconclusive results to the small sample size, which hindered the statistical analysis, leading to the study's dismissal (44).

4.8 POLYUNSATURATED FATTY ACIDS

PUFAs are essential fatty acids that serve as crucial components in maintaining cell membrane structural integrity. Research has demonstrated their importance in reducing inflammation, supporting mitochondrial function, decreasing the production of ROS, and reducing stress on the endoplasmic reticulum (68). However, PUFAs have a significant vulnerability – their carbon–hydrogen bonds can be easily damaged by ROS. When this occurs, it creates additional ROS, initiating a destructive cycle that leads to increased lipid peroxidation and cellular membrane damage (69).

Research conducted by Cotticelli et al. examined FRDA cell models and showed promising results using modified fatty acids, specifically linoleic acid and a deuterated form of α-linolenic acid. The modification process, which replaces standard hydrogen atoms (^1H) with deuterium (^2H), creates more chemically stable molecules. This stability enhancement reduces their toxic potential and makes them more resistant to lipid peroxidation damage (70). Studies using fibroblasts from KIKO and YG8R mouse models have shown that deuterated PUFAs (dPUFAs) can effectively prevent both lipid peroxidation and cellular death (71).

Based on these encouraging findings, clinical trials were initiated. One such trial examined RT001, a deuterated version of linoleic acid developed by Retrotope. This compound was evaluated in a 28-day clinical study that included 18 participants with FRDA, using a double-blind, placebo-controlled design. The results indicated that RT001 was both safe and well-tolerated. Participants in the RT001 group showed improved performance in cardiopulmonary exercise testing compared to the placebo group, along with modest trends toward neurological recovery (50).

These encouraging findings have prompted further exploration. A phase III, randomized, double-blind, placebo-controlled trial is currently recruiting FRDA patients to evaluate RT001. Cardiopulmonary exercise testing and modified Friedreich's Ataxia Rating Scale (mFARS) scores will be evaluated as primary outcomes (51).

4.9 RESVERATROL

Resveratrol is a polyphenolic phytoalexin present in various plants, known for its anti-inflammatory, antioxidant, neuroprotective, and anti-apoptotic properties. It plays a role in multiple cellular processes, such as activating NRF2, NAD-dependent deacetylase, sirtuin-1 (SIRT1), and PGC-1α, which contribute to improved mitochondrial biogenesis and enhanced antioxidant defences (72).

Studies have shown that resveratrol can effectively increase frataxin expression, a critical target in FRDA treatment. Research has revealed varied effects of resveratrol on frataxin expression across different cell types. Studies by Li et al. demonstrated promising results, showing increased frataxin levels after resveratrol treatment in multiple models: HeLa cells, lymphoblasts, brain tissue from YG8R FRDA mice, and obtained from FRDA patients (73). Additional research with FRDA patient fibroblasts produced similar positive outcomes (74). However, the effectiveness of resveratrol showed limitations when tested in more specialized cell types. When researchers used patient-derived induced pluripotent stem cells (iPSCs) and differentiated them into mesenchymal and neuronal cells, the response to resveratrol treatment was minimal or absent (74). This variation in cellular response suggests that resveratrol alone may not be sufficient as a primary treatment option for FRDA, given its inconsistent effectiveness across different cell types.

Clinical trials have produced inconsistent outcomes. A 12-week open-label study involving 27 participants tested two dose levels of resveratrol (low and high) but found no significant changes in frataxin levels. Notably, the high-dose group exhibited a decrease in the lipid peroxidation biomarker F2-isoprostane. Improvements in neurological symptoms were also noted, as measured by FARS and ICARS scores. However, high doses were associated with frequent gastrointestinal side effects, highlighting the need to improve the compound's tolerability and bioavailability (49).

To overcome these challenges, a randomized double-blind placebo-controlled crossover clinical trial is currently investigating the effectiveness of modified micronized resveratrol. The primary outcome measure is the mFARS scale, which evaluates disease progression (47). This research aims to maximize resveratrol's therapeutic benefits for FRDA while reducing adverse effects.

4.10 CATECHINS

Catechins, a class of polyphenolic flavonoids present in various fruits, red wine, and cocoa, are known for their antioxidant properties. These compounds function through multiple mechanisms, including scavenging ROS, chelating metal ions, enhancing the activity of antioxidant enzymes, and inhibiting pro-oxidant enzymes. In vitro studies have demonstrated the neuroprotective potential of catechins by reducing oxidative stress and promoting mitochondrial function, highlighting their possible role in mitigating neurodegenerative processes (75).

One specific catechin, (+)-epicatechin, has been investigated for its therapeutic potential in FRDA. A 24-week open-label study involving ten participants evaluated its safety and efficacy. The primary outcomes assessed included changes in the FARS and ventricular hypertrophy. Recent findings indicated that (+)-epicatechin was well-tolerated and led to improvements in cardiac function, as measured by an increase in left ventricular ejection fraction. No significant changes were observed in FARS or modified FARS (mFARS) scores (45).

4.11 CARNITINE FOR FRDA

L-Carnitine, also referred to as β-hydroxy-γ-trimethylaminobutyric acid, is a naturally occurring compound found in most mammalian tissues. Its main role is to assist in the transportation of long-chain fatty acids into mitochondria, where they undergo β-oxidation to produce ATP (76). Both L-carnitine and its acetylated derivative, acetyl-L-carnitine (ALCAR), are present in numerous human tissues, such as the brain and peripheral blood. ALCAR is often regarded as a more bioavailable version of L-carnitine because the acetyl group improves its water solubility and cellular absorption (Figure 4.1). As a mitochondrial cofactor, ALCAR has demonstrated potential as a neuroprotective agent, supported by its inclusion in multiple clinical trials. It aids energy metabolism by supplying critical chemical groups and displays anti-inflammatory and antioxidant properties (77).

Recent studies have highlighted the potential therapeutic applications of L-carnitine and acetyl-L-carnitine (ALCAR) in treating nervous system injuries and neurodegenerative conditions (78–80). A placebo-controlled, triple-phase crossover trial involving individuals with FRDA investigated the effects of these compounds on mitochondrial ATP production and neurological function over a four-month period (48). Phosphocreatine recovery was utilized as an indicator of mitochondrial ATP synthesis. Participants were given either 3 g/day of L-carnitine or 6.75 g/day of creatine. The findings demonstrated that L-carnitine significantly improved mitochondrial ATP production, whereas creatine had no measurable impact, as reflected by unchanged ICARS scores and

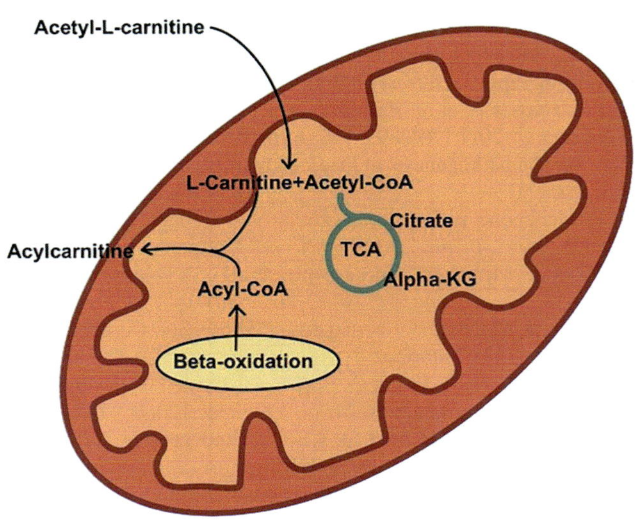

FIGURE 4.1 ALCAR metabolism in the brain.

echocardiographic data. These results suggest that L-carnitine could be a viable treatment option for FRDA. Additionally, a study conducted by the University of Florida examined the effects of ALCAR on cardiac function and gait ataxia symptoms in FRDA patients. Participants were administered 2 g/day of ALCAR over 24 months, with the study anticipated to conclude in 2017 (49). However, the outcomes of this research have yet to be published.

4.12 CONCLUSION

The clinical and therapeutic landscape of FRDA underscores the complexity of this progressive disorder and the challenges inherent in its management. The clinical manifestations, ranging from neurological impairments and cardiomyopathy to metabolic and orthopedic complications, highlight the multifaceted nature of FRDA. Advancements in clinical assessments, such as the development of mFARS, have improved our ability to monitor disease progression and therapeutic outcomes. Despite these strides, effective treatments remain elusive, with current therapeutic approaches offering limited success. The exploration of natural antioxidants, including coenzyme Q10, idebenone, EGb-761, PUFAs, resveratrol, and catechins, demonstrates promise in mitigating oxidative stress and improving mitochondrial function, yet these efforts are hindered by challenges in bioavailability, systemic delivery, and inconsistent clinical outcomes. Moving forward, a combination of enhanced delivery systems, advanced biomarker discovery, and robust clinical trials will be pivotal in translating these promising findings into viable treatment options for FRDA, offering hope for improved patient outcomes.

REFERENCES

1. Koeppen AH, Mazurkiewicz JE. Friedreich ataxia: Neuropathology revised. *J Neuropathol Exp Neurol*. 2013;72(2):78–90. https://doi.org/10.1097/NEN.0b013e31827e5762
2. Delatycki MB, Corben LA. Clinical features of Friedreich ataxia. *J Child Neurol*. 2012 Sep;27(9):1133–7.
3. Campuzano V, Montermini L, Moltò MD, Pianese L, Cossée M, Cavalcanti F, et al. Friedreich's ataxia: Autosomal recessive disease caused by an intronic GAA triplet repeat expansion. *Science*. 1996 Mar 8;271(5254):1423–7.
4. Abeti R, Baccaro A, Esteras N, Giunti P. Novel Nrf2-inducer prevents mitochondrial defects and oxidative stress in Friedreich's ataxia models. *Front Cell Neurosci*. 2018;12:188.
5. Cook A, Giunti P. Friedreich's ataxia: Clinical features, pathogenesis and management. *Br Med Bull*. 2017 Dec 1;124(1):19–30.
6. Lynch DR, Farmer JM, Balcer LJ, Wilson RB. Friedreich ataxia: Effects of genetic understanding on clinical evaluation and therapy. *Arch Neurol*. 2002 May;59(5):743–7.
7. Strawser C, Schadt K, Hauser L, et al. Pharmacological therapeutics in Friedreich ataxia: The present state. *Expert Rev Neurother*. 2017;17(9):895–907. https://doi.org/10.1080/14737175.2017.1356721
8. Pandolfo M, Pastore A. The pathogenesis of Friedreich ataxia and the structure and function of frataxin. *J Neurol*. 2009 Mar 1;256(1):9–17.
9. Schiff M, Rustin P. Idebenone in Friedreich ataxia and Leber's hereditary optic neuropathy: Close mechanisms, similar therapy?. *Brain*. 2016;139(Pt 7):e39. https://doi.org/10.1093/brain/aww085
10. González-Cabo P, Palau F. Mitochondrial pathophysiology in Friedreich's ataxia. *J Neurochem*. 2013 Aug;126 Suppl 1:53–64.
11. Bürk K. Friedreich Ataxia: Current status and future prospects. *Cerebellum Ataxias*. 2017 Apr 7;4:4. https://doi.org/10.1186/s40673-017-0062-x
12. Schöls L, Amoiridis G, Przuntek H, Frank G, Epplen JT, Epplen C. Friedreich's ataxia. Revision of the phenotype according to molecular genetics. *Brain*. 1997 Dec;120(Pt 12):2131–40.
13. Delatycki MB, Paris DB, Gardner RJ, et al. Clinical and genetic study of Friedreich ataxia in an Australian population. *Am J Med Genet*. 1999;87(2):168–74.
14. Lecocq C, Charles P, Azulay JP, et al. Delayed-onset Friedreich's ataxia revisited. *Mov Disord*. 2016;31(1):62–9. https://doi.org/10.1002/mds.26382
15. Tsou AY, Paulsen EK, Lagedrost SJ, et al. Mortality in Friedreich ataxia. *J Neurol Sci*. 2011;307(1–2):46–9. https://doi.org/10.1016/j.jns.2011.05.023

16. Trouillas P, Takayanagi T, Hallett M, Currier RD, Subramony SH, Wessel K, et al. International cooperative ataxia rating scale for pharmacological assessment of the cerebellar syndrome. The Ataxia Neuropharmacology Committee of the world federation of neurology. *J Neurol Sci*. 1997 Feb 12;145(2):205–11.
17. Cano SJ, Hobart JC, Hart PE, Korlipara LV, Schapira AH, Cooper JM. International cooperative ataxia rating scale (ICARS): Appropriate for studies of Friedreich's ataxia?. *Mov Disord*. 2005;20(12):1585–91. https://doi.org/10.1002/mds.20651
18. Subramony SH, May W, Lynch D, et al. Measuring Friedreich ataxia: Interrater reliability of a neurologic rating scale. *Neurology*. 2005;64(7):1261–2. https://doi.org/10.1212/01.WNL.0000156802.15466.79
19. Schmitz-Hübsch T, du Montcel ST, Baliko L, et al. Scale for the assessment and rating of ataxia: Development of a new clinical scale. *Neurology*. 2006;66(11):1717–20. https://doi.org/10.1212/01.wnl.0000219042.60538.92
20. Patel M, Isaacs CJ, Seyer L, et al. Progression of Friedreich ataxia: quantitative characterization over 5 years. *Ann Clin Transl Neurol*. 2016;3(9):684–94. https://doi.org/10.1002/acn3.332
21. Rummey C, Corben LA, Delatycki MB, et al. Psychometric properties of the Friedreich Ataxia Rating Scale. *Neurol Genet*. 2019;5(6):371. https://doi.org/10.1212/NXG.0000000000000371
22. Ackroyd RS, Finnegan JA, Green SH. Friedreich's ataxia. A clinical review with neurophysiological and echocardiographic findings. *Arch Dis Child*. 1984 Mar;59(3):217–21.
23. Morral JA, Davis AN, Qian J, Gelman BB, Koeppen AH. Pathology and pathogenesis of sensory neuropathy in Friedreich's ataxia. *Acta Neuropathol*. 2010 Jul;120(1):97–108.
24. Koeppen AH, Ramirez RL, Becker AB, Mazurkiewicz JE. Dorsal root ganglia in Friedreich ataxia: Satellite cell proliferation and inflammation. *Acta Neuropathol Commun*. 2016;4(1):46. https://doi.org/10.1186/s40478-016-0288-5
25. Nachbauer W, Boesch S, Reindl M, et al. Skeletal muscle involvement in friedreich ataxia and potential effects of recombinant human erythropoietin administration on muscle regeneration and neovascularization. *J Neuropathol Exp Neurol*. 2012;71(8):708–15. https://doi.org/10.1097/NEN.0b013e31825fed76
26. Koeppen AH, Davis AN, Morral JA. The cerebellar component of Friedreich's ataxia. *Acta Neuropathol*. 2011 Sep;122(3):323–30. https://doi.org/10.1007/s00401-011-0844-9
27. França MC, D'Abreu A, Yasuda CL, Bonadia LC, Santos da Silva M, Nucci A, et al. A combined voxel-based morphometry and 1H-MRS study in patients with Friedreich's ataxia. *J Neurol*. 2009 Jul;256(7):1114–20.
28. Folker J, Murdoch B, Cahill L, Delatycki M, Corben L, Vogel A. Dysarthria in Friedreich's ataxia: A perceptual analysis. *Folia Phoniatr Logop*. 2010;62(3):97–103. https://doi.org/10.1159/000287207
29. Fortuna F, Barboni P, Liguori R, et al. Visual system involvement in patients with Friedreich's ataxia. *Brain*. 2009;132(Pt 1):116–23. https://doi.org/10.1093/brain/awn269
30. Rance G, Fava R, Baldock H, et al. Speech perception ability in individuals with Friedreich ataxia. *Brain*. 2008;131(Pt 8):2002–12. https://doi.org/10.1093/brain/awn104
31. Keage MJ, Delatycki MB, Gupta I, Corben LA, Vogel AP. Dysphagia in Friedreich Ataxia. *Dysphagia*. 2017;32(5):626–35. https://doi.org/10.1007/s00455-017-9804-4
32. Weidemann F, Rummey C, Bijnens B, et al. The heart in Friedreich ataxia: Definition of cardiomyopathy, disease severity, and correlation with neurological symptoms. *Circulation*. 2012;125(13):1626–34. https://doi.org/10.1161/CIRCULATIONAHA.111.059477
33. Koeppen AH, Ramirez RL, Becker AB, Bjork ST, Levi S, Santambrogio P, et al. The pathogenesis of cardiomyopathy in Friedreich ataxia. *PLoS One*. 2015;10(3):e0116396.
34. Rajagopalan B, Francis JM, Cooke F, et al. Analysis of the factors influencing the cardiac phenotype in Friedreich's ataxia. *Mov Disord*. 2010;25(7):846–52. https://doi.org/10.1002/mds.22864
35. Schadt KA, Friedman LS, Regner SR, Mark GE, Lynch DR, Lin KY. Cross-sectional analysis of electrocardiograms in a large heterogeneous cohort of Friedreich ataxia subjects. *J Child Neurol*. 2012 Sep;27(9):1187–92.
36. Bourke T, Keane D. Friedreich's Ataxia: A review from a cardiology perspective. *Ir J Med Sci*. 2011;180(4):799–805. https://doi.org/10.1007/s11845-011-0744-y
37. Finocchiaro G, Baio G, Micossi P, Pozza G, di Donato S. Glucose metabolism alterations in Friedreich's ataxia. *Neurology*. 1988;38(8):1292–96. https://doi.org/10.1212/wnl.38.8.1292
38. Cnop M, Mulder H, Igoillo-Esteve M. Diabetes in Friedreich ataxia. *J Neurochem*. 2013;126 Suppl 1:94–102. https://doi.org/10.1111/jnc.12216
39. Coppola G, Marmolino D, Lu D, et al. Functional genomic analysis of frataxin deficiency reveals tissue-specific alterations and identifies the PPARgamma pathway as a therapeutic target in Friedreich's ataxia. *Hum Mol Genet*. 2009;18(13):2452–61. https://doi.org/10.1093/hmg/ddp183

40. Raman SV, Phatak K, Hoyle JC, Pennell ML, McCarthy B, Tran T, et al. Impaired myocardial perfusion reserve and fibrosis in Friedreich ataxia: A mitochondrial cardiomyopathy with metabolic syndrome. *Eur Heart J*. 2011 Mar;32(5):561–7.
41. Delatycki MB, Holian A, Corben L, Rawicki HB, Blackburn C, Hoare B, et al. Surgery for equinovarus deformity in Friedreich's ataxia improves mobility and independence. *Clin Orthop Relat Res*. 2005 Jan;(430):138–41.
42. Co-enzyme Q10 and idebenone use in Friedreich's ataxia – Parkinson – 2013 – Journal of Neurochemistry – Wiley Online Library [Internet]. [cited 2024 Dec 26]. Available from: https://onlinelibrary.wiley.com/doi/full/10.1111/jnc.12322
43. Lynch DR, Willi SM, Wilson RB, et al. A0001 in Friedreich ataxia: Biochemical characterization and effects in a clinical trial. *Movement Disord*. 2012 Jul;27(8):1026–33. https://doi.org/10.1002/mds.25058.
44. Burness CB. Efficacy of EGb761 in patients suffering from Friedreich ataxia. ctv.veeva.com [Internet]. Published 2024. Accessed December 25, 2024. Available from: https://ctv.veeva.com/study/efficacy-of-egb761-in-patients-suffering-from-friedreich-ataxia
45. Qureshi MY, Patterson MC, Clark V, et al. Safety and efficacy of (+)-epicatechin in subjects with Friedreich's ataxia: A phase II, open-label, prospective study. *J Inherit Metab Dis*. 2021;44(2):502–14. https://doi.org/10.1002/jimd.12285
46. Yiu EM, Tai G, Peverill RE, et al. An open-label trial in Friedreich ataxia suggests clinical benefit with high-dose resveratrol, without effect on frataxin levels. *J Neurol*. 2015;262(5):1344–53. https://doi.org/10.1007/s00415-015-7719-2
47. Delatycki MB. Micronised resveratrol as a treatment for Friedreich ataxia: A randomized, placebo-controlled, double-blinded crossover trial. ClinicalTrials.gov Identifier: NCT03933163. Published 2024. Accessed December 22, 2024. https://ctv.veeva.com/study/micronised-resveratrol-as-a-treatment-for-friedreich-ataxia
48. Shungu DC, Weiduschat N, Duvvuri V, et al. L-carnitine and creatine in Friedreich's ataxia: A randomized, placebo-controlled crossover trial. *J Neurol Sci*. 2005;237(1–2):49–57. Accessed December 22, 2024. https://pubmed.ncbi.nlm.nih.gov/15480852/
49. University of Florida. An Open-label Study of the Effects of Acetyl-L-Carnitine on Cardiovascular Outcomes in Friedreich's Ataxia. ClinicalTrials.gov Identifier: NCT01921868. Published 2016. Accessed September 16, 2025. https://clinicaltrials.gov/study/NCT01921868
50. Zesiewicz T, Heerinckx F, De Jager R, et al. Randomized, clinical trial of RT001: Early signals of efficacy in Friedreich's ataxia. *Mov Disord*. 2018;33(6):1000–5. https://doi.org/10.1002/mds.27353
51. Burness CB. A study to assess efficacy, long term safety and tolerability of RT001 in subjects with Friedreich's ataxia. ctv.veeva.com [Internet]. Published 2024. Accessed December 25, 2024. Available from: https://ctv.veeva.com/study/a-study-to-assess-efficacy-long-term-safety-and-tolerability-of-rt001-in-subjects-with-friedreichs
52. Orsucci D, Mancuso M, Ienco EC, LoGerfo A, Siciliano G. Targeting mitochondrial dysfunction and neurodegeneration by means of coenzyme Q10 and its analogues. *Curr Med Chem*. 2011;18(26):4053–64. https://doi.org/10.2174/092986711796957257
53. Bentinger M, Tekle M, Dallner G. Coenzyme Q-biosynthesis and functions. *Biochem Biophys Res Commun*. 2010;396(1):74–9. https://doi.org/10.1016/j.bbrc.2010.02.147
54. Jauslin ML, Meier T, Smith RA, Murphy MP. Mitochondria-targeted antioxidants protect Friedreich Ataxia fibroblasts from endogenous oxidative stress more effectively than untargeted antioxidants. *FASEB J*. 2003;17(13):1972–4. https://doi.org/10.1096/fj.03-0240fje
55. Lodi R, Hart PE, Rajagopalan B, Taylor DJ, Crilley JG, Bradley JL, et al. Antioxidant treatment improves in vivo cardiac and skeletal muscle bioenergetics in patients with Friedreich's ataxia. *Ann Neurol*. 2001 May;49(5):590–6.
56. Hart PE, Lodi R, Rajagopalan B, Bradley JL, Crilley JG, Turner C, et al. Antioxidant treatment of patients with Friedreich ataxia: Four-year follow-up. *Arch Neurol*. 2005 Apr;62(4):621–6.
57. Coenzyme Q10 and vitamin E deficiency in Friedreich's ataxia: Predictor of efficacy of vitamin E and coenzyme Q10 therapy – Cooper – 2008 – European Journal of Neurology – Wiley Online Library [Internet]. [cited 2024 Dec 25]. Available from: https://onlinelibrary.wiley.com/doi/10.1111/j.1468-1331.2008.02318.x
58. Gillis JC, Benefield P, McTavish D. Idebenone. A review of its pharmacodynamic and pharmacokinetic properties, and therapeutic use in age-related cognitive disorders. *Drugs Aging*. 1994;5(2):133–52. https://doi.org/10.2165/00002512-199405020-00007
59. Soriano S, Llorens JV, Blanco-Sobero L, et al. Deferiprone and idebenone rescue frataxin depletion phenotypes in a Drosophila model of Friedreich's ataxia. *Gene*. 2013;521(2):274–81. https://doi.org/10.1016/j.gene.2013.02.049

60. Seznec H, Simon D, Monassier L, et al. Idebenone delays the onset of cardiac functional alteration without correction of Fe-S enzymes deficit in a mouse model for Friedreich ataxia. *Hum Mol Genet.* 2004;13(10):1017–24. https://doi.org/10.1093/hmg/ddh114
61. Rustin P, von Kleist-Retzow JC, Chantrel-Groussard K, Sidi D, Munnich A, Rötig A. Effect of idebenone on cardiomyopathy in Friedreich's ataxia: A preliminary study. *Lancet.* 1999 Aug 7;354(9177):477–9.
62. Di Prospero NA, Baker A, Jeffries N, Fischbeck KH. Neurological effects of high-dose idebenone in patients with Friedreich's ataxia: A randomised, placebo-controlled trial. *Lancet Neurol.* 2007 Oct;6(10):878–86.
63. Lynch DR, Perlman SL, Meier T. A phase 3, double-blind, placebo-controlled trial of idebenone in friedreich ataxia. *Arch Neurol.* 2010 Aug;67(8):941–7.
64. Lagedrost SJ, Sutton MSJ, Cohen MS, Satou GM, Kaufman BD, Perlman SL, et al. Idebenone in Friedreich ataxia cardiomyopathy-results from a 6-month phase III study (IONIA). *Am Heart J.* 2011 Mar;161(3):639–45.e1.
65. Lynch DR. A study of efficacy, safety and tolerability of idebenone in the treatment of Friedreich's ataxia (FRDA) patients. ctv.veeva.com [Internet]. Published 2024. Accessed December 25, 2024. Available from: https://ctv.veeva.com/study/a-study-of-efficacy-safety-and-tolerability-of-idebenone-in-the-treatment-of-friedreichs-ataxia-f
66. Cook A, Boesch S, Heck S, Brunt E, Klockgether T, Schöls L, et al. Patient-reported outcomes in Friedreich's ataxia after withdrawal from idebenone. *Acta Neurol Scand.* 2019 Jun;139(6):533–9.
67. Singh SK, Srivastav S, Castellani RJ, Plascencia-Villa G, Perry G. Neuroprotective and antioxidant effect of *Ginkgo biloba* extract against AD and other neurological disorders. *Neurotherapeutics.* 2019 Jul;16(3):666–74.
68. Lepretti M, Martucciello S, Burgos Aceves MA, Putti R, Lionetti L. Omega-3 fatty acids and insulin resistance: Focus on the regulation of mitochondria and endoplasmic reticulum stress. *Nutrients.* 2018 Mar 14;10(3):350.
69. Nicolson GL. Metabolic syndrome and mitochondrial function: Molecular replacement and antioxidant supplements to prevent membrane peroxidation and restore mitochondrial function. *J Cell Biochem.* 2007 Apr 15;100(6):1352–69.
70. Cotticelli MG, Crabbe AM, Wilson RB, Shchepinov MS. Insights into the role of oxidative stress in the pathology of Friedreich ataxia using peroxidation resistant polyunsaturated fatty acids. *Redox Biol.* 2013;1(1):398–404.
71. Abeti R, Uzun E, Renganathan I, Honda T, Pook MA, Giunti P. Targeting lipid peroxidation and mitochondrial imbalance in Friedreich's ataxia. *Pharmacol Res.* 2015 Sep;99:344–50.
72. Pannu N, Bhatnagar A. Resveratrol: From enhanced biosynthesis and bioavailability to multitargeting chronic diseases. *Biomed Pharmacother.* 2019 Jan 1;109:2237–51.
73. Li L, Voullaire L, Sandi C, Pook MA, Ioannou PA, Delatycki MB, et al. Pharmacological screening using an FXN-EGFP cellular genomic reporter assay for the therapy of Friedreich ataxia. *PLOS ONE.* 2013 Feb 13;8(2):e55940.
74. Georges P, Boza-Moran MG, Gide J, Pêche GA, Forêt B, Bayot A, et al. Induced pluripotent stem cells-derived neurons from patients with Friedreich ataxia exhibit differential sensitivity to resveratrol and nicotinamide. *Sci Rep.* 2019 Oct 10;9(1):14568.
75. Bernatoniene J, Kopustinskiene DM. The role of catechins in cellular responses to oxidative stress. *Molecules.* 2018 Apr 20;23(4):965.
76. Ribas GS, Vargas CR, Wajner M. L-carnitine supplementation as a potential antioxidant therapy for inherited neurometabolic disorders. *Gene.* 2014 Jan 10;533(2):469–76.
77. Ferreira GC, McKenna MC. L-carnitine and acetyl-L-carnitine roles and neuroprotection in developing brain. *Neurochem Res.* 2017 Jun;42(6):1661–75.
78. Zanelli SA, Solenski NJ, Rosenthal RE, Fiskum G. Mechanisms of ischemic neuroprotection by acetyl-L-carnitine. *Ann N Y Acad Sci.* 2005 Aug;1053:153–61.
79. Alves E, Binienda Z, Carvalho F, Alves CJ, Fernandes E, de Lourdes Bastos M, et al. Acetyl-L-carnitine provides effective in vivo neuroprotection over 3,4-methylenedioximethamphetamine-induced mitochondrial neurotoxicity in the adolescent rat brain. *Neuroscience.* 2009 Jan 23;158(2):514–23.
80. Zhang R, Zhang H, Zhang Z, et al. Neuroprotective effects of pre-treatment with l-carnitine and acetyl-L-carnitine on ischemic injury in vivo and in vitro. *Int J Mol Sci.* 2012;13(2):2078–90. https://doi.org/10.3390/ijms13022078

5 The Effect of Natural Ingredients and Their Combination for Huntington's Disease

Nagasen Dasari,[1] *Sankhanil Panda,*[2]
Sunanda Kaligithi,[2] *and Sai Kiran S.S. Pindiprolu*[1]
[1]Department of Pharmacy, Aditya Pharmacy College, Surampalem, Andhra Pradesh, India
[2]Department of Pharm. D, Aditya Pharmacy College, Surampalem, Andhra Pradesh, India

5.1 INTRODUCTION

Huntington's disease (HD) is a genetic disorder that causes the neurons in specific brain regions to gradually deteriorate and die. The disease also affects other parts of the brain, including those that control voluntary (directed) movement. Strange body postures, uncontrollable chorea (dance-like movements), and behavioural, emotional, cognitive, and personality problems are all common in people with HD. For instance, the individual's fingers, feet, face, or torso move erratically. There are indications of chorea in these motions. When an individual experiences anxiety or distraction, they may become more acute; as HD worsens, the person may exhibit more pronounced and dramatic movements.

Two subtypes of HD exist:

Adult-onset HD: The most prevalent type of HD is this one. People between the ages of 30 and 40 are typically the ones who experience the symptoms.

Early-onset HD: Very rarely, the illness will strike youngsters or teenagers. Children who have the illness frequently exhibit symptoms that are quite similar to Parkinson's disease and frequently encounter sudden challenges with their schooling.

An equal number of males and females are impacted, and symptoms usually appear in their 30s or 40s of life. Any age between 1 and 90 years old can experience the onset of HD symptoms. Of those cases, 5% to 10% have juvenile onset, meaning that symptoms start before the age of 20. Since most juvenile instances are inherited paternally, children may show more indications of Parkinson's disease than chorea. Bradykinesia, dystonia, tremors, and cognitive deficiency are some of these symptoms. The Westphal form of HD is identified in patients who exhibit more hypokinetic symptoms, such as dystonia and bradykinesia, than hyperkinetic symptoms, such as chorea (1, 2).

5.2 AETIOLOGY AND PATHOPHYSIOLOGY OF HD

HD is brought on by a genetic disorder. It is transmitted from parents to offspring. A child's likelihood of developing HD is 50% if one of the parents has the condition. The child will not transmit

DOI: 10.1201/9781003478652-6

the sickness to their offspring if they do not contract it. It is never possible to determine a family history of HD in 1% to 3% of cases. The aetiology of HD is unclear and complex, despite the fact that the disease has a clearly documented genetic basis. The degenerative process of HD is explained by a variety of processes, including molecular genetics, which results in oxidative stress, metabolic failure, and mitochondrial malfunction. Early clinical manifestation diagnosis and careful management are therefore essential (3, 4).

HD is a degenerative brain condition brought on by sudden changes in the HD protein huntingtin (htt). The htt mutation that was caused by an expansion along the CAG repeat tract resulted in longer polyglutamine (polyQ) chains in the protein that was encoded (4–6). While alleles (wild-type) in unaffected groups can contain up to 35 CAG repeats, HD alleles require 36 or more. The total number of CAG repeats and the stage at which symptoms appear are inversely correlated; a later stage of onset is linked to larger CAG repeat extensions (7, 8).

A 348-kDa multidomain protein with a polymorphism glutamine/proline-rich domain at the amino terminus is encoded by the htt gene. Currently, genetic testing is available for both diagnosis and prognosis of HD, the most often reported inherited neurodegenerative disorder. Gene-targeted medicine could soon be available. One of the earliest conditions to be assessed genetically before delivery was HD. Undoubtedly, neuroimaging methods have made it possible to do predictive and diagnostic genetic screening for illness identification and its consequences for ill individuals and their families. This has been made possible by the implementation of specialized facilities and genetic testing protocols (9).

Two different neurodevelopmental findings in HD are stroke-related degeneration and neuronal loss, especially in the caudate nuclei that target the cerebral cortex, pallidum, thalamus, brainstem, and cerebellum. In the early stages of HD, neuroinflammation and microglia activation are characteristic features. Compared to IL-18, HD plasma had significantly higher concentrations of TGF-1, VEGF, matrix metallopeptidase 9, and IL–6. Plasma IL-6 was inversely correlated with HD prevalence. HD is a severe neurological illness with a late onset that is autosomal dominant. It causes persistent and irreversible motor impairments, which can cause mobility and coordination issues as well as psychiatric and behavioural disorders. A consequence of HD is visual impairment (10–12).

5.3 ONSET OF HD INVOLVES THE FOLLOWING MOLECULAR PATHWAYS

HD is caused by an aberrant expansion of the glutamine-encoding CAG repeats in exon 1 of the HTT gene (4p16.3). Since this mutation causes the polyglutamine tract of the "huntingtin" gene product to extend, HD is classified as one of the so-called polyglutamine illnesses. Genetic instability is a characteristic of the pathologically extended CAG tract in both somatic tissues and gametogenesis, particularly in spermatogenesis. The magnitude of the expanded HTT allele has a substantial impact on the polymorphism of the clinical symptoms associated with HD. The quantity of CAG repeats in the mutant allele is inversely connected with the age at which a disease first appears and the age at which it dies (13–15). The amount of CAG repeats and the pace at which HD advances are directly correlated, as Huang et al. were the first to show. Other authors later confirmed this conclusion. However, the mutation only explains about 66% of the variation in HD, highlighting the significance of other moderating variables in this illness. It is proposed that replication "slippage" aids in the development of trinucleotide repeat expansion. DNA polymerase, which copies DNA molecules and travels down the template chain during meiosis, loses nucleotide direction and restarts synthesis from the wrong location when a "loop" forms on the complementary chain. This kind of "slippage" is much more likely to occur during the replication of DNA sections that contain repeated nucleotide sequences. A similar outcome could be caused by abnormal activity of the enzymes (MSH2, OGG1) involved in single-stranded DNA break repair. For instance, inhibition of these enzymes is linked to a considerable reduction in CAG repeat instability in transgenic HD mice (6, 16, 17) (see Figure 5.1).

The discovery of homologs for the HTT gene in numerous evolutionary distant animals, starting with amoeba, suggests that huntingtin has a very conservative role in a cell. According to theory,

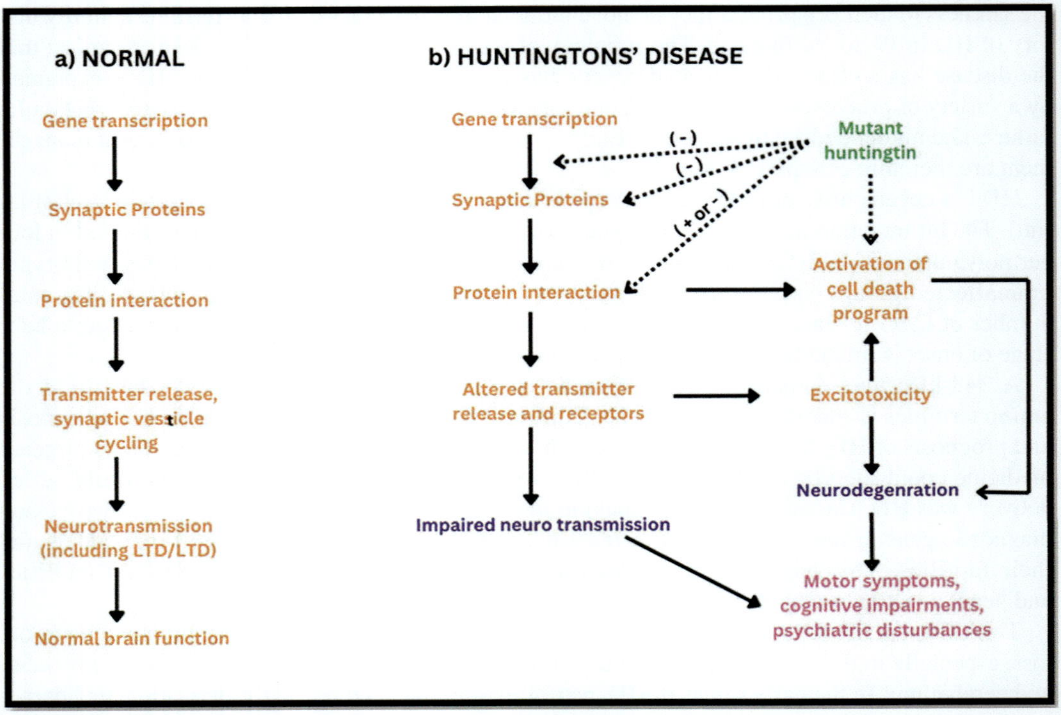

FIGURE 5.1 Pathophysiological progression of Huntington's disease.

huntingtin functions as a molecular "hub" that either forms complexes with other proteins or, conversely, encourages their dissociation to allow for the precise synchronization of various cellular functions. Huntingtin is connected to cell membranes and plays a part in autophagy regulation, synaptic transmission, and the dynamics of vesicular transport. HD is thought to play a crucial role in the nervous system's growth and maturation as well as in the control of cell adhesion, signal transmission, and neuron survival. Regarding the precise biological function of huntingtin, a final judgement is still pending and needs further studies to be conducted for a conclusive answer (3, 7).

5.4 CLINICAL CHARACTERISTICS OF HD

Unwanted, involuntary motions are the hallmark of motor alterations. At first, the movements usually start in the tiny muscles of the face as well as the distal extremities, including the fingers and toes. These muscular twitches are frequently undetectable to onlookers or may be mistaken for anxiousness. Walking becomes unsteady in normal life, and the person may appear to be somewhat intoxicated. From distal to more proximal and axial muscles, the undesirable motions gradually extend to every other muscle. Whenever the patient is awake, there are choreatic motions. Facial muscles may move continuously as a result of facial choreatic motions, even when there isn't a regular pattern. For instance, the head might be twisted or bent, the eye closed, the eyebrow lifted, or the tongue sticking out while the lips pout. The lengthy back muscles' extension motions are the most noticeable. For certain people, choking can occur at any time as speaking and swallowing become increasingly difficult (see Figure 5.2). Later on, the sufferer even loses their ability to speak. For instance, the head might be twisted or bent, the eye closed, the eyebrow lifted, or the tongue sticking out while the lips pout. The lengthy back muscles' extension motions are the most noticeable. For certain people, choking can occur at any time as speaking and swallowing become

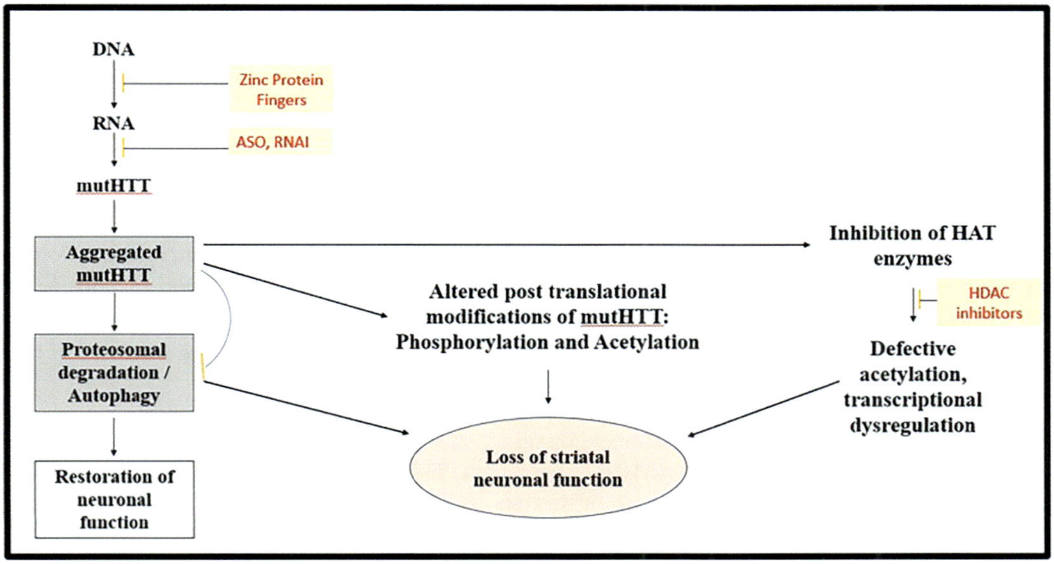

FIGURE 5.2 Possible molecular pathological events along with the possible therapeutic interventions.

increasingly difficult. The person eventually even loses the capacity to talk. There are two extremes: the old patient who is bedridden, has flexion contractures and rigidity in their extremities, and has been severely affected in the last stages of the disease for a long time, and the younger patient who has overwhelming rigidity (Westphal variation). Slower movements with elevated muscle tone, such as torticollis, as well as trunk or limb rotation, are some of the symptoms of dystonia. The initial motor symptom of HD may be dystonia, such as torticollis. Similar to Tourette syndrome, tics are another type of unpleasant movement, although they are not very common. Cerebellar signals can show up intermittently, much like hypometria and hypermetria. Walking is frequently compared to being "drunk" or having "cerebellar ataxia." It's really hard to tell the difference between ataxic and choreatic walking. Incidentally, there are pyramidal signs, sometimes known as Babinski signs. Over time, the impact of motor disturbance on day-to-day activities increases. When hyperkinesia and hypokinesia are present, walking and standing become difficult, and ataxic gait and frequent falls are common outcomes. Additionally, everyday tasks like dressing, toileting, getting out of bed, showering, cleaning the house, cooking, and eating get harder and harder. Even if mental and cognitive changes are still occurring, motor symptoms will eventually affect performance, depending on the type of work the patient conducts. In case of behavioural psychiatric symptoms, depression is the most common indication. Since inactivity, apathy, and weight loss are also symptoms of HD, the diagnosis is challenging. Typically, there is worry, guilt, and low self-esteem. In contrast to anxiety and sadness, apathy is correlated with the stage of the disease. Premanifest gene carriers and those with early symptoms are more likely to commit suicide. The period between the gene test and the reduction in independence is when suicide is most likely to happen. Another common occurrence is anxiety (34–61%), which can occasionally be linked to ambiguity over the onset and/or progression of the illness. Compulsions and obsessions can cause agitation and violence in addition to disrupting the patient's life. In hindsight, irritability is frequently the initial symptom, although it actually manifests at every stage of the illness. Irritability can manifest itself in a wide range of ways, from heated arguments to physical violence. Apathy syndrome is characterized by a decline in interest and an increase in sluggish behaviour. Distinguishing between depression and apathy can be challenging. Pyrosis may appear, usually in the later stages of the illness. Cognitive decline usually coexists with this. The overall clinical appearance is similar to the paranoid symptoms and auditory hallucinations associated with schizophrenia. Hypersexuality can lead to significant issues in a relationship in its early stages. Hyposexuality is the norm in later phases (8).

5.5 DIAGNOSIS OF HD

In general, a series or a combination of tests is utilized to confirm the diagnosis of a patient as HD. These include the following tests:

Medical History and Neurological Examination: To exclude other disorders and gather the patient's medical and family history, a neurologist will question them in detail. During neurological and physical examinations, reflexes, balance, mobility, muscle tone, hearing, walking, and mental state may all be evaluated. For specialized care and/or to support diagnosis, laboratory testing may also be done, and people with HD may be sent to experts such as speech pathologists, clinical neuropsychologists, genetic counsellors, or psychiatrists.

Imaging Techniques: Sometimes, especially when a patient's family history and genetic tests are unclear, a doctor may recommend brain imaging, such as computed tomography (CT) or, more often, magnetic resonance imaging (MRI). These scans usually show swelling of the brain's fluid-filled ventricles and shrinking of certain brain regions as the disease worsens. Since these alterations can happen in other conditions, they do not always signify HD. Even if a person has early HD symptoms, their CT or MRI scan results may be normal.

Testing for Genetic Conditions or Susceptibilities: Genetic testing can assist in identifying an individual's risk of having or passing on a genetic ailment, as well as confirm or rule out a suspected genetic condition. It is now able to predict with more precision whether someone will acquire HD, thanks to genetic testing. Using DNA extracted from a blood sample, the direct genetic test, the most reliable and accurate way to screen for HD, counts the amount of CAG repeats in the HD gene. A diagnosis of HD is supported when 36 or more repetitions are found. HD is ruled out by a test result of 26 or fewer repetitions. For those who have a family history of HD and are worried about passing the condition on to their unborn child, prenatal testing is an option (3).

5.6 MANAGEMENT OF HD PHARMACOLOGICALLY

Both men and women are impacted in equal measure, and the onset of symptoms usually occurs in the middle to late decades of life. Anybody between the ages of1 and 99 can experience symptoms of HD. Five to ten percent of those instances have juvenile onset, which indicates that the symptoms begin prior to the 20s (18, 19). Children may exhibit more Parkinsonian symptoms than chorea, including bradykinesia, dystonia, tremors, and cognitive deficits, because the majority of juvenile cases are inherited paternally. The Westphal form of HD is identified in patients who exhibit more hypokinetic symptoms, such as dystonia and bradykinesia, than hyperkinetic symptoms, such as chorea (8, 20). HD is diagnosed when certain motor symptoms are present and there is a family history of the disorder. Depending on behavioural and cognitive symptoms, personality changes, and other circumstances, there may be additional HD manifestations that are evident at the time of presentation or before diagnosis. In symptomatic people, a DNA test that looks for aberrant expansion of CAG in the htt gene can be utilized to confirm the diagnosis (21, 22). If a patient requests it and receives appropriate genetic counselling, DNA analysis can be carried out on those who are at risk of getting HD under the guidance of licenced medical specialists.

Pharmacological treatment for HD includes the following.

5.7 CHOREA

The characteristic movements of cholera are atypical, uncontrollable, spontaneous, uncontrollable, irregular, intermittent, non-rhythmic, and aimless, affecting the face, limbs, and trunk. Patients experiencing discomfort or distress connected to chorea may find that medication treatment is beneficial. Unless the patient has untreated depression or suicidal thoughts, tetrabenazine is one of the first-line therapies for this symptom. First-line treatment for this symptom involves second-generation neuroleptics, especially when patients also have co-occurring behavioural, psychotic, or personality disorders. Monotherapy is recommended for the treatment of chorea since combination therapy may make managing non-motor symptoms more difficult and increase the possibility of adverse effects. In case of disruptive chorea, appropriate protective measures should be implemented to prevent serious damage or choking, especially at mealtimes and when doing instrumental activities of daily life. The right placement techniques and assistive technology devices can be determined with the help of rehabilitation specialists (9, 10, 13).

5.8 RIGIDITY

Rigidity is the term for an increase in muscular tone that causes resistance to passive movement. Patients may experience discomfort as a result of joint stiffness and limited range of motion. Rigidity may be brought on or made worse by tetrabenazine or other neuroleptics. When considering whether to cut the dosage or cease taking neuroleptics and/or tetrabenazine, the overall benefit on chorea and/or behavioural symptoms versus the degree of stiffness should be taken into account if this compromises the patient's capacity to function. Levodopa, especially in younger individuals, can temporarily improve the akinetic-rigid symptoms of HD. The recommended method of starting levodopa treatment is gradually, and the daily dosage is usually less than that of Parkinson's disease. It is advised to have physical treatment to preserve or increase mobility and stop the onset of joint deformities and contractures (9, 10, 23).

5.9 PSYCHIATRIC SYMPTOMS

The side effect of antipsychotic medications like fluphenazine and haloperidol is movement suppression. As a result, they might help in chorea treatment. These medications, however, may exacerbate sleepiness, restlessness, and dystonia – involuntary spasms. While some medications, such as aripiprazole and olanzapine, may have fewer adverse effects, they should still be used cautiously as they have the potential to exacerbate symptoms. Mood-stabilizing medications, such as anticonvulsants like lamotrigine, carbamazepine, and divalproex, can help avoid the highs and lows linked to bipolar illness (9, 10, 24).

5.10 SLEEP DISORDERS

Approximately two-thirds of patients with HD experience sleep disturbances. There are various aetiologies. A sleep disorder may be present in certain patients during a depressed episode. For these patients, sedative antidepressants with no discernible anticholinergic adverse effects should be administered at night (25, 26).

A day-night reversal in sleep problems can occasionally result from antihyperkinetic medication's sedative side effects during the day. If the main dosage is given at night, in this case, it may be useful to leverage the sedative side effects of these medications to promote sleep. Treating hyperkinesia that causes trouble at night can also benefit from this dose schedule. This pharmacological treatment is also followed by regular physiotherapy sessions to help improve the motor functions and motor symptoms such as bradykinesia, dystonia, and/or rigidity and various other supportive measures which help in improving the symptomatic conditions of the person, thus easing the lifestyle they lead. But the main drawbacks which accompany pharmacological treatment options for HD are that a number of side effects have been observed. Even though the efficacy of the drugs and the medications shows benefits for certain conditions, these drawbacks, which come in the form of side effects, do not always appear to improve the patient's general and overall health and quality of life in day-to-day experiences. The medications that are most frequently used to treat this illness are listed in Table 5.1, along with any side effects that have been noted and any negative consequences on the patient (9, 10, 27).

TABLE 5.1
Pharmacological Agents Used for Treatment of Huntington's Disease and Their Adverse Effects

S. No	Name of Drug	Mechanism of Action	Side Effects	Adverse Effects	References
1	Tetrabenazine	Inhibits vesicular monoamine transporter 2 (VMAT), reducing dopamine levels in the brain	Sedation, fatigue, depression, akathisia	Parkinsonism, depression, suicidal ideas, and tendencies	(28)
2	Deutetrabenazine	VMAT 2 inhibitor	Somnolence, diarrhoea, dry mouth, insomnia	QT prolongation, serotonin syndrome, akathisia	(28)
3	Olanzapine	Dopamine D2 and serotonin 5-HT 2 receptor antagonist	Weight gain, sedation, orthostatic hypotension	Higher mortality rates among elderly people experiencing psychosis associated with dementia	(28)
4	Aripiprazole	Serotonin 5-HT1A receptor partial agonist, serotonin 5-HT2A receptor antagonist, and partial dopamine D2 receptor agonist	Anxiety, insomnia, akathisia, constipation	Neuroleptic malignant syndrome, tardive dyskinesia	(29)
5	Riluzole	Glutamate modulator, inhibits glutamate release	Nausea, dizziness, weakness, and elevated liver enzymes	Liver failure, neutropenia, and interstitial lung disease	(30)
6	Memantine	NMDA receptor antagonist, which reduces glutamate toxicity	Dizziness, headache, constipation, and confusion	Hallucinations, seizures	(31)

5.11 NATURAL INGREDIENTS AS TREATMENT OPTIONS FOR HD

5.11.1 List of Plants as Natural Sources for Treatment of HD

But, in recent times, various alternative approaches have also been found, which have also shown promising results in the treatment of this disease. These alternatives have been derived from natural sources, which mostly include plant extracts and isolated phytoconstituents which have shown the potential of being used as an alternative treatment for HD (11). A few of them have been described as follows and also in Table 5.2.

5.11.1.1 *Centella asiatica*

The ancient herb, commonly known as jal brahmi, is used to cure a variety of pathological conditions, such as psoriasis, leprosy, oxidative stress, memory loss, ulcers, anxiety, depression, and fever. The neuroprotective qualities of *C. asiatica* aqueous leaf extracts containing 500–1000 mg/kg were evaluated. Its significance in treating HD was explained by higher levels of antioxidants like glutathione (GSH) and catalase (CAT) and lower levels of oxidative stress indicators, including malondialdehyde (MDA), reactive oxygen species (ROS), and hydroperoxide. It was also helpful in the treatment of neurodegenerative illnesses by reducing the oxidative stress that 3-nitropropionic acid (3-NP) created in the brain's mitochondria. Aspartic acid and madecassic acid were the two

TABLE 5.2
Natural Agents Used for Treatment of Huntington's Disease and Their Pharmaceutical Constituents

S. No.	Plant Name	Phytochemical Constituent	Therapeutic Applications	References
1	*Centella asiatica*	Asiaticoside	Neuroprotective, anti-inflammatory, antioxidant	(42)
2	*Tripterygium wilfordi*	Triptolide	Neuroprotective, anti-inflammatory, reduction of protein aggregation, regulation of mitochondrial function	(42)
3	*Sesamum indicum*	Sesamin	Antioxidant, anti-inflammatory, modulation of mitochondrial function	(22)
4	*Camellia sinensis*	Epigallocatechin-3-gallate	Antioxidant, reduction of protein aggregation, anti-inflammatory, modulation of neurotrophic factors	(43)
5	*Momordica charantia*	Charantin	Neuroprotective, anti-inflammatory, antioxidant, modulation of cellular pathways	(43)
6	*Punica grantum*	Punicalogin	Neuroprotective, anti-inflammatory, antioxidant, reduction of protein aggregation	(42)
7	*Psoralea corylifolia*	Bakcuchiol, psoralen, flavonoids	Neuroprotective, anti-inflammatory, antioxidant, mitochondrial function	(42)
8	*Cannabis sativa*	Cannabidiol	Antioxidant, anti-inflammatory effects, neuroprotective, modulation of endocannabinoid systems, reduction of excitotoxicity	(42)
9	*Veeratrum grandiflorum*	Veratridine	Activation of sodium channels	(42)
10	*Elderberry*	Anthocyanins and flavonoids	Antioxidant, anti-inflammatory, enhancement of cognitive function	(42)
11	*Bacopa monnieri*	Bacoside-A	Antioxidant, anti-inflammatory effects, neuroprotective, reduction of apoptosis	(44)
12	*Calendula officinalis*	Triterpenoids, flavonoids, carotenoids	Antioxidant, anti-inflammatory effects, neuroprotective	(45)
13	*Centella asiatica*	Asiaticoside, modecassoside, asiatic acid	Antioxidant, anti-inflammatory effects, reduction of apoptosis, cognitive enhancement	(42)
14	*Convolvulus pluricaulis* choisy	Alkaloids, flavonoids, glycosides	Antioxidant, anti-inflammatory effects, neuroprotective, reduction of anxiety and stress	(46)
15	*Garcinia kola*	Kolaviron, flavonoids, tannins	Antioxidant, anti-inflammatory effects, potential benefits for mood and anxiety, cognitive enhancement	(42)
16	*Gastrodia elata* Blume	Gastrodin	Neuroprotective, antioxidant, anti-inflammatory, cognitive enhancement, and regulation of neurotransmission	(47)
17	*Gingko biloba*	Flavonoids, terpenoids	Antioxidant, neuroprotective effects, anti-inflammatory effect, improvement of cerebral blood flow	(42)
18	*Olea europaea*	Oleuropein, hydroxytyrosol, polyphenols	Neuroprotective, antioxidant, anti-inflammatory, cognitive enhancement	(48)
19	*Withania somnifera*	Withanolide D, sitoindosides	Antioxidant, anti-inflammatory, cognitive enhancement, adaptogenic, and stress-relieving effects, potential to reduce protein aggregation	(42)

most common ingredients in the CA extract. These substances were found to be in charge of neuron differentiation by triggering the signalling pathways associated with mitogen-activated protein kinase. But when it comes to stimulating nerve differentiation, the combination of madecassic acid and asiatic acid works better than either ingredient alone. The outcomes were also similar when nerve growth therapy was used (12, 32).

5.11.1.2 *Tripterygium wilfordi*

Many ailments, including cancer, autoimmune diseases, neurological conditions, and inflammatory diseases, are treated using Chinese herbal medicine, *T. wilfordi* Hook F. (TWHF), sometimes referred to as Thunder God Vine or Celastraceae.

The principal active biological principle of TWHF, quinone methide triterpenoid, is what gives it its therapeutic efficacy. But the primary active ingredient in TWHF that gives it its biological activity is celastrol. Celastrol (3 mg/kg) administered intraperitoneally in the 3-NP (neurotoxin) model of HD shows an anti-Huntington effect. Celastrol lowered both TNFa and nuclear factor kappa B (NF-kB), which dramatically reduced the extent of striatal lesions. This implies that celastrol might be a neuroprotective drug with promise for HD patients. Celastrol may be able to prevent polyglutamine expansion diseases, including Parkinson's disease and heart disease, because of its ability to upregulate heat shock protein (HSP) by increasing the quantity of inducible hsp gene expression, heat shock factor (HSF)-1. An estimate was made of how well celastrol inhibited the negative effects of polyglutamine repeat proteins. Celastrol has the potential to cure HD because it can correct the mutant HdhQ111/Q111 striatal cells' abnormal cellular localization of the huntingtin protein. This estimation's findings imply that celastrol prevents polyglutamine-mediated aggregation (12, 33, 34).

5.11.1.3 *Sesamum indicum*

Sesame (*S. indicum* L.), a member of the Pedaliaceae family, is known as the "queen of oilseeds" and is indigenous to Africa, China, and India. The plant is utilized as a nutritional supplement to cure a wide range of ailments, such as discomfort, skin lesions, oxidative stress, ageing, atherosclerosis, hypertension, and cataracts. Many active substances, including acteoside, pedaliin, luteolin, cistanoside, sesamoside, and sesamol, are found in the plant. The most active phenolic component produced by *S. indicum L.* is sesamol. Analysis was done on the potential use of sesamol as a remedy for 3-NP-induced neurotoxicity. Sesamol pretreatment at a dose of 10 mg/kg resulted in notable gains in body weight, enhanced locomotor activity, and a reduction in oxidative damage as determined by CAT, SOD, and lipid peroxidation (LPO) in the rats' brains. Therefore, this could be a helpful managerial tactic (12, 35).

5.11.1.4 *Camellia sinensis*

The most popular drink in the world is black tea, or *C. sinensis (L.) Kuntze*, a member of the Theaceae family. The phytochemical examination identified a number of active components, including rutin, epicatechin, gallic acid, catechin, and epigallocatechin. These constituents are responsible for the plant's therapeutic properties against a range of illnesses, including oxidative stress, microbial infection, cancer, ulcers, inflammation, arthritis, atherosclerosis, and inflammation (36).

Black tea's potential for memory impairment was suggested by a correlation between its consumption and improved attention and psychomotor speed. The electroencephalogram showed enhanced theta, alpha, and beta activity following oral administration of 300 mg/kg of epigallocatechin-3-gallate (EGCG). Thus, anxiety, memory loss, and attention deficit problems can all be treated with these medications. It was demonstrated that photoreceptor degeneration and huntingtin exon 1 protein overexpression were superior to the pathogenic polyglutamine (polyQ)-containing protein (12).

5.11.1.5 *Momordica charantia*

M. charantia L. is a plant that is believed to provide health advantages. It is sometimes referred to as bitter melon, bitter gourd, or karela. It belongs to the family Cucurbitaceae. Accordingly the plant has a long history of usage and is used to cure a variety of conditions, such as cancer, diabetes, obesity, hypertension, and viral infections. The bitter gourd's physiologically active ingredients include momordin, charantin, cucurbitaine, cucurbitacin, and momordin. The study evaluated the impact of bitter gourd on cognitive abilities by employing scopolamine-induced amnesia (12, 37, 38).

5.11.1.6 *Punica granatum*

Pharmaceutically active ingredients found in *P. granatum* L. include gallic acid, ellagic acid, luteolin, apigenin, punicalin, and punicafolin, which all contribute to the plant's biological activity. *P. granatum* L. juice has the unique ability to prevent memory loss due to the presence of polyphenols that successfully offset the negative effects of scopolamine on the brain. The downregulation of LPO, ROS, NO, and LDH and the upregulation of enzymatic and nonenzymatic antioxidants explained its capacity to protect against neuronal injury (12, 39).

5.11.1.7 *Psoralea corylifolia*

Mostly indigenous to China, India, and Korea, *P. corylifolia* Linn. (Leguminosae) is a widely known traditional medicinal plant that has long been used to treat a variety of illnesses. Numerous pharmacological effects, including anticancer, antibacterial, aphrodisiac, laxative, anthelmintic, and anti-inflammatory properties, have been documented for the plant. The plant contains bakuchiol, angelicin, psoralen, isopsoralen, dadzein, genistein, and bavachin as active phytoconstituents. PC12 cells have been used to assess the biological effectiveness of *P. corylifolia* Linn. seed extract against 3-NP-induced neurotoxicity. In contrast, pretreatment with *P. corylifolia* Linn. increased bioenergetic Jyotishmati, also known as *Celastras paniculatus* Willd., a plant in the Celastraceae family used in Ayurvedic medicine. PC12-treated cells showed decreased cellular consumption of oxygen, ATP generation, and mitochondrial membrane potential (MP). It is highly effective against a variety of ailments, including inflammation, asthma, emesis, coughing, cardiac debility, nephropathy, and skin and skin problems. Numerous phytoconstituents, such as celapanin, celapanigin, celastrine, and paniculatine energy, are abundant in the plant (12, 40).

5.11.1.8 *Cannabis sativa*

The oldest psychoactive medicinal plant in the family Cannabaceae, *C. sativa* L. has been utilized for decades in both agriculture and industry. The plant is said to have a broad variety of medicinal potential against conditions like inflammation, ulcers, neurodegeneration, and skin homeostasis. Treatment with the combination extract reduced the loss of Nissl-stained neurons and the GABA deficit caused by 3-NP. Furthermore, it increased NO synthase activity while downregulating the expression of the CB1 receptor and IGF, indicating that CBD's antioxidant properties may play a part in its ability to treat HD. The patients were given 10 mg/kg/day of oral CBD for a duration of six weeks. The findings demonstrated that when CBD was given to HD patients at a level of 700 mg/kg, neither toxic effects nor therapeutic improvement was seen. Further research is necessary to prove the benefits for HD patients (12).

5.11.1.9 *Veratrum grandiflorum*

The original source of resveratrol isolation is the white hellbore or *V. grandiflorum* O. Loes. The plant has garnered attention because of its extraordinarily therapeutic benefits against a wide range of ailments, such as microbial infections, cancer, diabetes, cardiovascular disease, neurological disorders, and metabolic disorders. In addition, it can be found in pomegranates, soybeans, dark chocolate, cranberries, blueberries, and mulberries. PPARg co-activator (PGC)-1a expression, central and peripheral pathologies, and motor performance were among the characteristics measured for the pharmacological assessment of resveratrol's effect on SIRT1 activation. Because it downregulated

NRF-1 and uncoupled protein-1 and stimulated PGC-1a expression, resveratrol supplementation was an effective method for managing HD (12).

5.11.1.10 Elderberry

The anti-inflammatory and antioxidant qualities of flavonoids, such as rutin and quercetin, and anthocyanins, which are abundant in elderberries, can account for their medicinal effects. It demonstrated that the antioxidant properties of the elderberry diet were linked to enhanced motor coordination and reduced cerebral infarcted tissue size. Elderberry powder contains quercetin, which has antioxidant qualities and can be taken consistently to cure neuropathies and motor coordination issues brought on by long-term drinking. Rutin guards against the long-term, erratic stress brought on by the loss of hippocampus neurons, which lessens anxiety and depressive symptoms while simultaneously enhancing motor coordination and locomotor function (41).

5.11.1.11 *Bacopa monnieri* – *Brahmi*

Its anti-inflammatory, antioxidative, and anti-apoptotic qualities have made it a promising pharmaceutical candidate. Alcoholic leaf extract from brahmi has been found to have a neuroprotective effect against 3-NPA-induced oxidative damage. By resuming the activities of crucial antioxidant enzymes such as superoxide dismutase, GSH reductase, GSH peroxidase, and thioredoxin reductase, it also enhances the antioxidant status of the striatal mitochondria (13).

5.11.1.12 *Calendula officinalis* L. – Marigold

Significant behavioural alterations and neuronal death in the brain's striatum were prevented by the *C. officinalis* flower extract containing 3-NP, an anti-inflammatory, antioxidant, and oestrogenic substance.

5.11.1.13 *Centella asiatica* – Gotu Kola

The plant's leaf powder and aqueous extract upregulated the activity of the plant that scavenges free radicals and reduced 3-NP-induced mitochondrial dysfunctions and oxidative stress in certain brain areas due to their antioxidant qualities (13).

5.11.1.14 *Convolvulus pluricaulis* Choisy – Shankhapushpi

Its biological relevance is attributed to the active ingredients, which include fatty acids, proteins, carbohydrates, alkaloids, coumarins, flavonoids, and glycosides. It has been suggested that it contains antioxidant, anti-apoptotic, and neuroprotective qualities.

5.11.1.15 *Garcinia kola* Heckel – Bitter Kola

It has been demonstrated that extracts of this species, including atrazine and kolaviron, protect human dopaminergic SH-SY5Y cells and PC12 cells.

5.11.1.16 *Gastrodia elata* Blume – Tian Ma

It prevents apoptosis in the dopaminergic neurons. Additionally, in PC-12 cell lines transfected transiently with mHtt, it inhibits the accumulation of mHtt by upregulating the activity of A(2A)-R, PKA, and proteasomes.

5.11.1.17 *Ginkgo biloba* L. – Maidenhair

Due to the variety of compounds it contains, such as flavanol glycosides and terpene lactones, the leaf extract has been used as a health supplement since ancient times. Antioxidative and anti-apoptotic properties are demonstrated by EGb 761.

5.11.1.18 *Olea europaea*

The component oleuropein comes from the leaf extract. It is a phenolic chemical, and the other products of its hydrolysis serve as antioxidants. Extra virgin olive oil blocks the neurotoxicity that 3-NP causes. By reducing LPO and halting GSH depletion, it acts as an antioxidant.

5.11.1.19 *Babchi*

The seed extracts from this plant exhibited exceptional protective properties; they increased ATP levels, decreased mitochondrial MP, and increased mitochondrial respiration. They also prevented 3-NP-induced cell death.

5.11.1.20 *Withania somnifera* L. – Ashwagandha

This plant possesses a multitude of significant pharmacological properties, including hemopoietic, immunomodulatory, antioxidant, and anticancer properties. It is made up of a class of bioactive substances called withanolides. By avoiding mitochondrial dysfunction, serving as an antioxidant, and altering behaviour, *W. somnifera* root extract served as a neuroprotective agent against the 3-NP challenge (13).

5.12 LIST OF PHYTOCONSTITUENTS AS OPTIONS FOR TREATMENT WITH NATURAL INGREDIENTS FOR HD

Now some of the phytoconstituents, which have been derived from various plant sources, and have been proven to be efficient in controlling the symptoms of HD have been listed out as follows.

5.12.1 Celastrol

The main active ingredient in the Chinese plant *T. wilfordi*, which is a member of the Celastraceae family, is celastrol. With several therapeutic actions against infections, inflammatory illnesses, cancer, diabetes, atherosclerosis, obesity, and mental disorders, it is a pentacyclic triterpenoid. Celastrol influences several antioxidant systems and cell signalling pathways, such as NF-kB, MAPK, JAK/STAT, and PI3K/Akt/mTOR signals. HSPs are a class of neuroprotective proteins that celastrol activates in a range of neurological circumstances. Celastrol boosted dopamine levels in the MPTP (1-methyl-4-phenyl-1,2,3,6-tetrahydropyridine) induced neurotoxicity paradigm via inducing HSP70 expression in dopaminergic neurons. Celastrol inhibited 3-NP-induced astrogliosis and striatal damage by inducing the heat shock response. This was accomplished by increasing the expression of the transcription factor HSF1 and dissolving PolyQ clumps in the cells. Moreover, several findings suggest that chaperones play a part in the early phases of HD pathogenesis by reducing polyglutamine toxicity. Interestingly, celastrol improves cognitive function by upregulating transcription factor EB expression, which is essential for autophagy and lysosomal-related genes, and prevents neurodegenerative diseases by preventing PTEN-induced kinase 1 (PIK1) from binding to the translocase mitochondrial protein TOM20. Therefore, celastrol may be utilized as a possible neuroprotective therapy for HD, according to the research (14, 49).

5.12.2 Curcumin

The main polyphenolic ingredient found in the rhizome of *Curcuma longa*, or turmeric, is curcumin. In India, it is most frequently used as a culinary element. The low-molecular-weight polyphenol curcumin gives the rhizome its yellow colour. Approximately 80% of all curcuminoids are made up of it. It has long been a mainstay in Ayurvedic therapy. Numerous medicinal benefits are evident in it, including hepatoprotective, analgesic, antioxidant, antibacterial, anti-fungal, anti-septic, anticarcinogenic, antiviral, anti-tumour, and anti-platelet actions. In the past few years, curcumin has been the focus of clinical research for a variety of illnesses, such as multiple myeloma, mastitis, diabetic nephropathy, Alzheimer's disease, pancreatic, colon and oral cancer, psoriasis, periodontal disease, recurring aphthous stomatitis, multiple myeloma, and precancerous lesions. It plays a significant function as a neuroprotective agent in HD at the moment. The exaggerated expression of the mHtt protein along with extended polyQ leads to a number of undesirable outcomes, such as cytotoxicity, inclusion body development, and cognitive and motor impairments. Supplementing with 10 μM curcumin dramatically improved motor dysfunction, decreased cytotoxicity, and prevented neuronal death in polyQ-mediated photoreceptor neurodegeneration. As a result, curcumin considerably reduced the onset and signs of HD in polyQ-induced neurodegeneration (14, 50, 51).

5.12.3 Epigallocatechin-3-Gallate

C. sinensis contains the polyphenolic flavonoid molecule EGCG, a primary catechin. It makes up between 50% and 80% of all catechins and is healthy. Its diverse therapeutic properties against cardiovascular disease, cancer, diabetes, and so on have been demonstrated in a number of recent studies. Additionally, it has antioxidant, antiviral, antibacterial, anti-inflammatory, and antifibrosis qualities. This nutraceutical has shown promise as a possible treatment for long-term neurological illnesses, including cerebral ischaemia, HD, Duchenne muscular dystrophy, and cognitive impairments. The pathogenic events linked to HD are induced by a variety of molecular processes, such as inflammation of the neurons and oxidative stress, excitotoxicity, mitochondrial toxicity, and other signalling pathways. All of the cellular processes mentioned above are successfully reversed by EGCG through its anti-inflammatory, anti-apoptotic, and antioxidant actions. In the hippocampus area of the HD brain, it effectively lowers MDA levels, returns normalcy to the functions of the enzymes like superoxide dismutase and GSH, and prevents the overexpression of caspase-3 (52, 53).

5.12.4 Ginsenoside

In *Panax ginseng* (ginseng, Araliaceae) and other *Panax* species, ginsenosides are the primary pharmacological component found in the root. Vasodilatory, immunomodulatory, anti-inflammatory, antioxidant, anticancer, antidiabetic, antiaging, and adaptogenic properties of this plant have been the subject of much investigation. The steroidal nucleus of triterpenoid glycosides, or ginsenosides, is attached to sugar moieties and forms their structural backbone. Plant parts containing over a hundred variants of ginsenosides have been identified and extracted (54).

5.12.5 Luteolin

Lutein (3,4,5,7-tetrahydroxyflavone), one of the significant flavonoids, is found in fruits, vegetables, and medicinal plants. Its antioxidant ability is the reason behind its anti-inflammatory, anticancer, antiviral, and protective characteristics.

According to a number of studies, luteolin can stop HD from progressing through changes in intracellular pathways, metal ion chelation, genetic modifications, and antioxidant and anti-inflammatory properties. Luteolin inhibits oxidative stress in neuronal cells via binding with many signalling components, including p53 and NF-kB, which are involved in intracellular pathways. Additionally, it improves mitochondrial dysfunction in neural cells and inhibits MAPK cascades to slow down NO generation in glial cells.

5.12.6 Rosmarinic Acid

The plant *Rosmarinus officinalis*, sometimes referred to as rosemary, belongs to the Laminaceae family and is the source of rosmarinic acid (RA), a polyphenolic compound used in herbal remedies. *Mellisa officinalis* (lemon balm), *Salvia lavandulifolia* (Spanish sage), *Ocimum tenuiflorum* (basil), *Salvia officinalis* (sage), and *Origanum majorana* (marjorum) are further noteworthy sources of rosmarinic acid. Anti-inflammatory, antibacterial, antiviral, depressive, anticarcinogenic, antioxidant, and chemotherapy prevention are just a few of the substances' many well-documented health benefits. There are already a plethora of herbal formulations and nutritional supplements on the market that include rhizome absorption, and all of them make health-promoting claims. Because of RA's potent antioxidant and preservation qualities, the food sector is very interested in this molecule.

To treat HD, solid lipid nanoparticles coated with rosmarinic acid have been created as an effective nasal medicine delivery system.

5.12.7 Alpha Mangostin

Cerebellar granule neurons produced in primary cultures with 3-NP showed concentration-dependent reductions in neuronal mortality and ROS production in response to α-mangostin.

5.12.8 Cannabigerol

It functions as a neuroprotective drug, normalizing the expression of several genes such as PPAR-gamma, IGF-1, and BDNF, as well as the set of genes associated with HD.

5.12.9 Epigallocatechin-Gallate

It has anti-apoptotic and anticancer qualities, because it has anti-beta-A triggered neuronal apoptotic properties. It enhances motor function, suppresses polyQ-mediated htt protein accumulation, and prevents the accumulation of mutated htt exon 1 (Httex1) in mice displaying HD-like symptoms (14).

5.13 CONCLUSION

Natural substitutes made from plants have demonstrated a great deal of promise in enhancing patients' quality of life through symptomatic or therapeutic remedies. Studies on these plants and their phytoconstituents demonstrate how well they can treat a range of health issues. By focusing on several physiological pathways, these natural remedies provide comprehensive advantages that may alleviate symptoms and promote general health. These treatments have a supporting or supplementary function in the management of chronic illnesses, especially those where traditional treatments might not be sufficient, by utilizing the special qualities of these bioactive substances.

When it comes to treating HD, these plant-based solutions may be a decent alternative to traditional pharmaceuticals. Natural therapies often provide a better degree of safety and tolerability than many manufactured pharmaceuticals, which frequently have a variety of side effects. They are an appealing option because of their capacity to change how the condition progresses or reduce symptoms without endangering others. As research into their mechanisms advances, these natural compounds might emerge as mainstream therapeutic options or competitive alternatives in the treatment of neurodegenerative disorders.

REFERENCES

1. Mayo Clinic. Huntington's disease – diagnosis and treatment – Mayo Clinic [Internet]. [cited 2024 Mar 29]. Available from: https://www.mayoclinic.org/diseases-conditions/huntingtons-disease/diagnosis-treatment/drc-20356122
2. Alzheimer's Association. Huntington's disease (HD) | symptoms & treatments | alz.org [Internet]. [cited 2024 Mar 18]. Available from: https://www.alz.org/alzheimers-dementia/what-is-dementia/types-of-dementia/huntington-s-disease
3. National Institute of Neurological Disorders and Stroke. Huntington's disease | National Institute of Neurological Disorders and Stroke [Internet]. [cited 2024 Mar 18]. Available from: https://www.ninds.nih.gov/health-information/disorders/huntingtons-disease
4. Ross CA, Tabrizi SJ. Huntington's disease: From molecular pathogenesis to clinical treatment. *The Lancet Neurology*. 2011 Jan 1;10(1):83–98.
5. Illarioshkin SN, Klyushnikov SA, Vigont VA, Seliverstov YuA, Kaznacheyeva EV. Molecular pathogenesis in Huntington's disease. *Biochemistry Moscow*. 2018 Sep 1;83(9):1030–9.

6. Huang WJ, Chen WW, Zhang X. Huntington's disease: Molecular basis of pathology and status of current therapeutic approaches (review). *Experimental and Therapeutic Medicine*. 2016 Oct 1;12(4):1951–6.
7. Landles C, Bates GP. Huntingtin and the molecular pathogenesis of Huntington's disease. *EMBO Reports*. 2004 Oct;5(10):958–63.
8. Roos RA. Huntington's disease: A clinical review. *Orphanet Journal of Rare Diseases*. 2010 Dec 20;5(1):40.
9. Frank S. Treatment of Huntington's disease. *Neurotherapeutics*. 2014 Jan;11(1):153–60.
10. Bachoud-Lévi AC, Ferreira J, Massart R, Youssov K, Rosser A, Busse M, Craufurd D, Reilmann R, De Michele G, Rae D, Squitieri F. International guidelines for the treatment of Huntington's disease. *Frontiers in Neurology* [Internet]. 2019 Jul 3 [cited 2024 Mar 29];10. Available from: https://www.frontiersin.org/journals/neurology/articles/10.3389/fneur.2019.00710/full
11. Irfan Z, Khanam S, Karmakar V, Firdous SM, Abou El Khier BSI, Khan I, Rehman MU, Khan A. Pathogenesis of Huntington's disease: An emphasis on molecular pathways and prevention by natural remedies. *Brain Sciences*. 2022 Oct 14;12(10):1389.
12. Gupta M, Sanjana, Singh N, Singh B, Alam P. Role of natural products in alleviation of Huntington's disease: An overview. *South African Journal of Botany*. 2022 Dec 1;151:263–276.
13. Banerjee S, Tudu CK, Nandy S, Pandey DK, Ghorai M, Shekhawat MS, Ghosh A, Nongdam P, Al-Tawaha AR, Bursal E, Batiha GE. Herbal remedies against Huntington's disease: Preclinical evidences and future directions. In: Sarwat M, Siddique H, editors. *Herbal Medicines* [Internet]. Academic Press; 2022 [cited 2024 Apr 15], Pages 37–69. Available from: https://www.sciencedirect.com/science/article/pii/B978032390572500010X.
14. Singh P, Mishra G, Molla M, Yimer YS, Sisay W, Andargie Y, Ewunetie A. Dietary and nutraceutical-based therapeutic approaches to combat the pathogenesis of Huntington's disease. *Journal of Functional Foods*. 2022 May 1;92:105047.
15. Chintamaneni PK, Nagasen D, Babu KC, Mourya A, Madan J, Srinivasarao DA, Ramachandra RK, Santhoshi PM, Pindiprolu SK. Engineered upconversion nanocarriers for synergistic breast cancer imaging and therapy: Current state of art. *Journal of Controlled Release*. 2022 Dec 1;352:652–672. https://doi.org/10.1016/j.jconrel.2022.10.056.
16. Pindiprolu SK, Madhan J, Srinivasarao DA, Kumar CS, Dasari N, Jyothi VG, Babu KC. Therapeutic targeting of aberrant sialylation for prevention of chemoresistance and metastasis in triple negative breast cancer. *Journal of Drug Delivery Science and Technology*. 2023 Jun 14:104617. https://doi.org/10.1016/j.jddst.2023.104617.
17. Dasari N, Balla S, Kondrapu P, Gummadi R, Surada N, Kondru U M, Pindiprolu SKSS. Targeting angiogenesis with fluphenazine-zinc oxide nanoconjugates: A potential mechanism for improving antipsychotic efficacy. 2023. https://doi.org/10.22159/ijap.2023v15i5.48317.
18. Aminabee S, Dasari SK, Rao KH, Sirisha V, Anitha V, Kumari ND, Sujiya B, Leelavati TS. Pharmacoeconomic analysis of biologic vs. biosimilar therapies in rheumatoid arthritis. *International Journal of Chemical and Biochemical Sciences*. 2023;24(4):395–400.
19. Dasari N, Guntuku GS, Pindiprolu SKSS. Targeting triple negative breast cancer stem cells using nanocarriers. *Discover Nano*. 2024;19:41. https://doi.org/10.1186/s11671-024-03985-y.
20. Dasari N, Phani Kumar CS, Gummadi RK, Pindiprolu SK. Application of novel natural sweetening agent-stevia in formulation, evaluation of nicardipine hydrochloride orodispersable for rapid absorption. *Indian Journal of Pharmaceutical Education and Research*. 2024;58(1s):s176–s186. https://doi.org/10.5530/ijper.58.1s.17.
21. Ajitkumar A, De Jesus O. Huntington disease. [Updated 2023 Aug 23]. In: *StatPearls* [Internet]. Treasure Island (FL): StatPearls Publishing; 2024 Jan-. Available from: https://www.ncbi.nlm.nih.gov/books/NBK559166/.
22. Choudhary S, Kumar P, Malik J. Plants and phytochemicals for Huntington's disease. *Pharmacognosy Reviews*. 2013 Jul;7(14):81–91. https://doi.org/10.4103/0973-7847.120505.
23. Gonzalez Rojas N, Cesarini M, Peker G, Da Prat G, Etcheverry J, Gatto E. Review of Huntington's disease: From basics to advances in diagnosis and treatment. *Journal of Neurology Research*. 2022;12(3):93–113.
24. Nanda B, Nandy S, Mukherjee A, Dey A. Plant-based natural products against Huntington's disease: Preclinical and clinical studies. 2019. https://doi.org/10.1201/9780429265204-7.
25. Hung CY, Zhu C, Kittur FS, He M, Arning E, Zhang J, Johnson AJ, Jawa GS, Thomas MD, Ding TT, Xie J. A plant-based mutant huntingtin model-driven discovery of impaired expression of GTPCH and DHFR. *Cellular and Molecular Life Sciences*. 2022;79:553. https://doi.org/10.1007/s00018-022-04587-6.
26. Kim A, Lalonde K, Truesdell A, Gomes Welter P, Brocardo PS, Rosenstock TR, Gil-Mohapel J. New avenues for the treatment of Huntington's disease. *International Journal of Molecular Sciences*. 2021; 22(16):8363. https://doi.org/10.3390/ijms22168363.

27. Lum PT, Sekar M, Gan SH, Bonam SR, Shaikh MF. Protectiveeffect of natural products against Huntington's disease: An overview of scientific evidence and understanding their mechanism of action. *ACS Chemical Neuroscience*. 2021;12(3):391–418. https://doi.org/10.1021/acschemneuro.0c00824
28. de Tommaso M, Serpino C, Sciruicchio V. Management of Huntington's disease: Role of tetrabenazine. *Therapeutics and Clinical Risk Management*. 2011;2011:123–129.
29. Paleacu D, Anca M, Giladi NJANS. Olanzapine in Huntington's disease. *Acta Neurologica Scandinavica*. 2002;105(6):441–444.
30. Ciammola A, Sassone J, Colciago C, Mencacci NE, Poletti B, Ciarmiello A, Squitieri F, Silani V. Aripiprazole in the treatment of Huntington's disease: A case series. *Neuropsychiatric Disease and Treatment*. 2009:1–4.
31. Landwehrmeyer GB, Dubois B, de Yébenes JG, Kremer B, Gaus W, Kraus PH, Przuntek H, Dib M, Doble A, Fischer W, Ludolph AC. Riluzole in Huntington's disease: A 3-year, randomized controlled study. *Annals of Neurology: Official Journal of the American Neurological Association and the Child Neurology Society*. 2007;62(3):262–272.
32. Sharifi-Rad M, Lankatillake C, Dias DA, Docea AO, Mahomoodally MF, Lobine D, Chazot PL, Kurt B, Boyunegmez Tumer T, Catarina Moreira A, Sharopov F. Impact of natural compounds on neurodegenerative disorders: From preclinical to pharmacotherapeutics. *Journal of Clinical Medicine*. 2020;9(4):1061. https://doi.org/10.3390/jcm9041061.
33. Pohl F, Kong Thoo Lin P. The potential use of plant natural products and plant extracts with antioxidant properties for the prevention/treatment of neurodegenerative diseases: In vitro, in vivo and clinical trials. *Molecules*. 2018;23(12):3283. https://doi.org/10.3390/molecules23123283.
34. Rahman MH, Bajgai J, Fadriquela A, Sharma S, Trinh TT, Akter R, Jeong YJ, Goh SH, Kim C-S, Lee K-J. Therapeutic potential of natural products in treating neurodegenerative disorders and their future prospects and challenges. *Molecules*. 2021;26(17):5327. https://doi.org/10.3390/molecules26175327.
35. Wahid M, Ali A, Saqib F, Aleem A, Bibi S, Afzal K, Ali A, Baig A, Ali Khan S, Bin Asad MHH. Pharmacological exploration of traditional plants for the treatment of neurodegenerative disorders. *Phytotherapy Research*. 2020;34:3089–3112. https://doi.org/10.1002/ptr.6742
36. Myers RH. Huntington's disease genetics. *Neurotherapeutics*. 2004;1:255–262. https://doi.org/10.1602/neurorx.1.2.255
37. Rosenblatt A. Neuropsychiatry of Huntington's disease. *Dialogues in Clinical Neuroscience*. 2007;9(2):191–197. https://doi.org/10.31887/DCNS.2007.9.2/arosenblatt
38. Paulsen JS, Ready RE, Hamilton JM, Mega MS, Cummings JL. Neuropsychiatric aspects of Huntington's disease. *Journal of Neurology, Neurosurgery & Psychiatry*. 2001;71:310–314.
39. Wyant KJ, Ridder AJ, Dayalu P. Huntington's disease – update on treatments. *Current Neurology and Neuroscience Reports*. 2017;17:33. https://doi.org/10.1007/s11910-017-0739-9
40. Bhat SA, Ahamad S, Dar NJ, Siddique YH, Nazir A. Theemerginglandscape of natural small-molecule therapeutics for Huntington's disease. *Current Neuropharmacology*. 2023;21(4):867–889. https://doi.org/10.2174/1570159X21666230216104621
41. Moghaddam MH, Bayat AH, Eskandari N, Abdollahifar MA, Fotouhi F, Forouzannia A, et al. Elderberry diet ameliorates motor function and prevents oxidative stress-induced cell death in rat models of Huntington disease. *Brain Research*. 2021 Jul 1;1762:147444.
42. Gupta M, Singh N, Singh B, Alam P. Role of natural products in alleviation of Huntington's disease: An overview. *South African Journal of Botany*. 2022;151:263–276.
43. Ehrnhoefer DE, Duennwald M, Markovic P, Wacker JL, Engemann S, Roark M, Legleiter J, Marsh JL, Thompson LM, Lindquist S, Muchowski PJ, Wanker EE. Green tea (–)-epigallocatechin-gallate modulates early events in huntingtin misfolding and reduces toxicity in Huntington's disease models. *Human Molecular Genetics*. 2006;15(18):2743–2751.
44. Shivasharan BD, Nagakannan P, Thippeswamy BS, Veerapur VP, Bansal P, Unnikrishnan MK. Protective effect of *Calendula officinalis* Linn. flowers against 3-nitropropionic acid induced experimental Huntington's disease in rats. *Drug and Chemical Toxicology*. 2013;36(4):466–473.
45. Gray NE, Magana AA, Lak P, Wright KM, Quinn J, Stevens JF, Maier CS, Soumyanath A. *Centella asiatica*: Phytochemistry and mechanisms of neuroprotection and cognitive enhancement. *Phytochemistry Reviews*. 2018;17:161–194.
46. Dey A, De JN. Neuroprotective therapeutics from botanicals and phytochemicals against Huntington's disease and related neurodegenerative disorders. *Journal of Herbal Medicine*. 2015;5(1):1–19.
47. Huang CL, Yang JM, Wang KC, Lee YC, Lin YL, Yang YC, Huang NK. Gastrodia elata prevents huntingtin aggregations through activation of the adenosine A2A receptor and ubiquitin proteasome system. *Journal of Ethnopharmacology*. 2011;138(1):162–168.

48. Huang ZN, Lee SY, Chen JM, Huang ZT, Her LS. Oleuropein enhances proteasomal activity and reduces mutant huntingtin-induced cytotoxicity. *Frontiers in Pharmacology*. 2024;15:1459909.
49. Silakari P, Yadav A, Arora A, Arora A, Gulsheen, Kaur P, Sahu SK. Investigating holistic natural strategies for the management of Huntington's disease. *BIO Web of Conferences*. 2024;86:01041.
50. Dey A. Naturalproducts against Huntington's disease (HD): Implications of neurotoxic animal models and transgenics in preclinical studies. In: Brahmachari G, editor. *Neuroprotective Natural Products*. 2017. https://doi.org/10.1002/9783527803781.ch8
51. Imarisio S, Carmichael J, Korolchuk V, Chen C-W, Saiki S, Rose C, Krishna G, Davies JE, Ttofi E, Underwood BR, Rubinsztein DC. Huntington's disease: From pathology and genetics to potential therapies. *Biochemical Journal*. 1 June 2008;412(2):191–209. https://doi.org/10.1042/BJ20071619
52. Borrell-Pages M, Zala D, Humbert S, Saudou F. Huntington's disease: From huntingtin function and dysfunction to therapeutic strategies. *Cellular and Molecular Life Sciences*. 2006;63(22):2642–2660. https://doi.org/10.1007/s00018-006-6242-0.
53. Gil JM, Rego AC. Mechanisms of neurodegeneration in Huntington's disease. *European Journal of Neuroscience*. 2008;27:2803–2820. https://doi.org/10.1111/j.1460-9568.2008.06310.x
54. Novak MJU, Tabrizi SJ. Huntington's disease: Clinical presentation and treatment. In: Brotchie J, Bezard E, Jenner P, editors. *International Review of Neurobiology*. Academic Press, Vol 98; 2011, Pages 297–323. ISSN 0074-7742, ISBN 9780123813282, https://doi.org/10.1016/B978-0-12-381328-2.00013-4.

6 Lewy Body Disease and Its Therapeutic Regimen Using Conventional Methods

Sagarkanta Mohanty,[1] *Srikanth Jupudi,*[1] *Neeru Dugar,*[1] *Swati Swagatika Swain,*[2] *and Nayudu Teja,*[2] *and Karri V. V. S. Narayana Reddy*[2]

[1]Department of Pharmaceutical Chemistry, JSS College of Pharmacy, JSS Academy of Higher Education & Research, Ooty, Tamil Nadu, India

[2]Department of Pharmaceutics, JSS College of Pharmacy, JSS Academy of Higher Education & Research, Ooty, Tamil Nadu, India

6.1 INTRODUCTION

Neurological disease is an emerging health concern as more than three billion people worldwide live with it, according to a study released by The Lancet Neurology in 2021 (Grisold, 2024). Neurodegenerative diseases are a group of progressive neurological diseases, which have various conditions characterized by selective disorder, which continues with loss of neurons, glial cells, and the neural networks in the brain and spinal cord (Taryn Berman, 2018). It has a considerable impact on worldwide health, such as the sixth leading cause of death in the United States; Alzheimer's disease (AD) and Parkinson's disease (PD) are the two most common causes of death in the United States, with dementia with Lewy bodies (DLB) coming in third with 52 cases per 100,000 (Kane et al., 2018). The prevalence of neurodegenerative disorder doubled from 4.0% (95% UI 3.2–5.0) in 1990 to 8.2% (6.6–10.2) in 2019 in India (Priya Kirushnakumar et al., 2024). α-Synucleinopathies are a disconcerting group of conditions that have long plagued neurodegenerative disorders. They are characterized by the abnormal accumulation and aggregation of the α-synuclein protein within the brain. LBD is defined by the atypical aggregation of α-synuclein, which is known as Lewy bodies and Lewy neurites in neurons and neuronal processes (Lewy-related pathology) among the α-synucleinopathies. Compared to clinically recognized cases, the true prevalence of LBD and the mixed disease of AD with LBD is probably much higher because symptoms of both are often similar, resulting in the misdiagnosis of LBD. LBD is a heterogeneous neurodegenerative condition that is the second-most frequent reason for neurodegenerative dementia after AD. PD, Parkinson's disease dementia (PDD), DLB, and several uncommon syndromes are all included in LBD, which is the most common synucleinopathy (Lin and Truong, 2019). LBD has symptoms such as a range of sleep, motor, autonomic, neuropsychiatric, and cognitive symptoms featuring DLB and PDD. LBD comprises 4–8% of patients with dementia, which is indicated by clinic-based studies (Prasad et al., 2023). There is a prevalence of about 80% of people having PD who can be diagnosed with LBD, according to the study of Hely and Reid et al. Lewy bodies were first identified in 1912 in the brainstem of PD patients, and their distribution has been found in the cortex and the autonomic nervous system. The DLB concept was first submitted by Dr. Kenji Kosaka through a case report in 1976. In 1980, the LBD term was put forward and divided into three classes based on the distribution of Lewy bodies: Type A (brainstem type), Type B (transitional type), and Type C (diffuse type). The cerebral type, another type of LBD, in which Lewy bodies were located in the cerebral cortex but

DOI: 10.1201/9781003478652-7

rarely found in the brainstem, was added in 1996 (Lin and Truong, 2019). The cerebral type has no Parkinsonian symptoms throughout the period of the disease, while the brainstem type is in tune with PD. LBD was mentioned by numerous names, including Lewy body disease (LBD), Lewy body dementia, diffuse LBD, and DLB. The clinicians were confused by the mixture of these terms until they reached a certain degree of concurrence. The aggregation of α-synuclein was identified to be the key factor of Lewy bodies in the 1990s.

6.2 SHORT HISTORICAL REVIEW OF THE LBD DIAGNOSTIC CRITERIA

The first international workshop on LBD was organized in 1995 by a group of LBD specialists, who issued consensus guidelines for clinical and pathological diagnosis of the disease in Newcastle upon Tyne. The first consensus guidelines of the diagnostic criteria for LBD, which include the three core characteristics of Parkinsonism – visual hallucinations, cognitive function fluctuations, and spontaneous motor symptoms – were specified in the workshop and were published in 1996 (Yamada et al., 2019). The second international workshop on DLB convened in 1998 in Amsterdam, and there was no revision of the diagnostic criteria in the second consensus report, which was published in 1999 (McKeith et al., 1999). After reviewing and modifying these criterias, in 2025 DLB Consortium symptoms and indicators like rapid eye movement (REM) sleep behaviour disorder, extreme neuroleptic sensitivity, and decreased striatal dopamine transporter activity on functional neuroimaging were given more diagnostic weight. The pathological criteria for diagnosis were upgraded which features the establishment of suggestive features such as REM sleep behavioural disorder, severe neuroleptic sensitivity, and low dopamine transporter uptake in basal ganglia demonstrated by single photon emission computed tomography (SPECT) or positron emission tomography (PET) imaging, such as 123iodine-labelled *N*-(3-fluoropropyl)-2β-carbomethoxy-3β-(4-iodophenyl) nortropane (123I-FP-CIT) SPECT and a temporal criterion that differentiates DLB from dementia in PD (; Yamada et al., 2019). The fourth International Workshop on LBD and PDD was held in Yokohama (McKeith et al., 2017). There was no revision of the criteria because it was just after the publication of the 2005 revised criteria. The International LBD Conference took place in Fort Lauderdale, Florida, in December 2015. The latest fourth consensus report of the DLB Consortium in 2017 was released combining the new diagnostic criteria for diagnosis of LBD. The most recent International Lewy Body Dementia Conference was held on June 15–17, 2022, at Newcastle upon Tyne, UK (Table 6.1). Stages of Lewy body dementia are shown in Figure 6.1.

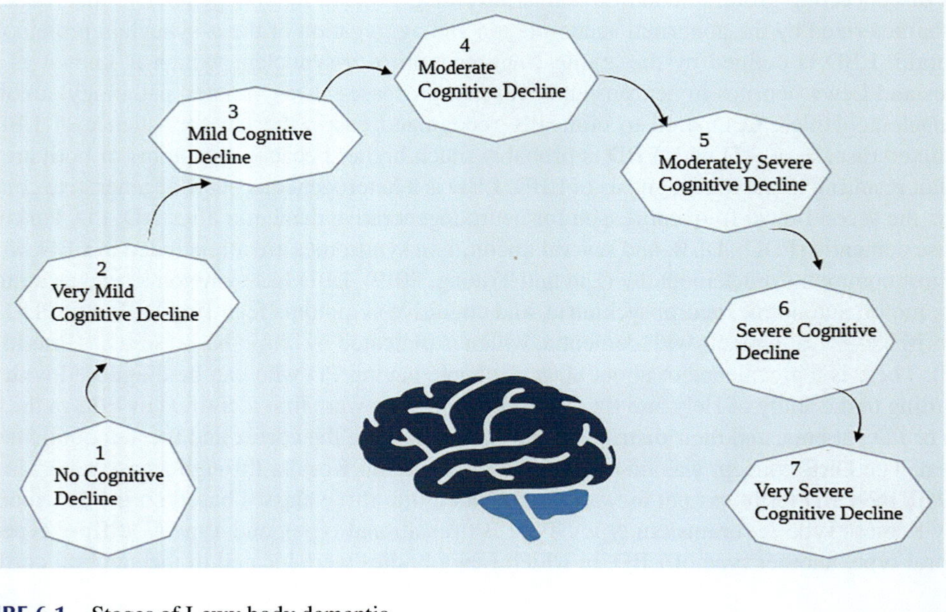

FIGURE 6.1 Stages of Lewy body dementia.

TABLE 6.1
Diagnostic Criteria for Lewy Body Disease

S. No	Criteria	Description
1	Central feature	• Essential for a diagnosis of LBD Dementia. • In the early stages, attention, executive function, and subperceptual impairment may be more noticeable than memory impairment.
2	Core clinical feature	The first three typically occur early and may persist throughout the course • Fluctuating cognition • Recurrent visual hallucinations • Rapid eye movement • Parkinsonism like bradykinesia, tremor, or rigidity
3	Supportive clinical feature	• Antipsychotic agents' sensitivity • Postural instability • Severe autonomic dysfunction • Hypersomnia • Hyposmia • Systematized delusions • Depression
4	Indicative biomarkers	• Reduced dopamine transporter uptake in the basal ganglia demonstrated by SPECT/PET • Abnormal (low uptake) 123I-MIBG myocardial scintigraphy • Polysomnographic confirmation of REM sleep without atonia
5	Supportive biomarkers	• preservation of medial temporal lobe structures on CT/MRI scan
6	Probable Lewy body disease	• Two or more core clinical features are present • Only one core clinical feature + one or more indicative biomarkers
7	Possible Lewy body disease	• Only one core clinical feature • One or more indicative biomarkers but no core clinical feature

6.3 STAGES OF LEWY BODY DEMENTIA

LBD can progress through mostly seven stages according to doctors and experts. The stages are:

Stage I: No cognitive decline:
In this stage the patient, doctors, and caretaker are not able to identify the symptoms.

Stage II: Very mild cognitive decline:
Appearance of slight changes, like forgetfulness

Stage III: Mild cognitive decline:
The loved ones/caretakers can notice the symptoms such as mild memory loss, forgetfulness, and concentration problems.

Stage IV: Moderate cognitive decline:
At this stage the individual with LBD has evidence of forgetfulness and difficulty performing daily tasks. The diagnosis of the LBD can be possible at stage IV.

Stage V: Moderately severe cognitive decline:
In this stage, individuals with LBD have remarkable memory issues and may require assistance with certain tasks.

Stage VI: Severe cognitive decline:
Memory loss at this stage is severe. Many individuals experience incontinence, psychological changes, and speech impairment. For daily life maintenance, they require assistance.

Stage VII: Very severe cognitive decline:

In this stage, patients are unable to communicate or move. For daily life maintenance, they need extensive assistance. This stage lasts 1.5–2.5 years (Susan Bernstein, 2024).

6.4 DIFFERENCE BETWEEN LBD AND PD

LBD and PD share several characteristics, like both are caused by protein deposition, which consists of α-synuclein called Lewy bodies. They tend to deteriorate over time because both are chronic and progressive. PD is a movement disorder, whereas LBD is a cognitive disorder; this is the key difference between both (Marcin, 2023; Borghammer et al., 2024). Specifically, both have differed in the following ways:

LBD: It is caused by α-synuclein protein deposition, called Lewy bodies in the brain, which affects cognition. It is a type of dementia that characterizes difficulty in thinking, language, and memory.

PD: It is caused by the demise of nerve cells in the basal ganglia, which are involved in various functions such as motor control, cognitive and emotional states, and learning. It causes erratic and ungovernable movements (tremors, shaking, etc.) and coordination and balancing issues. The dementia will develop over time (Marder, 2010).

The symptoms of LBD and PD may overlap as each progresses, which creates difficulty in diagnosis. To overcome this problem, the doctors use the one-year rule, which demonstrates (Yamada et al., 2019) (see Table 6.2):

LBD: Dementia symptoms should commence within one year of symptoms.

PD: Dementia symptoms typically start after one year of symptoms.

TABLE 6.2
Symptoms of LBD and PD

Lewy Body Disease	Parkinson's Disease
• Fluctuations in thinking and memory	• Muscle stiffness
• Hallucinations	• Difficulty in movement
• Confusion	• Tremor
• Sleeping issues	• Balance and coordination issues
	• Cognitive issues (later onset)

6.5 DIFFERENCE BETWEEN LBD AND AD

LBD: It is caused by α-synuclein protein deposition, called Lewy bodies in the brain, which affects cognition. It is a type of dementia which characterizes difficulty in thinking, language, and memory.

AD: It is the most common cause of dementia. It starts with the arrival of the protein buildup in the form of amyloid plaques and neurofibrillary tangles in the entorhinal cortex (EC) and hippocampus of the brain. Over time, these consequences in the death of brain cells, the shrinkage of the brain, and memory loss (see Table 6.3).

LBD and AD can seem similar in terms of symptoms at the early stage of both; it is quite common for a person with LBD to be mistakenly diagnosed with AD. Here are the main differences between LBD and AD (Heerema, 2024):

TABLE 6.3
Symptoms of LBD and AD

Lewy Body Disease	Alzheimer's Disease
• Age: Appears around 50 but can be earlier • Cause: Deposition of α-synuclein protein, called Lewy bodies, in the brain stem, the olfactory bulb, or Auerbach's plexus of the lower oesophagus. • Risk factors: Age, Family history of LBD, genetics, male gender (slightly more than female), viral infections, anxiety, and depression • Symptoms: Visual hallucinations, movement disorders, cognitive problems, varying attention, depression, apathy • Treatment: It has no cure. Medication can help manage symptoms for a period of time. • Prognosis: Five to eight years is the average survival time after diagnosis	• Age: Appears around mid-60s but can be earlier • Cause: Arrival of the protein buildup in the form of amyloid plaques and neurofibrillary tangles in the entorhinal cortex (EC) and hippocampus of the brain. • Risk factors: Age, family history of Alzheimer's condition, genetics, female gender (slightly more than male), medical conditions, lifestyle factors • Symptoms: Memory loss, confusion, hallucinations, and delusions, Language problems • Treatment: It has no cure. Medication and targeted treatments can slow the cognitive decline • Prognosis: Varies from person to person depending upon factors like age, rate of progression, and degree of impairment

6.6 PATHOGENESIS OF LBD

LBD is a complex and debilitating neurological condition that involves the gradual deterioration and dysfunction of various brain regions due to the deposition of aberrant protein aggregates known as Lewy bodies. The primary hallmark of this disease is the loss of dopaminergic neurons in the substantia nigra, which leads to the characteristic motor symptoms linked with PD. However, as the pathological process progresses, Lewy bodies can also spread to other brain areas, contributing to the development of non-motor symptoms, such as cognitive impairment, dementia, and autonomic dysfunction (Prasad et al., 2023; Tanei et al., 2021).

The pathogenesis of LBD is believed to involve the misfolding and aggregation of the protein α-synuclein, which is normally found in the presynaptic terminals of neurons. Mutations in the gene encoding α-synuclein or its overproduction can trigger the formation of these pathogenic protein aggregates, which then propagate throughout the brain, corrupting and killing healthy cells in a prion-like manner. Emerging evidence suggests that extracellular vesicles, such as exosomes, may play a crucial role in the intercellular transfer of α-synuclein (Jellinger, 2009), facilitating the spread of Lewy body pathology and contributing to the progressive nature of the disease. The α-synuclein protein contains 140 amino acids encoded by the SNCA gene on 4q21. Also, this protein has two other spliced variants: 126-AA, lacking exon 3, and 112-AA, lacking exon 5, which is separate from the 140-AA segment. The protein is exhibited throughout the nuclei and presynaptic areas. The substantia nigra, locus ceruleus, hypothalamus, nucleus basalis, cranial nerve motor nuclei, and the central and peripheral divisions of the autonomic nervous system are just a few of its numerous locations. Lewy bodies are present in the cerebral cortex neurons, especially in the limbic system's deep layers (V and VI). The α-synuclein protein is composed of three constitutional regions: (1) amphipathic N-terminal region, which contains 11 AA repeating, includes the sequence KTKEGV, which are the key factors for tetramerization of α-synuclein and are useful in the formation of α helix. (2) Central hydrophobic region, which contains the non-amyloid component region that is useful in the aggregation of protein. (3) C-terminal region, which is acidic in nature and rich in proline (Prasad et al., 2023). Abnormal α-synuclein proteins undergo phosphorylation, truncation, and nitration and also have abnormal solubility, which leads to the formation of oligomeric species that aggregate into fibrils and are ubiquitinated, are shown in Figure 6.2.

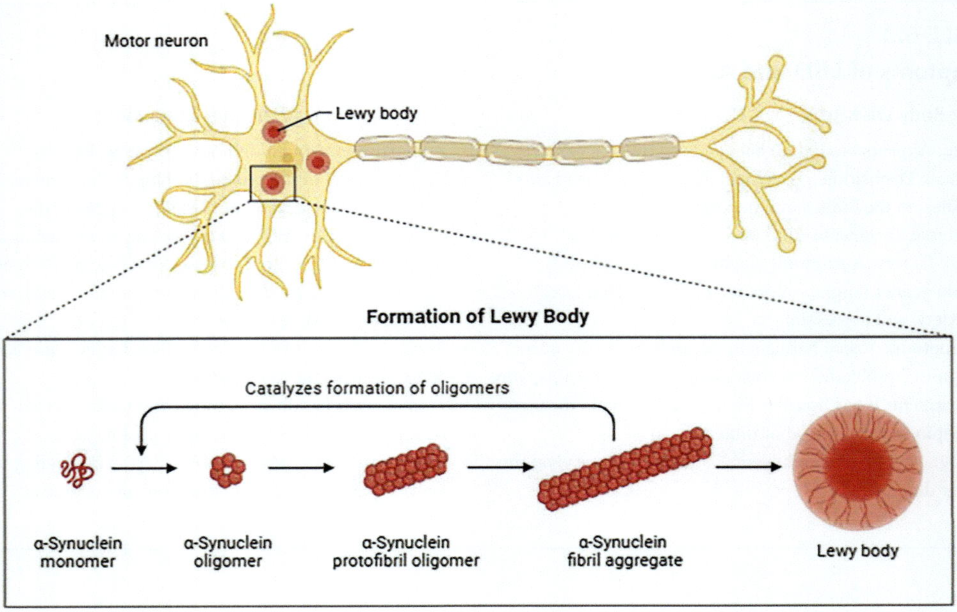

FIGURE 6.2 Formation of Lewy Body

Abnormal α-synuclein proteins undergo phosphorylation, truncation, and nitration and also have abnormal solubility, which leads to the formation of oligomeric species that aggregate into fibrils and are ubiquitinated. This leads to dysfunction and or loss of neurons, which leads to significant dopamine dysfunction, marked cholinergic pathology, and dysfunction of other neurotransmitters. The deterioration of memory and learning is led by the loss of acetylcholine-producing neurons, whereas the deterioration of sleep, behaviour, cognition, movement, and mood is led by the death of dopamine-producing neurons. The primary diagnostic criteria for LBD are associated with the following symptoms and the destruction of these neurons. As per a recent study, microglia can be activated by α-synuclein in different conformations. In LBD, microglial activation is observed in vivo. Studies on pathology and biomarkers provide more proof of inflammation. The cerebrospinal fluid biomarkers associated with LBD are neurofilament light chain, amyloid-β and tau proteins (Ferreira et al., 2020), α-synuclein, glucocerebrosidase activity, neuron-specific enolase, contactin-1, chromogranin A, neurosecretory protein VGF, and YKL-40 protein, whereas serum and plasma biomarkers are α-synuclein, which is present in serum and plasma and serum tyrosine (Armstrong et al., 2021). The Lewy body pathology associated with LBD can be further subdivided into three entities: pure autonomic failure, DLB, and sporadic PD. While these subtypes are nearly indistinguishable at a neuropathological level, there is a growing consensus that, clinically, they are not closely related, if not identical.

6.7 SYMPTOMS OF LBD

Dementia is identified as a progressive condition that is clinically significant enough to disrupt or impact social relationships and/or professional employment or the activities of daily living. For the diagnosis of LBD, the presence of dementia is required, which is an essential criterion. The clinical features of LBD overlap with other diseases of dementia onset, in particular, AD. Nevertheless, it is important to note the major differences in the occurrence of certain cognitive deficits among various pathological processes. In the course of research by Auning et al., where the presenting symptoms of DLB and AD were compared, the most common presenting symptom in patients with DLB was memory loss at 57%. This was then followed by 44% of the patients who had visual hallucinations, 34% who were depressed, 33% who were unable to solve problems, 28% who had disturbances in gait, and 25% of them had tremor and rigidity. However, in the AD group, 99% of participants had memory loss, explaining that it was the most common symptom, then problem-solving difficulties with 28% and depression with 24%. According to the result of this study, the memory impairment is very often as dominant symptom of the LBD, but not as normally seen in the case of AD (Taylor et al., 2020).

Further research sought to determine the chronicle of manner of symptoms in LBD and establish a precise distinction which subsist within the groups of men and women. For men, sleep disturbances such as REM sleep behaviour disorder (RBD) were considered first and then a crisis of cognitive deficits and Parkinsonian symptoms, fluctuation, and visual hallucination occurred. In women, RBD and cognitive symptoms appeared at almost the same intermediate ages. A noticeable difference was found because, even though other core features of DLB were similar, cognitive symptoms were reported to develop at an older age. Women presented with a significantly older median age of the first visit than men and had a lower mean score on cognitive testing (Prasad et al., 2023).

The most common symptoms found in LBD are neuropsychiatric symptoms, motor symptoms, movement disorders, poor regulation of body functions, cognitive problems, sleep disturbances such as RBD, varying attention, depression, falls, autonomic dysfunction, REM sleep behaviour disorder, hallucinations in other modalities, and systematized delusions (Taylor et al., 2020).

- **Neuropsychiatric Symptoms:** Patients with LBD have a variety of neuropsychiatric symptoms such as visual hallucinations and hallucinations in other sensory modalities, systematized delusions, apathy, aggression, anxiety, and depression. Depression occurs in about one-third of patients with LBD who have depression and is often accompanied by anxiety. Other neuropsychiatric symptoms are:
  - Auditory hallucinations
  - Delusions
  - Apathy
  - Agitation and psychosis
- **Motor Symptoms:**
  About 85% of patients with LBD have the motor symptoms, which include one of the core features of LBD in diagnostic criteria, which is an umbrella term for brain disorders that result in problems with movement, such as bradykinesia or slowed movements, rigidity or stiffness, and tremors. Significant movement problems may not appear for years in some LBD patients, but they may appear early in others. At first, motor symptoms may be very indefinite and easy to miss. Other motor symptoms of LBD include:
  - Imbalance, shakiness
  - Stumbling walk
  - Impaired swallowing
  - Hypomimia
  - Frequent falls
  - Ataxia
  - Smaller handwriting than what's normal for the person
- **Cognitive Problems/Impairment:**
  The cognitive problems associated with LBD are often severe and can significantly impact an individual's quality of life, presenting a significant challenge for clinicians and caregivers. Cognitive dysfunction in LBD is thought to arise from the widespread distribution of

Lewy body pathology throughout the cortex and subcortical regions of the brain. The presence of these Lewy bodies can disrupt normal neuronal function and lead to a decline in various cognitive domains, including attention, executive function, visuospatial abilities, and memory. It is a main characteristic and a core identifying symptom of LBD. The other cognitive symptoms included in LBD are the following:

- Disorganized thinking
- Poor judgement
- Confusion
- Difficulty with language and numbers
- Inability to multitask
- Trouble with depth perception
- Misidentifying objects

- **Autonomic Dysfunction:**
  LBD is often accompanied by a range of autonomic dysfunctions that can profoundly impact the quality of life of the affected individuals. These autonomic disturbances, which can precede the onset of motor symptoms, are thought to arise from the widespread distribution of Lewy body pathology throughout the central and peripheral nervous systems. Autonomic dysfunction in LBD encompasses a wide range of symptoms that include:
  - Orthostatic hypotension
  - Gastrointestinal issues such as constipation and dysphagia
  - Urinary incontinence
  - Sweating abnormalities
  - Sexual dysfunction
- **REM (Rapid Eye Movement) Sleep Behaviour Disorder (RBD):**
  RBD is a common LBD symptom, which entails acting out one's dreams, which can cause sleep loss and even harm to both the person and their sleep partner. Sleep problems may induce confusion and behavioural problems in people with LBD, which creates complications for the caregiver's burden. This can include violent actions like falling out of bed or hitting someone. It is a core clinical feature of LBD diagnosis. RBD is not as strongly associated with LBD in women, but it's a strong indicator of LBD in men with dementia.
- **Hallucinations in Other Modalities:**
  About 35% of LBD patients experience auditory hallucinations, which are less common than visual hallucinations. The most commonly reported hallucinations have been characterized as "verbal," which includes hearing voices, footsteps, doorbells, and conversations. Women experiencing musical hallucinations in DLB are higher identified by Tsunoda et al.
- **Conventional Therapeutic Regimens Used in the Management of LBD:**
  As it concerns LBD, although its pathophysiology was studied in detail, the therapy still is mostly palliative in nature, implying a thorough focus on the patient's quality of life rather than attempting to stop the disease process. Traditional symptomatic LBD treatment programmes are centred on both pharmacotherapy and non-pharmacological management strategies, as LBD manifests poly-symptomatically. These are drugs for motor development utilizing dopaminergic agents, for cognitive deterioration using cholinesterase inhibitors, and for psychiatric features using antipsychotics. However, curative benefits of these treatments are not very high, and complications tend to worsen other symptoms, in view of the fact that LBD is highly sensitive to any alterations in neurochemical balance. The spectrum of symptoms of LBD overlaps with other neurodegenerative diseases due to the coexistence of symptoms, which makes diagnosis and treatment management more difficult (interference of neuropsychiatric symptoms exacerbates LBD motor symptoms and vice versa). These factors frequently impact neurologists and clinical psychiatrists to make the difficult decision between psychosis and mobility, and they must exercise extreme caution when treating patients with LBD in any way (Khan et al., 2024).

6.8 PHARMACOLOGICAL APPROACHES

Pharmacological treatment of LBD aims to target and relieve symptoms affecting the domains of motor, cognitive, psychiatric, and autonomic symptoms. LBD is a complex disorder comprising PDD and DLB, which makes the therapeutic efficacy versus potential adverse effect balancing act considerably more complex; the vast majority of patients with LBD are very sensitive to many medications. The primary pharmacological approaches to LBD are listed in the following text (Stinton et al., 2015).

- **Dopaminergic Therapy:**
  Dopaminergic therapy has an important role in the treatment of LBD by addressing motor symptoms (including PDD and DLB). Dopaminergic neuron degeneration in the substantia nigra underlies motor symptoms that appear in LBD cases, including bradykinesia, rigidity, and tremor. The aim of dopaminergic therapy is functional, to reestablish dopamine levels and motor function, but its role in the lesser-known LBD is more complicated than in PD, as LBD patients are highly sensitive to the medication. They ameliorate motor features like bradykinesia and rigidity. However, their use is weighed down by the possibility of worsening psychiatric symptoms, such as hallucinations and delusions, which are common in LBD. It is important to titrate the dose carefully to achieve a balance between motor improvement and neuropsychiatric side effects (Goldman et al., 2008).
  Levodopa is the primary dopaminergic agent, which is combined with the peripheral decarboxylase inhibitor (e.g., carbidopa or benserazide), prescribed for the dopaminergic treatment as it converts into dopamine in the brain resulting in alleviation of the motor symptoms.
  Dopamine agonists (e.g., pramipexole, ropinirole) are generally less prescribed in LBD than in PD because of the increased prospect of psychiatric side effects such as confusion and hallucinations. Hence, they have less usage in LBD treatment and are generally avoided in favour of levodopa. Adjunctive therapies, such as monoamine oxidase-B (MAO-B) inhibitors (e.g., selegiline, rasagiline), are used to increase dopaminergic activity. Their benefits in LBD are more equivocal, and they can potentiate psychiatric symptoms.
  There are some practical considerations for dopaminergic therapy, such as:
  - Careful Dose Titration: The dosing of the levodopa should be prescribed in a careful manner for the achievement of motor symptoms improvement without triggering psychiatric side effects.
  - Combination with Cholinesterase Inhibitors: The dopaminergic therapy is often combined with cholinesterase inhibitors such as rivastigmine to maintain balance between motor and cognitive symptoms.
- **Cholinesterase Inhibitors:**
  Cholinesterase inhibitors inhibit the acetylcholinesterase, the enzyme which is responsible for the breakdown of acetylcholine and makes it available in synaptic clefts. By increasing acetylcholine availability, cholinesterase inhibitors enhance cholinergic neurotransmission, which is critically impaired in LBD due to neuronal loss in the basal forebrain. This results in restoration of cholinergic activity is believed to underlie the symptomatic benefits seen in LBD patients. Commonly used inhibitors of cholinesterase are rivastigmine, donepezil (Ikeda et al., 2013), and galantamine (Edwards et al., 2004). The clinical efficacy of them is cognitive improvement, reduction in neuropsychiatric symptoms, and enhanced sleep–wake regulation. They have significant cognitive improvement in domains such as memory, attention, and visuospatial ability, which are distressed in LBD, providing an enhanced lifestyle. They provide a reduction in neuropsychiatric symptoms such as hallucinations and delusions, which are the hallmarks of LBD symptoms. They reduce apathy, anxiety, and agitation, which contribute to more stable behaviour and fewer caregiver burdens. This medication may help normalize sleep patterns and reduce disturbances

associated with RBD, a common symptom in LBD, by improving cholinergic neurotransmission. Cholinesterase inhibitors play a vital role in the welfare of the patients and their caregivers, making them an integral part of LBD management strategies.

- **Atypical Antipsychotics:**
Atypical antipsychotics are also known as second-generation antipsychotics, which exert antagonism toward dopamine D2 receptors and serotonin receptors, which provide a significant use in LBD treatment. Quetiapine and clozapine, atypical antipsychotics, are commonly used in LBD for their ability to reduce psychotic symptoms without significantly exacerbating motor dysfunction. Nevertheless, their use still requires careful contemplation due to the potential for sedation, metabolic side effects, and, in some cases, worsening of cognitive function.
- **Antidepressants:**
Antidepressants are crucial for the management of depressive symptoms in LBD, affecting both PDD and DLB. Treatment of depression in LBD is complex due to the intrinsic neurochemical changes, hence the need to minimize side effects, and the high sensitivity of LBD patients to psychotropic medications.
Antidepressant medications commonly used in LBD are selective serotonin reuptake inhibitors (SSRIs), serotonin-norepinephrine reuptake inhibitors (SNRIs), tricyclic antidepressants (TCAs), and atypical antidepressants. SSRIs, sertraline, citalopram, and escitalopram increase serotonin availability by the inhibition of its reuptake in the synaptic cleft. They are normally well-tolerated and effective for the treatment of depression and anxiety in LBD. SNRIs include venlafaxine and duloxetine and enhance serotonergic and noradrenergic transmission, effective for the treatment of comorbid anxiety or pain symptoms. The common TCAs used in the management of LBD are nortriptyline and desipramine, which block serotonin and norepinephrine reuptake. These are recommended in case of severe depression if SSRIs or SNRIs have failed. The atypical antidepressants, which are used in LBD treatment, are mirtazapine and bupropion. Mirtazapine is used to treat insomnia and weight loss and also enhances the activity of serotonin and norepinephrine. Bupropion is not used in general because it exacerbates the motor symptoms, which are a norepinephrine-dopamine reuptake inhibitor. The clinical efficacies of antidepressants in LBD are mood improvement, apathy management, and sleep and anxiety.
- **Autonomic Dysfunction Medications:**
Autonomic dysfunctions, such as orthostatic hypotension, gastrointestinal issues such as constipation and dysphagia, urinary dysfunction, and excessive sweating and sexual dysfunction, are the prominent and debilitating symptoms of LBD, which significantly affect the patient's daily life. By addressing these complications, the patient's daily life will be normal.
Orthostatic hypotension: The orthostatic hypotension can be treated by using midodrine, fludrocortisone, and droxidopa. Midodrine helps to avert blood pressure drops by increasing vascular tone, as it is an alpha-1 adrenergic agonist. Fludrocortisone is a mineralocorticoid, which helps increase sodium retention and expand the plasma volume. Droxidopa is a norepinephrine prodrug and increases peripheral vascular resistance, which improves blood pressure.
Gastrointestinal Issues such as Constipation and Dysphagia: The constipation complication can be addressed by using osmotic laxatives (e.g., polyethylene glycol (PEG), lactulose), stimulant laxatives (e.g., bisacodyl, senna), and prokinetic agents (e.g., Prucalopride).
Urinary Dysfunction: The urinary dysfunction can be managed through antimuscarinics (e.g., Oxybutynin, Solifenacin, and Tolterodine), beta-3 adrenergic agonists (e.g., Mirabegron), and alpha-blockers (e.g., Tamsulosin, alfuzosin).
Excessive Sweating and Thermoregulatory Dysfunction: These complications in LBD can be treated with topical agents such as aluminium chloride antiperspirants and systemic medications such as glycopyrrolate and oxybutynin.

Sexual Dysfunction: Patients with LBD face erectile dysfunction and other sexual problems. To address this condition phosphodiesterase-5 (PDE5) inhibitor is used, which include sildenafil and tadalafil.

- **Sleep Disorder Management:**
Sleep disturbances are a general symptom of LBD, which includes mainly RBD, excessive daytime sleepiness, and insomnia, which significantly affects patients' daily life and caregivers. To treat the RBD the medications such as clonazepam reduce muscle activity during REM sleep and melatonin, by regulating the sleep–wake cycle and reducing RBD episodes. The excessive daytime sleepiness can be managed by Modafinil or Armodafinil and Methylphenidate, which promote wakefulness by modulating dopamine and norepinephrine systems. Restless legs syndrome and periodic limb movements of sleep are common symptoms associated with LBD, which can be treated by medications such as dopamine agonists (e.g., pramipexole and ropinirole) and gabapentin or pregabalin (Jung & St. Louis, 2016).

6.9 NON-PHARMACOLOGICAL APPROACHES

Non-pharmacological approaches play a significant role in the management of LBD, as the pharmacological treatments are not sufficient, and the symptoms are hard to manage via medications. Longer use of medications shows side effects and worsens the symptoms. These approaches focus on the enhancement of life quality and limitation of symptom severity and aid both patients and caregivers in dealing with the intricacies of the illness. These are cognitive and behavioural therapies, physical therapy and exercise programmes, occupational therapy, speech and swallowing therapy, sleep hygiene interventions, dietary modifications, caregiver education and support, music and art therapy, and environmental modifications. Therapies that aim to modulate the inflammatory response, such as dietary interventions and physical activity, have shown potential benefits in improving cognitive and motor function in LBD patients. Moreover, non-pharmacological interventions can also target the non-motor symptoms associated with LBD, which can significantly impact the quality of life of patients and their caregivers. Cognitive-behavioural therapies, mindfulness-based techniques, and various forms of occupational and physical therapy have been explored as potential complementary approaches to address issues such as depression, anxiety, and sleep disturbances. Sleep hygiene interventions include establishing consistent sleep routines, optimizing bedroom environments for rest, and using relaxation techniques such as meditation or breathing exercises, which can address the RBD. Caregiver has a magnificent role in the management of LBD symptoms. They should train about the LBD progression, symptoms, and management strategy, which results reduction of caregiver burnout and improvement of patient care. The environment and surroundings should be calm and comfortable to the patients by which can minimize confusion and agitation. While the efficacy of these non-pharmacological interventions remains an area of ongoing research, the integration of these approaches with traditional pharmacological therapies may offer a more comprehensive and personalized management strategy for individuals with LBD

6.10 CHALLENGES IN CONVENTIONAL THERAPEUTIC REGIMENS

The conventional therapeutic regimens present numerous challenges due to the disease's complex and multifaceted nature, which creates a setback in managing LBD through this. LBD is indicated by common neuronal dysfunction and involves dopaminergic, cholinergic, and serotonergic systems, along with severe autonomic dysfunctions and cognitive impairments. Conventional pharmacological approaches are intended to alleviate particular symptoms; however, their limited effectiveness, potential for side effects, and incapacity to address the underlying mechanisms of disease highlight the necessity of careful clinical consideration. The currently available medications have limited specificity and efficacy, which presents a significant challenge. Levodopa is primarily used in dopaminergic therapy, partially alleviate Parkinsonian motor symptoms, but it worsens the neuropsychiatric symptoms, including delusions and hallucinations, which are sensitive to LBD patients. For managing hallucinations and behavioural disturbances, antipsychotic medications are used, but they can worsen motor symptoms, which leads to severe adverse reactions like neuroleptic sensitivity syndrome. This paired challenge complicates symptom management and necessitates careful dose titration and observation.

Cholinesterase inhibitors such as rivastigmine are often used to treat cognitive function and manage psychiatric symptoms, but they produce side effects such as gastrointestinal complications and bradycardia. The medications like melatonin and clonazepam are used in the treatment of RBD and provide some relief, but they do not target the root causes and also have limited benefit in severe cases. Antidepressants used to treat mood disorders and anxiety show complications like drug sensitivity.

Conventional therapies don't provide a remedy or decelerate the progression of LBD, because they do not address the fundamental pathophysiological mechanisms, including α-synuclein aggregation and neuroinflammation. This highlights an emerging need for disease-modifying therapies that can alter the disease trajectory. In the management of LBD, polypharmacy is involved, which creates an increased risk of drug-drug interactions, adverse therapeutic effects, and poor adherence. Another complication related to this is the lack of biomarkers which can guide the treatment and difficult to monitor disease progression.

Other challenges in conventional therapeutic regimens are the psychosocial and economic burden on patients and caregivers, limited knowledge on LBD, and delayed diagnosis. Addressing these challenges requires an integrated, multidisciplinary approach, combining pharmacological treatments with non-pharmacological interventions, caregiver support, and ongoing research into novel therapeutic targets.

6.11 FUTURE PROSPECTS AND CONCLUSION

LBD presents a significant therapeutic challenge due to its complex pathological condition and the overlapping of the symptoms with other neurodegenerative diseases. The conventional therapeutic approaches provide symptomatic relief and limitation in targeting disease progression. Future therapeutic approaches should focus on the root cause of LBD, such as α-synuclein aggregation and neuroinflammation. The earlier diagnosis and more precise therapeutic interventions by the advancement of molecular biology and neuroimaging can provide a better strategy toward the management of LBD. The emerging research on disease-modifying treatment, which includes immunotherapy and small molecules to target α-synuclein, could revolutionize treatment paradigms. The efficacy of the existing medication can be enhanced by employing novel drug delivery systems, like nanoparticle-based methods. To alleviate genetic risk factors, gene therapy approaches such as CRISPR-mediated editing or RNA-based therapies can be used. Non-pharmacological therapies, including brain stimulation techniques and digital cognitive training tools, can complement conventional methods, offering holistic management of motor and cognitive symptoms in LBD. The artificial intelligence-driven monitoring systems can be used to manage the LBD medication.

Conventional therapies remain the cornerstone of LBD management, and a multidisciplinary approach integrating pharmacological, non-pharmacological, and emerging therapies will be critical for improving patient outcomes. Collaborative research efforts and clinical trials are essential to translating these advances into practical solutions. A deeper understanding of LBD's pathology and personalized medicine approaches heralds a new era of hope for patients and caregivers alike.

ABBREVIATIONS

AD Alzheimer's disease
DLB Dementia with Lewy bodies
LBD Lewy body disease
PD Parkinson's disease
PDD Parkinson's disease dementia
RBD Rapid eye movement sleep behaviour disorder

REFERENCES

Armstrong, M. J., Irwin, D. J., Leverenz, J. B., Gamez, N., Taylor, A., & Galvin, J. E. (2021). Biomarker use for dementia with Lewy body diagnosis: Survey of U.S. experts. *Alzheimer Disease and Associated Disorders*, *35*(1), 55. https://doi.org/10.1097/WAD.0000000000000414

Borghammer, P., Okkels, N., & Weintraub, D. (2024). Parkinson's disease and dementia with Lewy bodies: One and the same. *Journal of Parkinson's Disease*, *14*(3), 383–397. https://doi.org/10.3233/JPD-240002

Edwards, K. R., Hershey, L., Wray, L., Bednarczyk, E. M., Lichter, D., Farlow, M., & Johnson, S. (2004). Efficacy and safety of galantamine in patients with dementia with Lewy bodies: A 12-week interim analysis. *Dementia and Geriatric Cognitive Disorders*, *17 Suppl 1*(SUPPL. 1), 40–48. https://doi.org/10.1159/000074681

Ferreira, D., Przybelski, S. A., Lesnick, T. G., Lemstra, A. W., Londos, E., Blanc, F., Nedelska, Z., Schwarz, C. G., Graff-Radford, J., Senjem, M. L., Fields, J. A., Knopman, D. S., Savica, R., Ferman, T. J., Graff-Radford, N. R., Lowe, V. J., Jack, C. R., Petersen, R. C., Mollenhauer, B., . . . Kantarci, K. (2020). β-Amyloid and tau biomarkers and clinical phenotype in dementia with Lewy bodies. *Neurology*, *95*(24), E3257–E3268. https://doi.org/10.1212/WNL.0000000000010943

Goldman, J. G., Goetz, C. G., Brandabur, M., Sanfilippo, M., & Stebbins, G. T. (2008). Effects of dopaminergic medications on psychosis and motor function in dementia with Lewy bodies. *Movement Disorders: Official Journal of the Movement Disorder Society*, *23*(15), 2248–2250. https://doi.org/10.1002/MDS.22322

Grisold, W. (2024). The expanding burden of neurological disorders. *The Lancet Neurology*, 23(4), pp. 326–327.

Heerema, E. (2024). *Difference between Alzheimer's and Lewy body dementia*, s.l. Verywell health.

Ikeda, M., Mori, E., Kosaka, K., Iseki, E., Hashimoto, M., Matsukawa, N., Matsuo, K., Nakagawa, M., Katayama, S., Higashi, Y., Yamada, T., Maruki, Y., Orimo, S., Yoshiiwa, A., Hanyu, H., Yokochi, M., Kimura, T., Mizoguchi, K., Nakanishi, A., . . . Kawase, Y. (2013). Long-term safety and efficacy of donepezil in patients with dementia with Lewy bodies: results from a 52-week, open-label, multicenter extension study. *Dementia and Geriatric Cognitive Disorders*, *36*(3–4), 229–241. https://doi.org/10.1159/000351672

Jellinger, K. A. (2009). A critical evaluation of current staging of α-synuclein pathology in Lewy body disorders. *Biochimica et Biophysica Acta - Molecular Basis of Disease*, *1792*(7), 730–740. https://doi.org/10.1016/j.bbadis.2008.07.006

Jung, Y., & St. Louis, E. K. (2016). Treatment of REM sleep behavior disorder. *Current Treatment Options in Neurology*, *18*(11). https://doi.org/10.1007/S11940-016-0433-2

Kane, J. P. M., Surendranathan, A., Bentley, A., Barker, S. A. H., Taylor, J. P., Thomas, A. J., Allan, L. M., McNally, R. J., James, P. W., McKeith, I. G., Burn, D. J., & O'Brien, J. T. (2018). Clinical prevalence of Lewy body dementia. *Alzheimer's Research & Therapy*, *10*(1). https://doi.org/10.1186/S13195-018-0350-6

Khan, S. A., Khan, S., Kausar, H., Shah, R., Luitel, A., Gautam, S., Parajuli, S. B., Rauniyar, V. K., & Khan, M. A. B. (2024). Insights into the management of Lewy body dementia: A scoping review. *Annals of Medicine and Surgery (2012)*, *86*(2), 930–942. https://doi.org/10.1097/MS9.0000000000001664

Lin, Y. W., & Truong, D. (2019). Diffuse Lewy body disease. *Journal of the Neurological Sciences*, *399*, 144–150. https://doi.org/10.1016/j.jns.2019.02.021

Marcin, A., 2023. *Lewy Body Dementia (LBD) and Parkinson's Disease: How Are They Different?* Healthline.

Marder, K. (2010). Cognitive impairment and dementia in Parkinson's disease. *Movement Disorders*, *25*(S1), S110–S116. https://doi.org/10.1002/MDS.22787

McKeith, I. G., Boeve, B. F., Dickson, D. W., Halliday, G., Taylor, J. P., Weintraub, D., Aarsland, D., Galvin, J., Attems, J., Ballard, C. G., Bayston, A., Beach, T. G., Blanc, F., Bohnen, N., Bonanni, L., Bras, J., Brundin, P., Burn, D., Chen-Plotkin, A., . . . Kosaka, K. (2017). Diagnosis and management of dementia with Lewy bodies: Fourth consensus report of the DLB Consortium. *Neurology*, *89*(1), 88–100. https://doi.org/10.1212/WNL.0000000000004058

McKeith, I. G., Perry, E. K., & Perry, R. H. (1999). Report of the second dementia with Lewy body international workshop: Diagnosis and treatment. Consortium on Dementia with Lewy Bodies. *Neurology*, *53*(5), 902–905. https://doi.org/10.1212/WNL.53.5.902

Prasad, S., Katta, M.R., Abhishek, S., Sridhar, R., Valisekka, S.S., Hameed, M., Kaur, J. and Walia, N., 2023. Recent advances in Lewy body dementia: A comprehensive review. *Disease-a-Month*, 69(5), p. 101441.

Priya Kirushnakumar, H., Mohan, N., & M Kane, J. P. (2024). *Incidence and prevalence of Lewy body dementia in India: A systematic review*. https://doi.org/10.36922/an.4098

Stinton, C., McKeith, I., Taylor, J. P., Lafortune, L., Mioshi, E., Mak, E., Cambridge, V., Mason, J., Thomas, A., & O'Brien, J. T. (2015). Pharmacological management of Lewy body dementia: A systematic review and meta-analysis. *The American Journal of Psychiatry*, *172*(8), 731–742. https://doi.org/10.1176/APPI.AJP.2015.14121582

Susan Bernstein, J.M., 2024. What is Lewy body dementia? [Online] Available at: https://www.webmd.com/alzheimers/dementia-lewy-bodies [Accessed 30 Nov 2024]

Tanei, Z.I., Saito, Y., Ito, S., Matsubara, T., Motoda, A., Yamazaki, M., Sakashita, Y., Kawakami, I., Ikemura, M., Tanaka, S. and Sengoku, R., 2021. Lewy pathology of the esophagus correlates with the progression of Lewy body disease: A Japanese cohort study of autopsy cases. *Acta Neuropathologica*, 141, pp. 25–37.

Taryn Berman, A.B., 2018. What are neurodegenerative diseases and how do they affect the brain? *Frontiers in Young Minds*, *6*, 70.

Taylor, J.P., McKeith, I.G., Burn, D.J., Boeve, B.F., Weintraub, D., Bamford, C., Allan, L.M., Thomas, A.J. and T O'Brien, J., 2020. New evidence on the management of Lewy body dementia. *The Lancet Neurology*, 19(2), pp. 157–169.

Yamada, M., Komatsu, J., Nakamura, K., Sakai, K., Samuraki-Yokohama, M., Nakajima, K. and Yoshita, M., 2019. Diagnostic criteria for dementia with Lewy bodies: Updates and future directions. *Journal of Movement Disorders*, 13(1), pp. 1–10.

7 Parkinson's Disease and Synergistic Effect of Natural and Synthetic Agents

K.M. Muhasina,[1] Mehrukh Zehravi,[2] N.P. Deepika,[3] Krishnaveni Nagappan,[4] Md Ekhtiar Uddin,[5] and Dhanabal S. Palaniswamy[6]

[1]Department of Pharmaceutical Analysis, Prime College of Pharmacy, Erattayal, Palakkad, Kerala, India

[2]Department of Clinical Pharmacy, College of Dentistry & Pharmacy, Buraydah Private Colleges, Buraydah, Kingdom of Saudi Arabia

[3]Gujarat Grassroots Innovations Augmentation Network (GIAN), Ahmedabad, Gujarat, India

[4]Department of Pharmaceutical Analysis, JSS College of Pharmacy, Ooty, Tamil Nadu, India

[5]Department of Pharmacy, Southeast University, Dhaka, Bangladesh

[6]Department of Pharmacognosy, JSS College of Pharmacy, Ooty, Tamil Nadu, India

7.1 INTRODUCTION

Parkinson's disease (PD) is a complex neurological condition characterized by movement symptoms and other behavioral manifestations (1) and cognitive abnormalities that profoundly affect the quality of life of patients (2). Behavioral symptoms, including impulse control difficulties, anxiety, sadness, and apathy, and cognitive impairments such as memory problems and attention deficiencies, provide significant obstacles to the normal life of patients with PD (3). Various phytoconstituents have promise in alleviating these deficits (Figure 7.1). Curcumin, a phytoconstituent from *Curcuma longa*, has antioxidant and anti-inflammatory characteristics (4) that may reduce cognitive deficits in PD by addressing neuroinflammation and oxidative stress (5). Green tea catechins, especially epigallocatechin gallate (EGCG), are linked to improvements in cognitive function in PD due to their neuroprotective characteristics (6). Omega-3 fatty acids may have anti-inflammatory benefits that might improve cognitive function and emotional regulation in individuals with PD (7). Resveratrol, a phytochemical derived from grapes and berries, has anti-inflammatory and antioxidant characteristics that may mitigate cognitive decline and improve behavioral symptoms in PD (8). Ginkgolides, from *Ginkgo biloba*, are recognized for their cognitive-enhancing properties (9).

7.2 MECHANISM OF ACTION OF PHYTOCONSTITUENTS IN THE TREATMENT OF NEURODEGENERATIVE DISORDERS

PD involves a complicated interaction of several delicate physiological systems. Natural chemicals may engage the relevant pathways via different pathways. Natural substances may affect the quantity and activity of neurotransmitters in the brain, which may be crucial for effective neural transmission. Certain medications might enhance the impact and synthesis of tyrosine hydroxylase (TH) activator

DOI: 10.1201/9781003478652-8

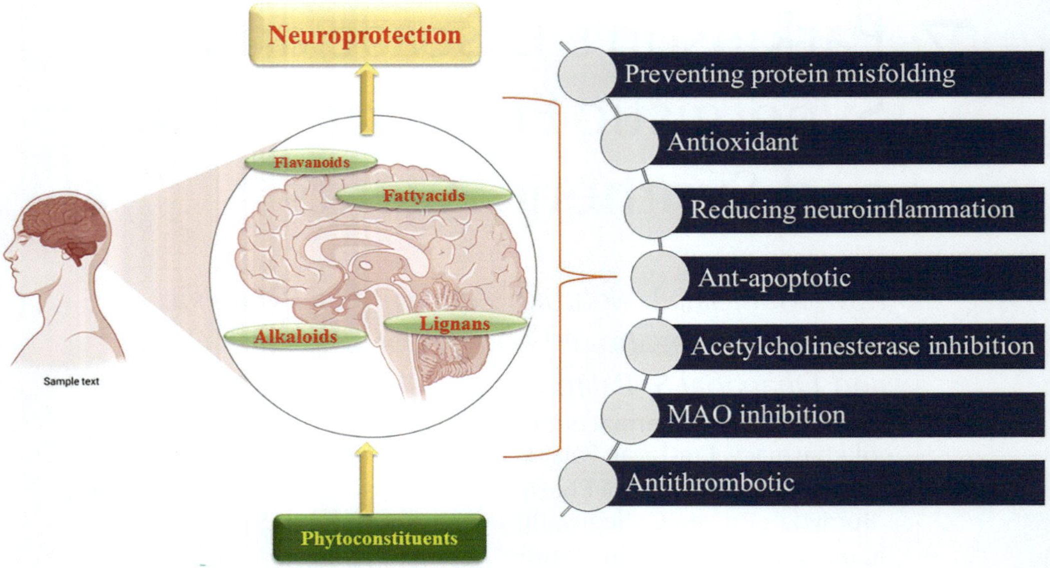

FIGURE 7.1 Phytoconstituents activity in the treatment of neurodegenerative disorders.

or release (MAO-B inhibitor) of dopamine (DA), the principal neurotransmitter diminished in PD and accountable for its motor symptoms, especially in the substantia nigra area (10). MAO comprises a collection of enzymes that participate in the metabolism of neurotransmitters, including DA, serotonin, and norepinephrine (NE). In PD, the two isoforms of monoamine oxidase have been associated with the disease's etiology. MAO-A is mostly accountable for the degradation of 5-HT and NE and is significant in mental disorders, including depression linked to PD. MAO-B is mostly responsible for the degradation of DA and is associated with the neurological illness of PD. In PD treatment, the inhibition of monoamine oxidases is a neuroprotective strategy used in disease treatment (11). Furthermore, TH is essential in the production of DA by facilitating the synthesis of L-dopa. PD may be seen as a deficit syndrome in the striatum (12). Conversely, COMT inhibitors may reduce the breakdown of levodopa (13). The adenosine A2A receptor is a G-protein-coupled receptor primarily located in the brain, especially in regions associated with motor control, such as the basal ganglia. Research indicates that this receptor interacts with other receptors, including DA D2, in a convoluted way inside the basal ganglia. Disruption in these connections, particularly the hyperactivation of adenosine A2A receptors, may cause an imbalance in the circuitry of the basal ganglia, leading to exceptional motor function and exacerbating the motor symptoms associated with PD. In light of this knowledge, the adenosine A2A receptor inhibition with antagonists seeks to re-establish equilibrium between adenosine A2A and DA D2 receptor signaling in the basal ganglia, possibly mitigating motor symptoms (14). Natural substances may have inhibitory effects on α-syn aggregation and fibrillation. These chemicals may safeguard neurons from injury and impede disease development by disrupting the creation of harmful protein aggregates (15). Moreover, natural substances may elevate the levels of microtubule-associated protein 1A/1B-light chain 3 (LC3-II), an autophagy indicator that is diminished in PD, resulting in the formation of aggregates of α-Syn (16). Phytochemicals may safeguard mitochondria against oxidative damage. They can augment mitochondrial activity, optimize energy metabolism, and preserve the integrity of the mitochondria. Natural compounds sustain the function of mitochondria, thereby ensuring optimal energy of neuronal supply and mitigating cellular stress through the regulation of lactate dehydrogenase and succinate dehydrogenase enzyme activities, which are integral to cellular metabolism and frequently employed as biomarkers in diagnostic assessments and research to evaluate damage to the tissue (17). Additionally, cyt *c* (cytochrome *c*), a protein integral to the mitochondrial electron transport chain, evolved from compromised mitochondria and may initiate apoptosis, hence contributing to the degeneration of dopaminergic neurons (18). Furthermore, PGC-1α, essential for the oxidative metabolism and control of mitochondrial biogenesis, is pivotal in sustaining cellular energy balance (19). Research indicates that PGC-1α expression is diminished in PD, perhaps leading to mitochondrial malfunction and impaired energy generation in dopaminergic neurons (20). Sirtuin 1 regulates cellular functions, including mitochondrial function, apoptosis, and DNA repair in PD (21). In PD, there is evidence of impairment in both the cholinergic system and the endocannabinoid system. These systems have often been examined for their involvement in the pathogenesis and management of PD (22). Phyto-origin cannabinoids have shown the ability to reverse modifications in the endocannabinoid system, including abnormalities in cannabinoid receptor expression and dysregulation of endocannabinoid signaling associated with PD. The cannabinoid receptor activation has been shown to have neuroprotective activity by diminishing inflammation, excitotoxicity, and oxidative stress in preclinical models of PD (23). The nicotinic acetylcholine receptors (nAChRs) and the cholinergic system are very important for brain and muscle functions. In PD, nicotinic receptors become less available and cholinergic neurones in the basal forebrain break down. Researchers have looked into whether nicotine, which selectively binds to nicotinic acetylcholine receptors, could help protect neurons in people with PD. Nicotine may control the release of DA, improve cognitive and attentional skills, and ease movement symptoms (24). Natural compounds neutralize harmful free radicals and mitigate oxidative stress by enhancing biomarkers like glutathione (GSH), superoxide dismutase (SOD), GSH peroxidase (GPx), and complex I, while reducing MDA levels and caspase-3 (Casp-3) activity (25). GSH is an essential antioxidant that prevents cell damage. In PD, GSH levels decrease, indicating an impairment in the antioxidant defense mechanism, where GCLC, GCLM, and GSR are essential enzymes for GSH synthesis and recycling (26). A kind of enzyme known as SOD is responsible for assisting in the process of neutralizing superoxide radicals. In addition, GPx1 is an enzyme that reduces GSH, which makes it easier for hydrogen peroxide and lipid peroxides to be eliminated from the body (27). Mitochondrial electron transport chain complex I inhibition increases oxidative stress and neuronal damage. PD is associated with elevated lipid peroxidation (LPO) and MDA. These pathways cause neuronal death and oxidative stress. Casp-3 is crucial to mitochondria-dependent and mitochondria-independent apoptosis. When activated, Casp-3 targets various proteins, including PARP, which is necessary for DNA repair (28). Natural compounds may suppress Casp-3, which promotes apoptosis. In addition, natural substances may increase nuclear factor erythroid 2-related factor 2 (Nrf2), a transcription factor that regulates the production of antioxidant and cytoprotective genes via the antioxidant response element (ARE), which is decreased in PD (29). Keap1, on the other hand, controls the cellular antioxidant defense. It regulates Nrf2 activity crucially. Deregulation of the Keap1/Nrf2 pathway in PD reduces antioxidant defenses and increases oxidative stress risk (30). Emerging evidence suggests that Nrf2-ARE, PI3K-Akt, JNK-ERK1/2, and TLR4/NF-kB pathways play key roles in PD pathogenesis (31). In experimental settings different than PD, phenolic phytoconstituent orientin modulates these pathways (32). Natural compounds may limit nitric oxide (NO) production and iNOS activity, which are important in neurotransmission and immunological defense, to reduce inflammation. Also, pro-inflammatory cytokines like IL-6 and TNF-α are elevated in PD. They may increase reactive oxygen species, glial cell activation, and apoptosis (33). Another novel inflammatory cytokine, IL-33, binds to the orphan receptor suppression of tumorigenicity 2 (ST2) to activate signaling pathways that kill neurons. PD activates NLRP3, an inflammasome protein complex that triggers inflammatory responses, leading to the release of pro-inflammatory cytokines like IL-1β, which can worsen neuronal damage and disease progression (34). PD activates microglia, central nervous system immune cells, which may release inflammatory chemicals, causing neuroinflammation and neuronal damage (35). In PD, neuroinflammation is common and linked to elevated levels of CRP, Cox-2, and other inflammatory mediators like ILs, TNF-α, and IFN-γ (36). Many important biological factors and pathways form a complex PD etiology network. Matrix metalloproteinases (MMPs) degrade the extracellular matrix. MMP activity is elevated in PD, suggesting a role in disease progression. MMPs may contribute to neuroinflammation, oxidative stress, and blood–brain barrier breakdown in PD (37). Furthermore, both JNK and p38 belong to the family of proteins known as mitogen-activated protein kinases. In patients with PD, the activation of JNK and p38 has been linked to the loss of neuronal cells and inflammation. It is possible that the deregulation of these kinases might result in the degeneration of dopaminergic neurons. These kinases participate in intracellular signaling networks that influence cellular responses to stress (38). Moreover, the signaling molecules Janus kinase 2 (JAK2) and signal transducer and activator of transcription 3 (STAT3) participate in several physiological processes, including immunological responses and inflammation. Research indicates that the JAK2/STAT3 signaling pathways are dysfunctional in PD. Abnormal phosphorylation of JAK2 and STAT3 has been seen in the brains of individuals with PD, indicating that these proteins may contribute to the inflammatory processes and neurodegeneration associated with the condition (39). Conversely, p53 is a transcription factor recognized for its function in regulating cell cycle progression, DNA repair, and apoptosis. Research suggests that the activation of p53 may contribute to the neurodegenerative processes linked with PD. This stimulation may induce cell death pathways and oxidative stress in dopaminergic neurons (40). The proteins Bax and Bcl-2 regulate apoptosis. Increased levels of pro-apoptotic Bax and lower levels of anti-apoptotic Bcl-2 have been associated with PD (41). Moreover, a link exists between neuronal cell death and the dysregulation of intracellular calcium levels. In PD, the concentrations of the calcium-binding protein calbindin D28K are diminished. This protein regulates calcium homeostasis in neurons and functions in the modulation of calcium levels (42). Natural substances may provide direct neuroprotection by inhibiting neuronal cell death and enhancing cell survival. They may impede apoptosis, mitigate neuronal damage caused by oxidative stress and inflammation, and enhance cellular defense systems. Moreover, natural substances may enhance the synthesis of neurotrophic factors, including brain-derived neurotrophic factor (BDNF) and nerve growth factor, which are recognized for their role in supporting neuronal survival and proliferation. These advantages may enhance neural function, facilitate neuroplasticity, safeguard against neurodegeneration, and alleviate behavioral abnormalities and cognitive deficiencies (43).

7.3 SYNERGISTIC ACTIVITY OF PHYTOCONSTITUENTS WITH SYNTHETIC MEDICATIONS

The combined effect of anti-Parkinsonism drugs and phytoconstituents signifies the enhanced therapeutic results obtained from the integration of conventional Parkinson's treatments with natural plant-derived compounds. PD is a neurodegenerative disorder characterized by the loss of DA-producing neurons in the brain, leading to motor symptoms including tremors, rigidity, bradykinesia, and postural instability. Phytoconstituents include flavonoids, polyphenols, alkaloids, and terpenoids, which demonstrate neuroprotective effects. These chemicals may mitigate oxidative stress, a major factor in the destruction of dopaminergic neurons in PD. Curcumin (derived from turmeric) (45) and resveratrol (extracted from grapes) possess antioxidant qualities that may safeguard brain cells from harm (46). Certain phytoconstituents may facilitate the enhancement of DA release or optimize its reuptake processes. Compounds such as apomorphine and *Mucuna pruriens*, a leguminous plant, include L-dopa, which directly restores DA levels. When administered with conventional dopaminergic therapies (e.g., levodopa), this may enhance symptom management (47).

Prolonged use of levodopa and other Parkinson's therapies may lead to motor fluctuations, dyskinesias (involuntary movements), and further adverse effects (Table 7.1). Certain phytoconstituents, such as those included in *Withania somnifera* (Ashwagandha) (48) and *Bacopa monnieri* (brahmi) (49), are proposed to mitigate these negative effects and enhance overall results by functioning as adaptogens and neuroprotective agents. Phytoconstituents may affect other neurotransmitter systems implicated in PD. Compounds found in *G. biloba* may enhance cerebral blood flow and have neuroprotective properties on the dopaminergic system. This, when integrated

TABLE 7.1
Mechanism of Action and Side Effects of the Conventional Treatment Options for Neurodegenerative Disorders

Drugs Used for PD	Type	Mechanism	Side Effects
L-Dopa and carbidopa	Dopamine precursor	Blocks aromatic amino acid decarboxylase	Insomnia, hypotension, nightmares, hallucinations, somnolence, vomiting, and nausea
Selegiline, rasagiline, safinamide	MAO-B inhibitors	Increase the dopamine synaptic availability. Safinamide prevents glutamate release	Dry mouth, vomiting, dizziness, nausea, persistent diarrhea, constipation, abnormal movements, and weight loss
Entacapone and tolcapone	COMT inhibitors	Decrease L-dopa metabolism and increase CNS L-DOPA bioavailability	Hypotension, dizziness, nausea, vomiting, dyskinesia, liver toxicity
Ropinirole, apomorphine, pramipexole, rotigotine	Dopamine agonists	At postsynaptic receptors, stimulate the actions of dopamine	Sleep stacks, orthostatic hypotension, confusion, constipation, angle edema, dyskinesia, hallucinations, and sleepiness
Benztropine and trihexyphenidyl	Anticholinergics	In the striatum, block acetylcholine	Urine retention, dry eyes, blurred vision, dry mouth, and confusion
Amantadine	NMDA-type glutamate receptor antagonist	Stimulate dopamine release	Insomnia, edema, hallucinations, and confusion

with conventional therapies, may enhance the overall therapeutic outcomes. Inflammation is an additional contributing factor to PD. Numerous phytoconstituents, including curcumin, green tea catechins (6), and omega-3 fatty acids (present in flaxseeds, chia seeds, etc.) (7), are recognized for their anti-inflammatory effects. Mitigating neuroinflammation may decelerate disease development and enhance the effectiveness of standard pharmacological treatments. Table 7.2 summarises the sources and mechanisms of phytoconstituents on neurodegenerative disorders.

Mitochondrial support is associated with the advancement of PD due to mitochondrial malfunction. Certain phytochemicals, including coenzyme Q10, have been studied for their ability to enhance mitochondrial activity and mitigate oxidative damage in the brain (50). When used with Parkinson's medications, this may aid in preserving dopaminergic neurons. The synergistic action of anti-Parkinsonism pharmaceuticals with phytoconstituents denotes the amplified therapeutic outcomes achieved by the amalgamation of standard Parkinson's medications with natural plant-derived chemicals.

Table 7.3 summarizes major phytoconstituents (plant-derived substances) and their possible synergistic effects when used with synthetic anti-Parkinsonism medications. The objective is to emphasize how these phytochemicals may augment or increase the therapeutic efficacy of traditional pharmaceuticals.

7.4 CONCLUSION

The integration of phytoconstituents with synthetic anti-Parkinsonism medications is a viable adjunctive therapy strategy. Plant-derived chemicals may augment the efficacy of traditional drugs by offering neuroprotective, anti-inflammatory, and antioxidant advantages, enhancing motor and cognitive capabilities, and perhaps mitigating adverse effects. Patients must contact healthcare professionals before integrating these natural treatments to guarantee their safe and successful use in conjunction with conventional medicines.

TABLE 7.2
Sources and Mechanism of Action of the Phytoconstituents for the Treatment of Neurodegenerative Disorders (44)

Source	Compound	Mechanism of Action
Ginkgo biloba	EGb761	Inhibited activity of MAO reduces oxidative stress and lipid peroxidation; anti-apoptotic activity
	Ginkgetin	Reduces reactive oxygen species generation
	Ginkgolide B and bilobalide	Anti-apoptotic activity and reduced α-synuclein
Ginseng	Ginsenosides RB1 and Rg1	Anti-apoptotic activity by increasing bcl-2, reducing Bax, caspase-3, and reducing TNF-α
	Ginsenosides Rd and Re and Rd	Improved mitochondrial function
Scutellaria baicalensis	Baicalein	Reduce lipid peroxidation, dopamine level, α-synuclein aggregation, and increase serotonin expression
Celery, broccoli, parsley, thyme, and olive oil	Luteolin	Decrease inflammation, decrease ROS, and decrease oxidative stress
Citrus fruit, onions, and grains	Quercetin	Decrease oxidative stress and increase SOD, GPx, ATPase, PKD, BDNF, Akt, and so on
Grapefruit	Kaempferol	Decrease lipid peroxidation, and increase autophagy, SOD, and GHS
Coffee, cereal grains, tea, and vegetables	Rutin	Decrease oxidative stress, lipid peroxidation, and increase SOD, GPx, and catalase
Mangifera indica and *Rheum nobile*	Isoquercetin	Decrease oxidative stress
Parsley, oranges, onions, chamomile, wheat sprouts, tea	Apigenin	Decrease in inflammation and reactive oxygen species
Coffee, cereal grains, tea, and vegetables	Troxerutin	Decrease lipid peroxidation, reactive oxygen species, and DNA fragmentation
oranges and lemons	Hesperidin	Antioxidant

TABLE 7.3
Sources and Mechanism of Action of the Phytoconstituents along with Conventional Medications to Enhance the Therapeutic Activity and Reduce Side Effects

Phytoconstituent	Source (Plant)	Synthetic Anti-Parkinsonism Drugs	Mechanism of Synergy	Reference
L-Dopa (levodopa)	Velvet beans (*Mucuna pruriens*)	Levodopa (L-dopa)	Enhances the synthesis of dopamine. *Mucuna pruriens* comprises natural L-dopa, which elevates dopamine concentrations in the brain. When used with synthetic levodopa, it may enhance motor symptoms and decrease levodopa dosage needs, hence mitigating adverse effects such as dyskinesia.	(47)
Curcumin	Turmeric (*Curcuma longa*)	Dopamine agonists, levodopa	Antioxidant and anti-inflammatory activity. Curcumin mitigates oxidative stress and neuroinflammation, possibly safeguarding dopaminergic neurons from injury and augmenting the neuroprotective efficacy of Parkinson's medications.	(45)

TABLE 7.3 (*Continued*)

Sources and Mechanism of Action of the Phytoconstituents along with Conventional Medications to Enhance the Therapeutic Activity and Reduce Side Effects

Phytoconstituent	Source (Plant)	Synthetic Anti-Parkinsonism Drugs	Mechanism of Synergy	Reference
Bacopa monnieri	Brahmi (*B. monnieri*)	Dopamine agonists, levodopa	Enhancement of cognitive function. Bacopa enhances cognitive function, perhaps augmenting mental clarity and attention in people with Parkinson's disease, so complementing the motor advantages of Parkinson's medications.	(49)
Withanolides	Ashwagandha (*Withania somnifera*)	Dopamine agonists, levodopa	Neuroprotective and adaptogenic activity. Ashwagandha mitigates stress, inflammation, and neurodegeneration, possibly augmenting the efficacy of Parkinson's medications and alleviating adverse effects including anxiety and sadness.	(48)
Ginkgolides	*Ginkgo biloba*	Dopamine agonists, levodopa, MAO-B inhibitors	Enhanced circulation and neuroprotection. It improves cerebral blood circulation, perhaps augmenting the transport of dopaminergic medications to the brain, hence increasing their efficacy. It has neuroprotective properties.	(51)
Resveratrol	Grapes (*Vitis vinifera*)	Levodopa, MAO-B inhibitors	Anti-inflammatory and antioxidant activity. Resveratrol has potent antioxidant and anti-inflammatory characteristics, potentially mitigating oxidative damage and enhancing the efficacy of Parkinson's medications.	(46)
Epigallocatechin gallate (EGCG)	Green tea (*Camellia sinensis*)	Dopamine agonists, levodopa, MAO-B inhibitors	Neuroprotective and antioxidant activity. EGCG mitigates oxidative stress and inflammation, offering neuroprotection that may decelerate disease development and enhance the efficacy of synthetic Parkinson's medications.	(52)
Silymarin (silibinin)	Milk thistle (*Silybum marianum*)	Dopamine agonists, levodopa	Antioxidant and liver protection. Silymarin safeguards the liver, is crucial to medication metabolism, and has antioxidant qualities, possibly augmenting the efficacy of Parkinson's therapies while diminishing drug toxicity.	(53)
Omega-3 fatty acids	Flaxseeds, fish oil	Dopamine agonists, levodopa	Anti-inflammatory and neuroprotective activity. Omega-3 fatty acids mitigate neuroinflammation and bolster cerebral function, possibly augmenting the efficacy of dopaminergic therapies and diminishing cognitive deterioration.	(54)
Anthocyanins	Blueberries, blackberries	Dopamine agonists, levodopa, MAO-B inhibitors	Anti-inflammatory and metabolic support. Berberine enhances metabolic function and reduces inflammation, hence augmenting the effectiveness of standard Parkinson's treatments.	(55)

(Continued)

TABLE 7.3 (*Continued*)

Sources and Mechanism of Action of the Phytoconstituents along with Conventional Medications to Enhance the Therapeutic Activity and Reduce Side Effects

Phytoconstituent	Source (Plant)	Synthetic Anti-Parkinsonism Drugs	Mechanism of Synergy	Reference
Ginsenosides	Ginseng (*Panax ginseng*)	Dopamine agonists, levodopa	Energy-boosting and neuroprotective activity. Ginseng can improve cognitive function, reduce fatigue, and provide neuroprotection, potentially enhancing the action of Parkinson's drugs.	(56)

REFERENCES

1. Deepika NP et al (2024) Neuroprotective role of coconut oil for the prevention and treatment of Parkinson's disease: Potential mechanisms of action. *Biotechnology and Genetic Engineering Reviews* 40(4):3346–3378.
2. Kalaba Ö, Güzeloğlu ÖMC (2024) Cognition and quality of life in Parkinson's disease. *Psikiyatride Güncel Yaklaşımlar* 16(4):604–616.
3. Maristany AJ et al (2024) Psychiatric manifestations of neurological diseases: A narrative review. *Cureus* 16(7):e64152.
4. Muhasina KM, Palaniswamy DS, Deepika NP (2021) Therapeutic importance of curcumin with a special emphasis on Mdr cancer cells and the factors influencing pharmacodynamics. *Indonesian Journal of Pharmacy/Majalah Farmasi Indonesia* 32(4):429–441.
5. Turer BY, Sanlier N (2024) Relationship of curcumin with aging and Alzheimer and Parkinson disease, the most prevalent age-related neurodegenerative diseases: A narrative review. *Nutrition Rev* nuae079. https://doi.org/10.1093/nutrit/nuae079.
6. Pandit N, Kulkarni S, Singhvi G (2024) Effect of green tea on human brain health. *Nutraceutical fruits and foods for neurodegenerative disorders*. Elsevier, pp 301–331.
7. Rao AS et al (2024) Impacts of omega-3 fatty acids, natural elixirs for neuronal health, on brain development and functions. *Neuroprotection: Method and protocols*. Springer, pp 209–229.
8. Jadidian F et al (2024) Pharmacotherapeutic potential of *Vitis vinifera* (grape) in age-related neurological diseases. *Boletín Latinoamericano y Del Caribe De Plantas Medicinales y Aromáticas* 23(3):349–370.
9. Ali D et al (2024) Implications of herbal components in the treatment of neurological disorders. *Curr Nutr Food Sci* 20(6):677–686.
10. Alosaimi F et al (2022) The role of neurotransmitter systems in mediating deep brain stimulation effects in Parkinson's disease. *Front Neurosci* 16:998932.
11. Tan Y-Y, Jenner P, Chen S-D (2022) Monoamine oxidase-B inhibitors for the treatment of Parkinson's disease: Past, present, and future. *J Parkinsons Dis* 12(2):477–493.
12. Nagatsu T et al (2019) Human tyrosine hydroxylase in Parkinson's disease and in related disorders. *J Neural Transm* 126:397–409.
13. Gershanik OS (2015) Improving L-dopa therapy: The development of enzyme inhibitors. *Mov Disord* 30(1):103–113.
14. Pinna A (2014) Adenosine A2A receptor antagonists in Parkinson's disease: Progress in clinical trials from the newly approved istradefylline to drugs in early development and those already discontinued. *CNS Drugs* 28(5):455–474.
15. Fields CR, Bengoa-Vergniory N, Wade-Martins R (2019) Targeting alpha-synuclein as a therapy for Parkinson's disease. *Front Mol Neurosci* 12:299.
16. Wang ZY et al (2017a) Neuroprotective natural products for the treatment of Parkinson's disease by targeting the autophagylysosome pathway: A systematic review. *Phytother Res* 31(8):1119–1127.
17. Grünewald A, Kumar KR, Sue CM (2019) New insights into the complex role of mitochondria in Parkinson's disease. *Prog Neurobiol* 177:73–93.

18. Meng H et al (2017) Loss of Parkinson's disease-associated protein CHCHD2 affects mitochondrial crista structure and destabilizes cytochrome c. *Nat Commun* 8(1):15500.
19. Scarpulla RC (2011) Metabolic control of mitochondrial biogenesis through the PGC-1 family regulatory network. *Biochimica et Biophysica Acta (BBA)-Mol Cell Res* 1813(7):1269–1278.
20. Bennett CF, Latorre-Muro P, Puigserver P (2022) Mechanisms of mitochondrial respiratory adaptation. *Nat Rev Mol Cell Biol* 23(12):817–835.
21. Li X et al (2020a) The critical role of SIRT1 in Parkinson's disease: Mechanism and therapeutic considerations. *Aging Dis* 11(6):1608.
22. Scherma M et al (2016) Interactions between the endocannabinoid and nicotinic cholinergic systems: Preclinical evidence and therapeutic perspectives. *Psychopharmacology* 233:1765–1777.
23. Lutz B (2022) Neurobiology of cannabinoid receptor signaling. *Dialogues Clin Neurosci* 22:207.
24. Quik M et al (2015) Alpha7 nicotinic receptors as therapeutic targets for Parkinson's disease. *Biochem Pharmacol* 97(4):399–407.
25. Essa M et al (2014) Review of natural products on Parkinson's disease pathology. *J Aging Res Clin Pract* 3(1):1–8.
26. Wu C-R et al (2015) Carnosic acid protects against 6-hydroxydopamine-induced neurotoxicity in in vivo and in vitro model of Parkinson's disease: Involvement of antioxidative enzymes induction. *Chem Biol Interact* 225:40–46.
27. Ruszkiewicz J, Albrecht J (2015) Changes in the mitochondrial antioxidant systems in neurodegenerative diseases and acute brain disorders. *Neurochem Int* 88:66–72.
28. Lu JY et al (2017) The neuroprotective effect of nicotine in Parkinson's disease models is associated with inhibiting PARP-1 and caspase-3 cleavage. *PeerJ* 5:e3933.
29. Aguiar AS et al (2016) Moderate-intensity physical exercise protects against experimental 6-hydroxydopamine-induced hemiparkinsonism through Nrf2-antioxidant response element pathway. *Neurochem Res* 41:64–72.
30. Mahapatra P (2018) Identification of natural inhibitors of proteins involved in the pathology of Parkinson's disease.
31. Chakkittukandiyil A et al (2022) The principal molecular mechanisms behind the activation of Keap1/Nrf2/ARE pathway leading to neuroprotective action in Parkinson's disease. *Neurochemistry International* 156:105325.
32. Sajini DV et al (2024) Orientin modulates Nrf2-ARE, PI3K/Akt, JNK-ERK1/2, and TLR4/NF-kB pathways to produce neuroprotective benefits in Parkinson's disease. *Neurochemical Research* 49(6):1577–1587.
33. Kempuraj D et al (2021) Neuroprotective effects of flavone luteolin in neuroinflammation and neurotrauma. *BioFactors* 47(2):190–197.
34. Khot M et al (2022) NLRP3 inflammasomes: A potential target to improve mitochondrial biogenesis in Parkinson's disease. *Eur J Pharmacol* 934:175300.
35. Ramirez AI et al (2017) The role of microglia in retinal neurodegeneration: Alzheimer's disease, Parkinson, and glaucoma. *Front Aging Neurosci* 9:214.
36. Mehta N et al (2023) C-reactive protein as the biomarker of choice to monitor the effects of exercise on inflammation in Parkinson's disease. *Front Immunol* 14:1178448.
37. Behl T et al (2021) Multifaceted role of matrix metalloproteinases in neurodegenerative diseases: Pathophysiological and therapeutic perspectives. *Int J Mol Sci* 22(3):1413.
38. Jha SK et al (2015) p38 MAPK and PI3K/AKT signalling cascades in Parkinson's disease. *Int J Mol Cell Med* 4(2):67.
39. Lashgari N-A et al (2021) The involvement of JAK/STAT signaling pathway in the treatment of Parkinson's disease. *J Neuroimmunol* 361:577758.
40. Luo Q et al (2022) Association of p53 with neurodegeneration in Parkinson's disease. *Parkinson's Disease* 2022. https://doi.org/10.1155/2022/6600944.
41. Liu J, Liu W, Yang H (2018) Balancing apoptosis and autophagy for Parkinson's disease therapy: Targeting BCL–2. *ACS Chem Neurosci* 10(2):792–802.
42. Ricke KM et al (2020) Mitochondrial dysfunction combined with high calcium load leads to impaired antioxidant defense underlying the selective loss of nigral dopaminergic neurons. *J Neurosci* 40(9):1975–1986.
43. Solayman M et al (2017) Natural products combating neurodegeneration: Parkinson's disease. *Curr Drug Metable* 18(1):50–61.
44. Corona JC (2018) Natural compounds for the management of Parkinson's disease and attention-deficit/hyperactivity disorder. *BioMed Research International* 2018(1):4067597.

45. Hu S et al (2015) Clinical development of curcumin in neurodegenerative disease. *Expert Review of Neurotherapeutics* 15(6):629–637.
46. Zhang Y, Guo H, Fu H (2024) Protective effect of resveratrol combined with levodopa against oxidative damage in dopaminergic neurons. *Cell Biochemistry and Biophysics* 82(2):817–826.
47. Rai SN et al (2020) Mucuna pruriens in Parkinson's and in some other diseases: Recent advancement and future prospective. *3 Biotech* 10(12):522.
48. *Withania somnifera* (Ashwagandha) and its potential therapeutic role in neurodegenerative diseases. *Journal of Neurosciences in Rural Practice* 4(2):198–201.
49. Pham HTN et al (2020) Bacopa monnieri (L.) Wettst. Extract improves memory performance via promotion of neurogenesis in the hippocampal dentate gyrus of adolescent mice. *International Journal of Molecular Sciences* 21(9):3365.
50. Beal MF (2004) Mitochondrial dysfunction and oxidative damage in Alzheimer's and Parkinson's diseases and coenzyme Q10 as a potential treatment. *Journal of Bioenergetics and Biomembranes* 36(4):381–386.
51. Mandel S, Moussa BHY (2004) Catechin polyphenols: Neurodegeneration and neuroprotection in neurodegenerative diseases. *Free Radical Biology and Medicine* 37(3):304–317.
52. Tang S et al (2025) Epigallocatechin-3-gallate inhibits oxidative stress through the Keapl/Nrf2 signaling pathway to improve Alzheimer disease. *Molecular Neurobiology* 62(3):3493–3507.
53. Devi KP et al (2017) A mini review on the chemistry and neuroprotective effects of silymarin. *Current Drug Targets* 18(13):1529–1536.
54. Avallone R, Vitale G, Bertolotti M (2019) Omega-3 fatty acids and neurodegenerative diseases: New evidence in clinical trials. *International Journal of Molecular Sciences* 20(17):4256.
55. Ullah R et al (2019) Natural antioxidant anthocyanins—A hidden therapeutic candidate in metabolic disorders with major focus in neurodegeneration. *Nutrients* 11(6):1195.
56. He Y-B et al (2021) Effect of ginsenoside-Rg1 on experimental Parkinson's disease: A systematic review and meta-analysis of animal studies. *Experimental and Therapeutic Medicine* 21(6):552.

8 Natural Extracts for Spinal Muscular Atrophy

Swati Swagatika Swain,[1] *Karri V. V. S. Narayana Reddy,*[1] *Gowthamarajan Kuppusamy,*[1] *Subhasri Bogadi,*[1] *Sagarkanta Mohanty,*[2] *Ashish Sriram Mishra,*[3] *and Vetriselvan Subramaniyan*[4]

[1]Department of Pharmaceutics, JSS College of Pharmacy, JSS Academy of Higher Education & Research, Ooty, Tamil Nadu, India

[2]Department of Pharmaceutical Chemistry, JSS College of Pharmacy, JSS Academy of Higher Education & Research, Ooty, Tamil Nadu, India

[3]Department of Pharmaceutics, SRM College of Pharmacy, SRM Institute of Science and Technology, Kattankulathur, Tamil Nadu, India

[4]Professor of Pharmacology, Department of Medical Sciences, School of Medical and Life Sciences, Sunway University

8.1 INTRODUCTION

A devastating genetic condition called spinal muscular atrophy (SMA) mostly affects motor neurons, resulting in muscle weakening and atrophy. It has an enormous impact on patient and their families and is the most prevalent genetic cause of death in infants. Due to the mutation in the Survival Motor Neuron 1 (SMN1) gene, SMA is caused, which leads to SMN protein deficiency, which is a critical molecule for the survival of motor neurons. These SMN neurons are located in the anterior horn of the spinal cord. These SMN proteins are essential for the movement of voluntary muscle. Motor neurons deteriorate in the absence of a sufficient amount of SMN protein, which results in an indication of muscular weakness and atrophy (Shababi et al., 2014). The variability largely depends on the number of copies of a related gene, SMN2, which produces limited functional SMN protein. Individuals with more SMN2 copies often exhibit milder symptoms due to partial compensation for the loss of SMN1 functionality. A fundamental basis for comprehending SMA and creating focused treatments is provided by the well-established genetic and molecular mechanisms underlying SMA. Carriers of SMA are usually unaffected but are able to pass their defective genes on to the next generation, since the disease has an autosomal recessive inheritance pattern. The amount of SMN protein determines the severity of SMA. Exon 7 is excluded by alternative splicing of the majority of SMN2 transcripts, resulting in a shortened, non-functional protein. On the other hand, the modest percentage of full-length SMN2 transcript is essential for altering the severity of the condition. The development of new therapeutic strategies helps in increasing SMN protein levels through genetic architecture. These genetic approaches include gene therapy for replacing the defective SMN1 gene (Ottesen et al., 2023). For many patients, the SMA became a manageable disease instead of a fatal condition, due to the development of such therapies. There are five main types of SMA based on motor abilities and age: types 0, 1, 2, 3, and 4. The most severe and rare form of SMA is called type 0, where prenatal fetal movements are decreased and severe difficulties in respiration, which often leads to the death of neonatal. Type 1, also known as Werdnig–Hoffman disease, is the most common and severe form in infancy. The characteristics of Type 1 SMA in infants show poor head control, profound hypotonia, and difficulty in feeding within the first six months of life. Due

DOI: 10.1201/9781003478652-9

to these respiratory issues, mostly, those children do not survive more than two years (Kolb and Kissel, 2015). In addition, those children may experience progressive muscle weakness and respiratory insufficiency and sometimes develop complications like scoliosis. People with type 3 SMA, also known as Kugelberg–Welander disease, can walk on their own at first but may eventually lose this ability as the illness worsens. Type 3 SMA starts after 18 months. The mildest form of SMA is known as type 4 SMA, which can affect in adulthood and can be characterized by gradual muscle weakness, primarily in proximal muscles, and minimal impact on life expectancy (Park et al., 2010). A few important cellular processes, including RNA splicing, axonal transport, and synaptic maintenance, have been disrupted due to this SMN protein deficiency. Motor neurons are susceptible due to their high metabolic demands and reliance on efficient RNA processing. Pre-mRNA splicing requires the formation of small nuclear ribonucleoproteins (snRNPs), which is hindered by dysfunctional splicing machinery. Additionally, reduced SMN protein levels disrupt axonal transport and neuromuscular junction integrity, further contributing to motor neuron degeneration and muscle atrophy. The progressive loss of motor neurons leads to muscle denervation, weakness, and eventual atrophy, severely affecting mobility and respiratory function. While motor neurons are the primary cells affected, emerging evidence suggests that other cell types and tissues may also contribute to SMA pathogenesis, offering new avenues for research and therapeutic intervention. The clinical presentation of SMA varies widely based on the disease type but generally includes progressive muscle weakness, hypotonia, and areflexia (Angilletta et al., 2023). In severe cases, respiratory insufficiency is a significant cause of morbidity and mortality due to weakness in intercostal muscles. Feeding difficulties, joint contractures, and scoliosis are common secondary complications that further impact quality of life. Diagnosis typically begins with clinical suspicion based on presenting symptoms, followed by confirmatory genetic testing to identify SMN1 gene deletions or mutations. Additional diagnostic tools, such as electromyography and muscle biopsies, may provide supportive evidence but are rarely needed in the presence of genetic confirmation. Carrier testing and prenatal screening are increasingly used to identify at-risk families, enabling early diagnosis and intervention. Historically, SMA management focused on supportive care, including physical therapy, respiratory support, nutritional management, and orthopedic interventions to address complications such as scoliosis. However, the treatment landscape has dramatically changed with the advent of disease-modifying therapies. Nusinersen (Spinraza), an antisense oligonucleotide, was the first FDA-approved treatment for SMA, designed to enhance the inclusion of exon 7 in SMN2 transcripts and thereby increase SMN protein production. Onasemnogene abeparvovec (Zolgensma), a gene replacement therapy, delivers a functional copy of the SMN1 gene through a viral vector, offering a potentially one-time treatment for SMA. Risdiplam (Evrysdi), an oral small molecule, also promotes exon 7 inclusion in SMN2 transcripts, providing a more accessible treatment option. These therapies have significantly improved survival and motor function, particularly when administered early, emphasizing the importance of newborn screening programs for SMA. Despite these advancements, challenges remain in ensuring equitable access to these therapies due to their high cost and complex administration requirements. Additionally, while these treatments address the underlying genetic defect, they do not reverse existing motor neuron damage or fully restore lost motor function. Research continues to explore combination therapies that could synergistically enhance treatment outcomes (Ponomarev et al., 2023). Advances in biomarker discovery are also underway, aiming to improve early diagnosis, monitor disease progression, and evaluate therapeutic efficacy. SMA is a testament to the power of translational research, where fundamental discoveries in molecular genetics have led to life-changing treatments. The disease remains a global health challenge, requiring ongoing efforts to improve patient outcomes and quality of life. Multidisciplinary care teams, patient advocacy, and continued investment in research are crucial to addressing the unmet needs of the SMA community. By fostering innovation and collaboration, the medical and scientific communities are making strides toward a future where SMA is no longer a life-limiting condition.

8.2 ETIOLOGY, EPIDEMIOLOGY, AND PATHOPHYSIOLOGY OF SMA

8.2.1 Etiology

In 95% of cases, SMA results from a homozygous deletion of SMN1 on chromosome 5q13; however, this does not explain how there can be significant clinical heterogeneity in phenotype. The answer lies in there being two versions of SMN telomeric version (SMN1) 2) centromeric version (SMN2), with individuals varying in the number of copies of SMN2 they possess (Lamadrid-González et al., 2024). SMN1 transcription produces a functionally complete mRNA that can then be translated to create the SMN protein. SMN2 transcription results in functionally complete mRNA 10 to 15% of the time, resulting in far fewer SMN protein being encoded than SMN1. SMN2 is identical to SMN1 except for a single C-T substitution in exon 7. This substitution promotes splicing 80 to 85% of the time during transcription and results in the removal of exon 7. This truncated mRNA causes similarly truncated non-functional proteins. Patients with SMA lack SMN1 and are therefore dependent on residual SMN2 production of functional SMN protein for alpha motor neuron function and subsequent survival. There is, therefore, a positive correlation seen between the number of copies of SMN2 and phenotype severity, with SMA type 1 typically having 1 to 2 copies of SMN2 and SMA type 4 having 3 to 5 SMN2 copies. There is no perfect correlation, as there are anomalies in phenotypic variation in SMN2, leading to varying amounts of functional SMN protein production in different individuals. Therefore, a low number of SMN2 copies with a milder clinical phenotype has been described (Lamadrid-González et al., 2024).

8.2.2 Epidemiology

SMA incidence has been estimated at 1 in 6,000 to 11,000, with a carrier frequency of 2% to 3% (1 in 40) of mutations in SMN1 in the general population. Incidence varies according to ethnicity, with the incidence reported as 8/100,000 for people of white ethnicity when compared with 0.89 in 100,000 for black ethnicity and 0.96/100,000 for those of mixed ethnicity. Extrapolating from carrier frequency in the general population, one should expect a higher incidence of live births with SMA. One hypothesis is that fetuses possessing 0 copies of SMN1 + SMN2 fail to develop, as is seen in other species with this genotype (Zhou et al., 2024).

8.2.3 Pathophysiology

Insufficient SMN protein, which is necessary for survival & upkeep of motor neurons, is the primary cause of SMA. The SMN protein is encoded by the SMN1 gene, which is responsible for producing the majority of SMN protein in the body. In SMA patients, mutations or deletions in the SMN1 gene reduce the levels of SMN protein, particularly affecting motor neurons, which are highly dependent on SMN for survival. Pathophysiology for SMA includes two important pathologies: (1) the role of the SMN protein and its deficiency, and (2) weakening of muscles and effects on motor neurons (Butchbach, 2021).

8.2.3.1 Role of SMN Protein and Its Deficiency

The SMN1 gene encodes the SMN protein, which is significant for several cellular processes. One of its important roles is in the assembly of spliceosomal small nuclear ribonucleoproteins (snRNPs), which are necessary components of RNA splicing – a procedure that allows cells to produce a diverse set of proteins from a single gene. Due to the lack of sufficient SMN protein, cells struggle to produce necessary proteins efficiently, which is specifically detrimental to motor neurons. Meanwhile, motor neurons have long axons that need a notable amount of cellular transport and maintenance, and they are mainly vulnerable to SMN deficiency. Research has shown that a deficiency of SMN disrupts cellular trafficking within neurons and interferes with the function and repair of motor neuron axons (Fallini et al., 2012).

8.2.3.2 Weakening of Muscles and Effects on Motor Neurons

Motor neurons are specific cells that send signals to muscles from the spinal cord, enabling coordination and movement. When the levels of SMN are inadequate, motor neurons become dysfunctional and start to degenerate. This degeneration leads to the progressive weakening of muscles, as they no longer receive proper signals from the damaged motor neurons. Over time, this weakening results in muscle atrophy, significantly impacting patients' ability to move, breathe, and perform daily activities. A 2023 study published in *Frontiers in Neurology* highlighted how SMN deficiency leads to axonal degeneration, disrupting neuromuscular junctions and progressively weakening muscle responses (Zhou et al., 2024). In addition to the SMN1 gene, humans possess a backup gene, SMN2, which also produces the SMN protein but in much smaller amounts. Due to a single nucleotide difference, SMN2 primarily produces a truncated and less functional form of the SMN protein. However, the number of SMN2 copies varies among individuals, and higher SMN2 copy numbers are often associated with milder SMA symptoms since they can produce more SMN protein, partially compensating for the SMN1 deficiency (Monani, 2005).

8.2.3.3 Progression and Symptoms

SMA is classified into different types based on the age of onset and severity of symptoms. Type 1, the most severe form, typically presents in infancy and is characterized by severe muscle weakness and respiratory difficulties. Type 2 usually appears in early childhood and is less severe than type 1. Type 3, also known as Kugelberg–Welander disease, starts in early childhood or adolescence and is milder than type 2. Type 4, the adult-onset form, presents in adulthood with milder symptoms. The progressive muscle weakness often starts in proximal muscles (those closer to the center of the body) and spreads over time. Symptoms commonly include difficulty swallowing, respiratory complications, and limited motor function. In severe cases, respiratory failure is a primary cause of mortality, as the muscles required for breathing also weaken. The pathophysiology of SMA highlights the need for therapies that can either increase SMN protein levels or support motor neuron health despite SMN deficiency (Kolb and Kissel, 2015, p. 1). Understanding these underlying mechanisms has led to innovative therapeutic approaches, including gene therapies that aim to replace or augment the missing SMN protein, thereby providing new hope for SMA management. Clinical symptoms of SMA include muscle atrophy, motor dysfunction, respiratory issues, and severity spectrum across different types.

- **Muscle Atrophy and Motor Dysfunction:**
  A primary symptom of SMA is muscle atrophy or the wasting away of muscles, resulting from the loss of signals from damaged motor neurons. This atrophy leads to muscle weakness, especially in muscles close to the body's core, such as the shoulders, hips, and back. As SMA progresses, patients experience increasing difficulty in performing voluntary movements, leading to motor dysfunction (Potgieser et al., 2014). Tasks such as sitting, standing, or walking become progressively challenging, and fine motor skills may also deteriorate.
  - **Type 1 SMA** (severe) often presents in infancy, with muscle weakness observable shortly after birth. Children with this type typically struggle to lift their heads or sit without support and may not achieve milestones like crawling or walking.
  - **Type 2 SMA** (intermediate) generally presents between 6 and 18 months. Individuals with this type may sit independently but often cannot walk without support.
  - **Type 3 SMA** (mild) presents later in childhood or adolescence, with affected individuals able to walk independently, although they may lose this ability over time as muscle weakness progresses.

- **Respiratory Issues**
  Respiratory complications are common in SMA, particularly in severe types, due to the weakening of intercostal muscles and the diaphragm, which are essential for breathing. As these muscles lose strength, individuals experience difficulty clearing respiratory secretions, increasing the risk of respiratory infections and reducing lung function. In severe cases, respiratory failure can occur and is often the primary cause of mortality in Type 1 SMA (Lagae et al., 2024)
  - **Type 1 SMA** patients frequently require respiratory support from infancy and may need ventilators for assistance.
  - **Type 2 SMA** patients may experience respiratory complications as the disease progresses, especially during infections, though these complications are generally milder than in Type 1.
  - **Type 3 and Type 4 SMA** individuals usually experience less severe respiratory symptoms but may still face reduced lung capacity and an increased vulnerability to infections.
- **Severity Spectrum across Different Types of SMA**
  The severity of SMA varies widely and is typically categorized into types based on the age of onset and achieved motor milestones:
  - **Type 1 SMA (Werdnig–Hoffman Disease)** is the most severe form, with symptoms appearing before six months of age. Most infants with Type 1 experience profound weakness and respiratory complications and have significantly shortened life expectancy without medical intervention.
  - **Type 2 SMA (Dubowitz Disease)** presents between 6 and 18 months and is of intermediate severity. These individuals can often sit independently but may not walk unassisted, and their life expectancy varies depending on respiratory health.
  - **Type 3 SMA (Kugelberg–Welander Disease)** typically appears in late childhood or adolescence. Individuals with Type 3 SMA are usually able to walk, though they may need mobility assistance over time.
  - **Type 4 SMA** is adult-onset and the least severe form, with symptoms usually emerging after age 21. Individuals with Type 4 SMA generally retain much of their motor function and have a normal life expectancy, though they may experience some progressive weakness.

Understanding the clinical spectrum of SMA symptoms is essential for effective care and targeted therapeutic development. This overview highlights the broad variability in SMA presentations and the need for individualized treatment approaches. So, further natural extracts have a significant effect on SMA, which is discussed in the following section.

8.3 PROMISING NATURAL EXTRACTS FOR SMA

Skeletal muscle (SkM) accounts for 30–40% of the total body weight. Maintaining crucial functions like contractibility, mobility, morbidity, mortality, and quality of life is essential. An imbalance between types of machinery involved in protein production and degradation is a hallmark of SMA, a significant consequence of SkM. Numerous clinical issues that significantly impair an individual's quality of life, such as sarcopenia, immobility, denervation, disuse, and cachexia, have been because of it. Furthermore, because of elevated oxidative stress, compromised mitochondria, and persistent inflammation, metabolic diseases and sedentary lifestyles result in necrosis-induced myopathies. Loss of body lean mass, a reflection of damaged muscle cell fibers and cross-sectional area regeneration, is the root cause of musculoskeletal disorders. The significance of SMA can be explained by the fact that patient death may be predicted only by appendicular lean mass. Comparing cancer patients with intensive care unit patients, it has been found that lower muscle mass is directly associated with a higher mortality rate. Approximately 1% of muscle mass is lost per year due to sarcopenia after the age of 45, resulting in SMA, which causes falls, fragility, and death. Consequently, the effects of SMA are noticeable health effects and major causes of morbidity and death. Multiple signaling pathways, maybe resulting from various circumstances, contribute to SMA. There had been, therefore, just a few medications with adverse effects that could be used to treat SMA. Exercise has been known to improve mass and muscular tone. Sirtuin1 (SIRT1), peroxisome proliferator-activated receptor-gamma coactivator-1 alpha (PGC-1α), and AMP-activated protein kinase (AMPK) are all activated by exercise-induced metabolic demand. By activating the AMPK/SIRT1/PGC-1α axis, insulin resistance caused by several reasons, including cachexia, age, sedentary lifestyle, and others, is overcome, and mitochondrial oxidation has been increased in SkM. However, older or catechistic people may find it difficult or impossible to engage in physical activity. To tackle SMA, there has been a desire to develop alternative treatments. While several FDA-approved drugs for SMA, like ezutromid, bimagrumab, domagrozumab, and others, can partially prevent muscle loss, most of them don't meet specific clinical objectives (Chaytow et al., 2021). To combat SkM mass loss during the early stages of atrophy, specific combinatorial techniques are needed. To control protein synthesis and degradation, as well as muscle regeneration, tissue engineering, nano-carrier delivery, natural products, cell treatments, and extracellular vesicles have all been created recently. Nonetheless, it had been impossible to overestimate the urgency of additional study in these areas, as it has been essential to the creation of successful SMA treatments. Since ancient times, dietary components and natural products have been sought-after sources for treating a wide range of illnesses and physical diseases because of their high functional, structural diversity and few negative effects on humans. In addition, having been used as model compounds to design and create medications to cure diseases, many natural chemicals have been developed as standard pharmaceuticals (Najmi et al., 2022). Many natural substances, such as polyphenols (curcumin, resveratrol, urolithin A), terpenoids (ursolic acid, tanshinone IIA, celastrol), flavonoids (apigenin, quercetin), alkaloids (magnoflorine, tomatidine), vitamin D, and others, are well known to increase muscle mass, strength, mitochondrial biogenesis, and muscle stem cell differentiation and to decrease inflammation and hydrogen peroxide production in SkMs. According to some data, herbal remedies and their refined constituents, such as alkaloids and polyphenols, are thought to have pleiotropic properties. Nonetheless, it has been demonstrated that certain substances have particular targets and regulate particular signaling pathways. For instance, to promote muscle myogenesis, ursolic acid, resveratrol, curcumin, berberine, and others increase the expression of AMPK-mediated PGC-1α and inhibit nuclear factor kappa-light-chain-enhancer of activated B cells (NF-κB) and activation of mitogen-activated protein kinase pathways of protein catabolism (A. Yadav et al., 2022). Nevertheless, their practice had been less successful than that of synthetic medications due to a dearth of studies on bioavailability, molecular processes, and clinical trials on natural products. The purpose of this review has been to provide an overview of current developments in our understanding of mechanisms, therapeutic potentials, and limitations of natural compounds that affect the strength and mass of SMA and SkM and that may be produced as adjuvant therapies, medicines, or dietary supplements. These natural extracts not only work individually but may have synergistic effects when combined or with existing SMA therapies (Wu et al., 2023). Their ability to simultaneously reduce oxidative stress, enhance mitochondrial health, modulate inflammation, and promote neuron regeneration provides a multifaceted approach to support motor neuron health in SMA. The mechanism of action of SMA can be understood by examining the effects of abnormality in the SMN1 gene. Motor neuron survival depends on the SMN protein, which is produced less frequently as a result of this mutation (Figure 8.1). As a result, motor neurons begin to degenerate, leading to neuronal degradation and ultimately causing muscle weakness and atrophy.

8.3.1 Curcumin

A neurological condition called SMA causes muscles to weaken and atrophy as a result of progressive loss of motor neurons. Although SMA is typically managed with supportive treatments, the

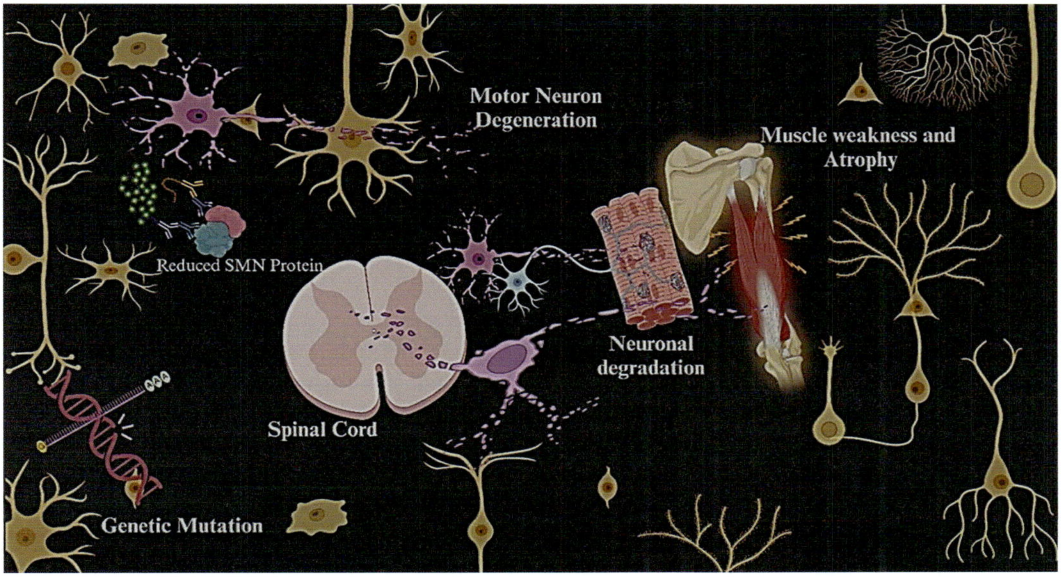

FIGURE 8.1 Mechanism of spinal muscular atrophy: SMN1 gene mutations reduce SMN protein levels, causing motor neuron degeneration, neuronal damage, and progressive muscle atrophy.

potential of natural extracts to enhance neuroprotection is a beacon of hope. Among these, curcumin an active compound in turmeric which has gained attention due to its strong antioxidant and anti-inflammatory.

8.3.1.1 Anti-Inflammatory and Antioxidant Properties

The development of SMA is significantly influenced by oxidative damage and inflammation. The heightened inflammatory response associated with SMA accelerates motor neuron degeneration. However, oxidative stress, which has been caused by an imbalance between antioxidants and free radicals, further harms nervous system cells. Curcumin has demonstrated the capacity to mitigate certain procedures effectively. Investigations have shown how curcumin inhibits inflammatory mediators, like COX-2 enzymes and cytokines, thus reducing inflammation that may worsen SMA symptoms (Peng et al., 2021). Additionally, curcumin acts as a strong antioxidant, neutralizing free radicals and boosting natural antioxidant defenses like superoxide dismutase (SOD) and glutathione. These actions create a protective environment within cells, potentially slowing down motor neuron damage in SMA.

8.3.1.2 Potential Role in Protecting Motor Neurons

SMA has been characterized by the loss of motor neurons, which carry messages from the brain and spinal cord to muscles. Protecting these cells from inflammatory and oxidative damage has been crucial for improving SMA outcomes. Preclinical studies have shown that curcumin is a promising candidate for offering such protection. Importantly, curcumin crosses blood–brain barrier, permitting it to directly reach and affect neural tissues. By reducing oxidative and inflammatory stress in the spinal cord, curcumin has the potential to significantly slow motor neuron degeneration and prolong cellular function, providing a ray of hope for SMA patients.

8.3.1.3 Preclinical Studies on Curcumin in Neurodegenerative Conditions

Curcumin's neuroprotective effects have been widely studied in various neurodegenerative diseases, suggesting potential applications for SMA. In animal models of amyotrophic lateral sclerosis (ALS), another motor neuron disease, curcumin treatment resulted in decreased spinal cord inflammation and oxidative stress (Wal et al., n.d.). Curcumin modulates critical pathways that may lessen neuronal damage, including NF-κB (nuclear factor kappa-light-chain-enhancer of activated B cells) pathway, a crucial regulator of inflammation. Curcumin has also been shown to elevate levels of brain-derived neurotrophic factor (BDNF) in animal models of neurodegeneration, a protein essential for neuron survival and health. BDNF's role in neurogenesis and neuron maintenance could be beneficial for motor neuron preservation in SMA. Additionally, curcumin appears to interact positively with AMPK/SIRT1/PGC-1α pathway, pathway that promotes mitochondrial health and resilience to stress. These interactions may help sustain cellular energy in motor neurons, improving their survival and function.

8.3.1.4 Potential as an Adjunct Therapy for SMA

While these preclinical findings are promising, clinical studies in humans are essential to confirm curcumin's safety and effectiveness in SMA treatment. It's important to note that curcumin, like many potential treatments, may have side effects. For instance, high doses of curcumin may cause gastrointestinal issues. Nonetheless, curcumin's unique combination of anti-inflammatory and antioxidant properties, along with its demonstrated support for motor neuron health, makes it a promising candidate for SMA research. Given progressive nature of SMA and limitations of existing treatments, curcumin may provide a complementary approach to slow neurodegeneration and improve motor neuron resilience. The preclinical evidence of curcumin's neuroprotective qualities and its anti-inflammatory and antioxidant actions make it a useful natural substance for SMA study. Future clinical studies could validate curcumin's role as an adjunct therapy in SMA, potentially offering new ways to protect motor neurons and improve lives of those who are impacted by this difficult illness.

8.3.2 Resveratrol

A polyphenolic substance that has been widely present in red wine, berries, and grapes is resveratrol gained attention in neurodegenerative research for its robust neuroprotective properties. Its potential benefits in SMA are currently being explored, particularly in improving mitochondrial function and reducing oxidative stress. As oxidative damage and mitochondrial dysfunction are central to SMA progression, resveratrol's protective effects may offer meaningful support for motor neuron health (E. Yadav et al., 2022).

8.3.2.1 Neuroprotective Properties and Mechanism

It is well known that resveratrol has neuroprotective properties primarily due to its ability to alter cellular stress responses and decrease oxidative injury. This molecule increases the overall capacity of mitochondria and boosts the cell's energy generation by activating a protein, SIRT1, that controls cell aging. This SIRT1 activation assists in preventing a frequent problem in SMA models, which is mitochondrial dysfunction, and thus maintains the energy equilibrium necessary for motor neuron sustenance. Furthermore, through its antioxidant properties, it prevents oxidative stress, which otherwise brings about cellular damage and speeds up motor neurons degeneration. Observations made in different neurodegenerative models show that resveratrol has the ability to decrease oxidative stress markers and thus could be useful in SMA.

8.3.2.2 Potential Role in Improving Mitochondrial Function and Reducing Oxidative Stress

Another major problem observed in SMA is thus, mitochondrial dysfunction that interferes with cellular energy homeostasis and leads to motor neuronal death. Resveratrol also increases the sirtuin-1 (SIRT1)/PGC-1α (Peroxisome proliferator-activated receptor-gamma coactivator 1-alpha) pathway which enhances mitochondrial synthesis, and increases protection of mitochondria under stress conditions. This superior mitochondrial function can be particularly advantageous in SMA models characterized by elevated energy needs as well as oxidative stress. Apart from lowering oxidative stress, resveratrol also protects cells from the destructive impact of free radicals. Investigations have revealed that resveratrol treatment has a preventive effect on lipid peroxidation as a sign of oxidative stress and enhanced the activities of antioxidants enzymes such as catalase and SOD in neural models. These effects design a healthier cellular environment, and may therefore halt or even reverse SMA by maintaining motor neuron function.

8.3.2.3 Preclinical Evidence for Resveratrol in Neurodegenerative Conditions

The discovery of resveratrol's neuroprotective possibilities is as vast as the list of neurodegenerative diseases, including SMA. Resveratrol increases motor function and reduces the loss of motor neurons in animal models of ALS, another motor neuron disease (Mancuso et al., 2014). This is partly attributable to its capacity to increase mitochondrial function and decrease oxidative stress. These findings are consistent with SMA models in which amelioration of mitochondrial dysfunction and decrease of oxidative stress could greatly affect the disease course. Furthermore, resveratrol has been found to upregulate BDNF in neurodegenerative disease models, promoting neuronal integrity and survival. This could be especially beneficial for SMA motor neuron preservation because neurotrophic support is minimal in this disease.

8.3.2.4 Potential as an Adjunct Therapy for SMA

While preclinical studies highlight resveratrol's neuroprotective effects, clinical research is essential to confirm its safety and efficacy in SMA patients. Nonetheless, resveratrol's combination of SIRT1 activation, mitochondrial enhancement, and antioxidant properties makes it an intriguing candidate for SMA research. Given the progressive nature of SMA and the lack of effective treatments, resveratrol may offer a complementary approach, potentially slowing motor neuron degeneration and supporting cellular health. The preclinical evidence supports resveratrol's ability to reduce oxidative stress and enhance mitochondrial function, positioning it as a promising natural compound for SMA research (de Oliveira et al., 2016). Ongoing and future studies could pave the way for resveratrol to serve as an adjunct therapy, improving outcomes for those affected by this challenging neurodegenerative condition.

8.3.3 Green Tea Polyphenols (EGCG)

Green tea's main polyphenol, epigallocatechin-3-gallate (EGCG), has drawn interest due to its treatment benefits for neurological conditions such as SMA. Its anti-inflammatory, neuroprotective, and antioxidant qualities point to a potential involvement in promoting the health and functionality of motor neurons (Mokra et al., 2022).

8.3.3.1 Antioxidant Effects and Neuroprotection

One important contributing element to the development of neurodegenerative illnesses such as SMA is oxidative stress. EGCG exhibits strong antioxidant properties via scavenging free radicals and boosting the activity of natural antioxidant enzymes like catalase and SOD. This action reduces oxidative damage in neuronal cells and might halt motor neuron degeneration. EGCG was also shown to regulate the oxidative stress signaling pathways and thus has neuroprotective effects in different models of neurodegenerative diseases, according to a scientific article in *Nutrients*. Likewise, a review in *Neurochemical Research*, a scholarly journal in the field, presents EGCG's ability to protect against inflammation in cerebral ischemia/reperfusion injury and its possible relevance to other neurodegenerative diseases (Wang et al., 2022).

8.3.3.2 Improved Muscle and Motor Function

Besides the neuroprotective potential, EGCG has been shown to have a positive impact on muscle and motor function. Studies show that EGCG can increase the number and efficiency of mitochondria, an organelle that generates energy within the cell. This leads to increased energy production in muscle cells. This enhancement assists muscle endurance as well as strength and is beneficial to those with SMA. As was stated in the study in the *International Journal of Molecular Sciences*, EGCG has helped improve the muscular function and reduce the level of muscle atrophy in animals, which could be beneficial for SMA patients.

8.3.3.3 Role in Reducing Inflammation

Chronic inflammation contributes to the progression of neurodegenerative diseases, including SMA. By blocking pro-inflammatory cytokines and altering signaling pathways like NF-κB, EGCG reduces inflammation. By decreasing inflammation, EGCG may help protect motor neurons from inflammatory damage. A review in *Neurochemical Research* discusses EGCG's role in attenuating inflammation in cerebral ischemia/reperfusion injury, a condition where blood flow to the brain is temporarily reduced or stopped, and then re-established, highlighting its potential applicability to other neurodegenerative conditions. In summary, EGCG is a promising natural substance for promoting motor neuron health and function in SMA because of its anti-inflammatory, antioxidant, and neuroprotective qualities. While additional investigation, including clinical studies, is required to confirm its effectiveness in SMA patients, current evidence strongly supports the potential of EGCG as an adjunctive therapy in managing this condition, providing reassurance about the validity of the research in this field.

8.3.4 Omega-3 Fatty Acids (Fish Oil)

Omega-3 fatty acids (fish oil) are a supportive agent for SMA. Omega-3 fatty acids are widely known for their neuroprotective and anti-inflammatory qualities, especially docosahexaenoic acid (DHA) and eicosapentaenoic acid (EPA), which are contained in fish oil. These essential fatty acids have been studied extensively in the context of neurodegenerative diseases and offer potential benefits for SMA by slowing disease progression and supporting muscle health. This section explores ways omega-3 fatty acids may contribute to SMA management, based on research findings from animal studies and neurodegenerative disease models (Avallone et al., 2019).

8.3.4.1 Anti-Inflammatory Benefits

One important element in the development of SMA is chronic inflammation, where inflammation can accelerate motor neuron damage. It has been demonstrated that omega-3 fatty acids, particularly EPA and DHA, block inflammatory pathways, notably by decreasing the NF-κB pathway, a key regulator of inflammation, and lowering pro-inflammatory cytokines. Research has demonstrated that DHA and EPA decrease the synthesis of inflammatory chemicals that are increased in neurodegenerative diseases, such as interleukin-6 (IL-6) and tumor necrosis factor-alpha (TNF-α). In a study published in *Behavior, Brain, and Immunity*, omega-3 fatty acids were found to lower markers of neuroinflammation and improve motor neuron survival in animal models of ALS, a disease that shares similarities with SMA. The reduction of neuroinflammatory mediators suggests that omega-3 fatty acids could help slow motor neuron degeneration by creating a more anti-inflammatory environment in the central nervous system.

8.3.4.2 Supporting Muscle Health and Disease Progression

Omega-3 fatty acids are also beneficial for muscle health, which is essential for managing SMA. Studies have shown that EPA and DHA improve muscle protein synthesis and reduce muscle atrophy by modulating the mTOR signaling pathway, which plays a vital part in muscle repair and growth. An *American Journal of Clinical Nutrition* highlighted that omega-3 fatty acid supplementation could enhance muscle strength and reduce muscle loss in older adults, suggesting similar benefits might be extended to SMA patients, where muscle atrophy is a central concern. Animal studies support these findings, demonstrating that omega-3 fatty acids reduce muscle wasting and improve muscle endurance. In a mouse model of muscular dystrophy, omega-3 supplementation helped preserve muscle function and slow disease progression, hinting at the possibility of similar effects in SMA. This preservation of muscle health may help counteract severe muscle atrophy characteristic of SMA, improving overall outcomes and quality of life.

8.3.4.3 Evidence from Neurodegenerative Disease Models

In SMA, the effects of omega-3 fatty acids have not been well investigated, but many studies have been published on the role of omega-3 fatty acids in neurodegenerative disease models. In models of Parkinson's and Alzheimer's diseases, omega-3 supplementation was found to improve mitochondrial biogenesis and decrease oxidative stress, which are critical for neuronal survival. The particular significance of SMA is the enhancement of mitochondrial function; motor neurons need a lot of energy to work and are prone to oxidative stress. Furthermore, the company has proved that DHA increases the BDNF, which is a protein that is essential for the health and flexibility of neurons. BDNF levels have been shown to rise in neurons, which might help support motor neuron integrity in SMA. This neuroprotective effect has been observed in both ALS and Alzheimer's disease models, suggesting that omega-3 fatty acids might offer similar support in the context of SMA (Miller et al., 2016).

8.3.4.4 Potential as an Adjunct Therapy for SMA

While evidence from preclinical studies and research in other neurodegenerative illnesses supports omega-3 fatty acids' potential function in managing SMA, further research is required to validate their effectiveness in this condition. It's crucial to remember that most of the proof that omega-3 fatty acids are beneficial in SMA management comes from preclinical studies, such as animal models. Clinical studies involving human subjects are still relatively limited. However, promising results from this preclinical research indicate that omega-3 fatty acids could be a valuable adjunct therapy for slowing disease progression and improving muscle and motor function in SMA. Among many advantages of omega-3 fatty acids are their ability to lower inflammation, offer neuroprotection, and enhance muscle health. Their effectiveness in preclinical models suggests that they may complement current SMA treatments, potentially leading to better outcomes for patients. Future clinical studies are essential to validate these findings and determine optimal dosing strategies for individuals affected by SMA, but the multifaceted benefits of omega-3 fatty acids provide reassurance about their potential in SMA management.

8.3.5 Ginseng

Ginseng, particularly *Panax ginseng*, has been traditionally valued for its various health benefits, particularly in neurological function. Recent studies highlight ginseng's potential to support nerve health and regeneration, making it an intriguing candidate for managing neurodegenerative diseases like SMA. Ginsenosides, active compounds in ginseng, have shown promise in enhancing nerve growth factor (NGF) expression and potentially exerting synergistic effects when combined with other SMA treatments (Zhang et al., 2024).

8.3.5.1 Benefits for Neurological Function

Ginsenosides have been studied for their neuroprotective properties, particularly in reducing inflammation and oxidative stress, both crucial in neurodegenerative diseases. A 2022 investigation that appeared in the *Journal of Ginseng Research* demonstrated that ginsenosides promote neuronal survival and reduce inflammation in neuronal cells, protecting them against oxidative damage, which is key to preserving motor neuron health in SMA (68,69). In another study done in *Nutrition* in 2023, the author was able to establish that ginseng has neuroprotective properties and also aids in boosting brain power due to its ability to improve mitochondrial health and decrease oxidative stress in neural tissues. This increased activity of mitochondria may be useful in providing the high energy requirement of motor neurons, thereby promoting cellular integrity in SMA. The ability of ginseng to promote cognitive performance suggests that there is much more to this plant with regard to neurological disorders.

8.3.5.2 Role in Enhancing NGF Expression

NGF is critical to neuron growth, maintenance, and survival and therefore holds significant value in neurodegenerative diseases. Literature reviews indicate that ginsenosides may upregulate the production of NGF, which may in turn improve neuronal health in SMA by conserving motor neurons. According to 2022 research conducted at the *Frontiers in Pharmacology*, ginsenoside Rb1 enhances the levels of NGF and enhances neurite outgrowth in vitro, implying the possibility of use in nerve support and regeneration for degenerative diseases. Neurite outgrowth is defined as the ability of nerve cells to grow their projections important in nerve repair and regeneration. Also, it was observed that ginsenosides offer protection to the neural cells against stress-induced damage, which is very important in SMA, where motor neurons are known to be sensitive to cellular stress. Cellular stress refers to the strain on cells caused by various factors, including oxidative damage and inflammation. By enhancing NGF production, ginseng may provide a nurturing environment for neuron health, possibly reducing the rate of motor neuron degeneration (Shi et al., 2022).

8.3.5.3 Possible Synergistic Effects with Existing SMA Treatments

Ginseng's role in managing SMA may be further enhanced when combined with other SMA treatments. Ginsenosides have shown potential to synergize with existing therapies by reducing inflammation and oxidative stress, which can complement the neuroprotective effects of SMA-targeted treatments. This synergistic potential refers to the combined effect of ginseng and other therapies being greater than the sum of their individual effects. A 2022 review in *Pharmaceuticals* discussed the synergistic potential of herbal supplements like ginseng with conventional neuroprotective therapies, noting that they may increase treatment efficacy by reducing cellular stress and enhancing overall neuroprotection.

8.4 MECHANISMS OF ACTION OF NATURAL EXTRACTS IN SMA

Natural extracts such as curcumin, resveratrol, EGCG (from green tea), omega-3 fatty acids, and ginseng are being researched for their therapeutic potential in SMA. These compounds hold promise in impacting SMA pathology through mechanisms such as reducing oxidative stress, improving mitochondrial function, modulating inflammatory pathways, and promoting motor neuron survival and regeneration. This research offers hope for the development of effective SMA treatments (Guo et al., 2024). However, it's important to note that the use of natural extracts in SMA treatment may also have potential risks or limitations, which will be discussed in the following sections. The following sections outline recent findings on these mechanisms, with supporting research articles.

8.4.1 Reducing Oxidative Stress

Neurodegeneration is the gradual decline in the structure or function of neurons in SMA, and oxidative stress, which results from the unequal ratio of free radicals and antioxidants, is the main cause. Natural extracts provide a source of antioxidants that help to fight oxidative stress and preserve motor neurons. Curcumin has been found to inhibit oxidation and decrease the action of free radicals by increasing the action of SOD and other antioxidants. A study published recently in *Antioxidants* (2023) showed that curcumin has the potential to prevent motor neuron damage by lowering the levels of oxidative stress in neuronal cells. Resveratrol acts as a potent antioxidant by activating the SIRT1 pathway that helps in improving mitochondrial tolerance to oxidative stress. A study in the *Journal of Neuroinflammation* (2023) revealed that resveratrol could reduce the oxidative markers in models of neurodegenerative diseases and, therefore, is likely to protect SMA.

8.4.2 Improving Mitochondrial Function

Another mechanism involved in SMA pathogenesis is mitochondrial dysfunction because motor neurons are known to need a lot of energy for their work. Natural extracts can enhance the function and production of mitochondria, which is a requirement in motor neurons. Green tea extract EGCG increases the number and effectiveness of mitochondria. A recent study in *Cellular Physiology & Biochemistry* (2023) demonstrated that EGCG enhances mitochondrial function by upregulating the PGC-1α pathway, the primary regulator of mitochondrial biogenesis. Omega-3 fatty acids have been shown to improve mitochondrial health by modulating lipid composition in mitochondrial membranes. A 2023 review in *Molecular Neurobiology* reported that omega-3 supplementation enhanced mitochondrial function and reduced oxidative damage in motor neurons (Chen et al., 2023).

8.4.3 Modulating Inflammatory Pathways

Chronic inflammation exacerbates SMA pathology by accelerating motor neuron degeneration. Natural extracts have anti-inflammatory effects that can reduce inflammation and protect neuronal health. Ginseng has demonstrated anti-inflammatory properties by inhibiting the NF-κB pathway, a key regulator of inflammation. A study in *Frontiers in Immunology* (2022) highlighted that ginsenosides in ginseng reduce inflammation markers like TNF-α and IL-6, which are implicated in neurodegeneration. Omega-3 fatty acids also modulate inflammation by reducing pro-inflammatory cytokines. Recent research published in *Cells* (2024) showed that omega-3s lower IL-1β and TNF-α levels in neurodegenerative disease models, suggesting their potential in SMA (24).

8.4.4 Promoting Motor Neuron Survival and Regeneration

Natural extracts also show potential in supporting motor neuron survival and promoting neurogenesis, which could help preserve muscle function in SMA. Ginseng, specifically ginsenoside Rb1, has been demonstrated to enhance NGF expression, promoting regeneration and health of neurons. A study in *Neurochemical Research* (2023) reported that ginseng treatment enhanced NGF levels and improved motor neuron survival in an SMA animal model (Schroth et al., 2024). Curcumin has been found to promote motor neuron survival by inhibiting apoptosis (cell death) pathways. Research in *Molecular Neurobiology* (2022) demonstrated that curcumin protects neurons by reducing caspase activity, which is involved in programmed cell death, potentially prolonging motor neuron function in SMA (Groulx-Boivin et al., 2024).

8.4.5 Synergistic Effects of Combined Mechanisms

These natural extracts not only work individually but may have synergistic effects when combined with each other or with existing SMA therapies. Their ability to simultaneously reduce oxidative stress, enhance mitochondrial health, modulate inflammation, and promote neuron regeneration provides a multifaceted approach to support motor neuron health in SMA. Recent studies support the potential of ginseng as a complementary therapy for SMA. By enhancing NGF expression, supporting mitochondrial function, and reducing oxidative stress, ginseng may bolster motor neuron health, complementing existing SMA treatments. The need for future clinical research is paramount to confirm these effects and determine optimal dosing, and your continued interest and support in this field are crucial. However, current findings indicate that ginseng holds promise as a supportive agent in the broader management of SMA. Natural compounds like resveratrol, green tea polyphenols (EGCG), curcumin, ginseng, and omega-3 fatty acids have demonstrated encouraging neuroprotective effects in SMA. These compounds help reduce oxidative stress, modulate inflammatory pathways, improve mitochondrial function, and promote motor neuron survival and regeneration, as depicted in Figure 8.2.

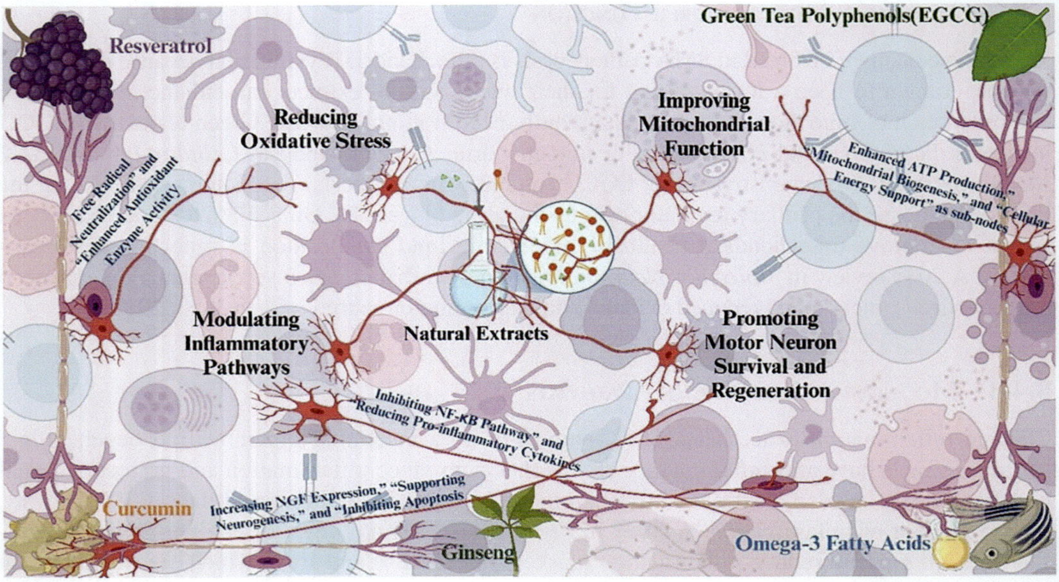

FIGURE 8.2 Neuroprotective effects of natural extracts in SMA.

8.5 CHALLENGES AND LIMITATIONS

Natural extracts hold potential as therapeutic agents for SMA due to their bioactive compounds, but significant challenges and limitations hinder their clinical application. The variability in the composition of natural extracts, influenced by factors like plant source, geographic location, and extraction methods, poses a major challenge to ensuring consistency and reproducibility. Additionally, the bioavailability and pharmacokinetics of these compounds are often suboptimal, requiring advanced delivery systems to enhance their therapeutic efficacy. Limited understanding of their precise mechanisms of action in SMA further complicates their development, as does the potential for off-target effects and toxicity at therapeutic doses. Regulatory hurdles and the lack of rigorous clinical trials on natural extracts for SMA impede their translation into approved treatments. Finally, the high cost and scalability of producing standardized extracts remain a concern, limiting their accessibility for broader application in SMA management.

8.6 CONCLUSION

SMA is a very common genetic illness that is one of the parts of the motor neuron diseases throughout the world today, which causes nerve cell injury, resulting in the loss of muscle use and further decrease in body functionality. Despite the progress of gene therapy in the management of SMA, such as the introduction of nusinersen and onasemnogene abeparvovec, which have brought great improvement, the latter approaches involve a high price and potential side effects and are limited in accessibility. This has evoked curiosity in alternative methods as well as complementary treatment, mainly the investigation of herbal extract, during your treatment course as a therapy. Natural extracts from these kinds of resources, plants, fungi, and marine organisms, are a promising option for SMA treatment. These bioactive compounds are excellent examples of rich sources of antioxidants, anti-inflammatory agents, and neuroprotective molecules that might fight oxidative stress, inflammation, and neuronal apoptosis in SMA pathology. The main phytochemicals, like flavonoids, alkaloids, and polyphenols, that have been discovered are known for their contribution to the activation of specific signaling pathways that promote motor neuron survival and muscle regeneration. For example, curcumin is a plant chemical that comes from *Curcuma longa*. It was found to be several fold higher in effect than both curcumin and dimethoxy curcumin and proved to be capable of decreasing oxidative stress and inflammation, which were the main causes of motor neuron degeneration. As well as resveratrol, a polyphenol from grapes and berries, which regulate the SIRT1 pathway, the pathway known for neuronal health protection, also offers protection for the mitochondria. Natural substances are further believed to increase the electrical connection between neurons and strengthen weakened muscles caused by SMA. One of the advantages of using the extract of ginseng, a plant commonly found in North America and Asia, is the fact that it has been found to induce such neurologic changes as increasing neuronal size and decreasing neuronal death. In addition, compounds like quercetin and EGCG are anti-apoptotic and anti-inflammatory, which can help decrease the duration of SMA treatment. Worth noting is that natural substances could be incorporated into the therapy as a part of a more comprehensive treatment encompassing gene therapies and other drug interventions, while possibly decreasing the required dosages and side effects. However, despite the fact that these extracts have demonstrated their effectiveness in preclinical research, the primary obstacle still stands in the advancement of moving these substances from the laboratory bench to the bedside. By supplementing gene treatments and other pharmacological choices, these natural medicines can play a critical part in more comprehensive treatment strategies by potentially lowering dosages and minimizing negative effects. However, there are significant obstacles in getting natural extracts from preclinical research to actual clinical application. The main problems are the low bioavailability, the difficulty in standardizing dosages, and the variation in natural product compositions. To overcome these obstacles, sophisticated formulation techniques may be used, such as encapsulating these bioactive substances in liposomes or nanoparticles, which may enhance their stability, targeted distribution, and therapeutic benefits. To verify the safety, effectiveness, and ideal dosages of these extracts for SMA patients, thorough clinical trials are also necessary. Furthermore, natural extracts have the potential to be accessible and affordable treatment choices, particularly in settings with limited resources where state-of-the-art gene treatments would not be practical. Combining conventional wisdom with contemporary pharmacological developments enables the development of cost-effective and long-lasting treatments that may help a wider range of patients. In conclusion, natural extracts are a potential but underutilized strategy for SMA management. Their promise is highlighted by their varied bioactive characteristics, neuroprotective advantages, and compatibility with current therapies. However, realizing this promise would necessitate a cooperative, interdisciplinary effort that combines clinical validation, creative delivery methods, and pharmacological research. With ongoing research and development, natural extracts could revolutionize the SMA treatment landscape, offering renewed hope to both patients and caregivers confronting this challenging condition.

REFERENCES

Angilletta, I., Ferrante, R., Giansante, R., Lombardi, L., Babore, A., Dell'Elice, A., Alessandrelli, E., Notarangelo, S., Ranaudo, M., Palmarini, C., De Laurenzi, V., Stuppia, L., Rossi, C., 2023. Spinal Muscular Atrophy: An Evolving Scenario through New Perspectives in Diagnosis and Advances in Therapies. *Int. J. Mol. Sci.* 24, 14873. https://doi.org/10.3390/ijms241914873

Avallone, R., Vitale, G., Bertolotti, M., 2019. Omega-3 Fatty Acids and Neurodegenerative Diseases: New Evidence in Clinical Trials. *Int. J. Mol. Sci.* 20, 4256. https://doi.org/10.3390/ijms20174256

Butchbach, M.E.R., 2021. Genomic Variability in the Survival Motor Neuron Genes (SMN1 and SMN2): Implications for Spinal Muscular Atrophy Phenotype and Therapeutics Development. *Int. J. Mol. Sci.* 22, 7896. https://doi.org/10.3390/ijms22157896

Chaytow, H., Faller, K.M.E., Huang, Y.-T., Gillingwater, T.H., 2021. Spinal Muscular Atrophy: From Approved Therapies to Future Therapeutic Targets for Personalized Medicine. *Cell Rep. Med.* 2, 100346. https://doi.org/10.1016/j.xcrm.2021.100346

Chen, Q., Ruan, D., Shi, J., Du, D., Bian, C., 2023. The Multifaceted Roles of Natural Products in Mitochondrial Dysfunction. *Front. Pharmacol.* 14. https://doi.org/10.3389/fphar.2023.1093038

de Oliveira, M.R., Nabavi, S.F., Manayi, A., Daglia, M., Hajheydari, Z., Nabavi, S.M., 2016. Resveratrol and the Mitochondria: From Triggering the Intrinsic Apoptotic Pathway to Inducing Mitochondrial Biogenesis, a Mechanistic View. *Biochim. Biophys. Acta BBA – Gen. Subj.* 1860, 727–745. https://doi.org/10.1016/j.bbagen.2016.01.017

Fallini, C., Bassell, G.J., Rossoll, W., 2012. Spinal Muscular Atrophy: The Role of SMN in Axonal mRNA Regulation. *Brain Res.* 1462, 81. https://doi.org/10.1016/j.brainres.2012.01.044

Groulx-Boivin, E., Osman, H., Chakraborty, P., et al. 2024. Variability in Newborn Screening Across Canada: Spinal Muscular Atrophy and Beyond. *Can J Neurol Sci.* 51, 203–209. https://doi.org/10.1017/cjn.2023.34

Guo, X., Wang, Y., Zhu, Z., Li, L., 2024. The Role of Plant Extracts in Enhancing Nutrition and Health for Dogs and Cats: Safety, Benefits, and Applications. *Vet. Sci.* 11, 426. https://doi.org/10.3390/vetsci11090426

Kolb, S.J., Kissel, J.T., 2015. Spinal Muscular Atrophy. *Neurol. Clin.* 33, 831–846. https://doi.org/10.1016/j.ncl.2015.07.004

Lagae, L., Proesmans, M., Van den Hauwe, M., Vermeulen, F., De Waele, L., Boon, M., 2024. Respiratory morbidity in patients with spinal muscular atrophy – a changing world in the light of disease-modifying therapies. *Front. Pediatr.* 12, 1366943. https://doi.org/10.3389/fped.2024.1366943

Lamadrid-González, J., Castellar-Leones, S., Contreras-Velásquez, J.C., Bermúdez, V., 2024. SMN2 Copy Number Association with Spinal Muscular Atrophy Severity: Insights from Colombian Patients. *J. Clin. Med.* 13, 6402. https://doi.org/10.3390/jcm13216402

Mancuso, R., del Valle, J., Modol, L., et al 2014. Resveratrol Improves Motoneuron Function and Extends Survival in SOD1G93A ALS Mice. *Neurotherapeutics* 11, 419–432. https://doi.org/10.1007/s13311-013-0253-y

Miller, N., Shi, H., Zelikovich, A.S., Ma, Y.-C., 2016. Motor Neuron Mitochondrial Dysfunction in Spinal Muscular Atrophy. *Hum. Mol. Genet.* 25, 3395–3406. https://doi.org/10.1093/hmg/ddw262

Mokra, D., Joskova, M., Mokry, J., 2022. Therapeutic Effects of Green Tea Polyphenol (–)-Epigallocatechin-3-Gallate (EGCG) in Relation to Molecular Pathways Controlling Inflammation, Oxidative Stress, and Apoptosis. *Int. J. Mol. Sci.* 24, 340. https://doi.org/10.3390/ijms24010340

Monani, U.R., 2005. Spinal Muscular Atrophy: A Deficiency in a Ubiquitous Protein; a Motor Neuron-Specific Disease. *Neuron* 48, 885–895. https://doi.org/10.1016/j.neuron.2005.12.001

Najmi, A., Javed, S.A., Al Bratty, M., Alhazmi, H.A., 2022. Modern Approaches in the Discovery and Development of Plant-Based Natural Products and Their Analogues as Potential Therapeutic Agents. *Molecules* 27, 349. https://doi.org/10.3390/molecules27020349

Ottesen, E.W., Singh, N.N., Luo, D., Kaas, B., Gillette, B.J., Seo, J., Jorgensen, H.J., Singh, R.N., 2023. Diverse Targets of SMN2-Directed Splicing-Modulating Small Molecule Therapeutics for Spinal Muscular Atrophy. *Nucleic Acids Res.* 51, 5948–5980. https://doi.org/10.1093/nar/gkad259

Park, G.-H., Maeno-Hikichi, Y., Awano, T., Landmesser, L.T., Monani, U.R., 2010. Reduced Survival of Motor Neuron (SMN) Protein in Motor Neuronal Progenitors Functions Cell Autonomously to Cause Spinal Muscular Atrophy in Model Mice Expressing the Human Centromeric (SMN2) Gene. *J. Neurosci. Off. J. Soc. Neurosci.* 30, 12005–12019. https://doi.org/10.1523/JNEUROSCI.2208-10.2010

Peng, Y., Ao, M., Dong, B., Jiang, Y., Yu, L., Chen, Z., Hu, C., Xu, R., 2021. Anti-Inflammatory Effects of Curcumin in the Inflammatory Diseases: Status, Limitations and Countermeasures. *Drug Des. Devel. Ther.* 15, 4503–4525. https://doi.org/10.2147/DDDT.S327378

Ponomarev, A.S., Chulpanova, D.S., Yanygina, L.M., Solovyeva, V.V., Rizvanov, A.A., 2023. Emerging Gene Therapy Approaches in the Management of Spinal Muscular Atrophy (SMA): An Overview of Clinical Trials and Patent Landscape. *Int. J. Mol. Sci.* 24, 13743. https://doi.org/10.3390/ijms241813743

Potgieser, A.R.E., de Jong, B.M., Wagemakers, M., Hoving, E.W., Groen, R.J.M., 2014. Insights from the Supplementary Motor Area Syndrome in Balancing Movement Initiation and Inhibition. *Front. Hum. Neurosci.* 8, 960. https://doi.org/10.3389/fnhum.2014.00960

Schroth, M., Deans, J., Arya, K., et al. 2024. Spinal Muscular Atrophy Update in Best Practices. *Neurology Clinical Practice* 14, e200310. https://doi.org/10.1212/CPJ.0000000000200310

Shababi, M., Lorson, C.L., Rudnik-Schöneborn, S.S., 2014. Spinal Muscular Atrophy: A Motor Neuron Disorder or a Multi-Organ Disease? J. Anat. 224, 15–28. https://doi.org/10.1111/joa.12083

Shi, Z., Chen, H., Zhou, X., Yang, W., Lin, Y., 2022. Pharmacological Effects of Natural Medicine Ginsenosides Against Alzheimer's Disease. *Front. Pharmacol.* 13. https://doi.org/10.3389/fphar.2022.952332

Wal, P., Saraswat, N., Pal, R.S., Wal, A., Chaubey, M., n.d. A Detailed Insight of the Anti-inflammatory Effects of Curcumin with the Assessment of Parameters, Sources of ROS and Associated Mechanisms. https://doi.org/10.2174/1874220301906010064

Wang, L., Dai, M., Ge, Y., Chen, J., Wang, C., Yao, C., Lin, Y., 2022. EGCG Protects the Mouse Brain Against Cerebral Ischemia/Reperfusion Injury by Suppressing Autophagy via the AKT/AMPK/mTOR Phosphorylation Pathway. *Front. Pharmacol.* 13, 921394. https://doi.org/10.3389/fphar.2022.921394

Wu, J., Li, Y., He, Q., Yang, X., 2023. Exploration of the Use of Natural Compounds in Combination with Chemotherapy Drugs for Tumor Treatment. *Molecules* 28, 1022. https://doi.org/10.3390/molecules28031022

Yadav, A., Yadav, S.S., Singh, S., Dabur, R., 2022. Natural Products: Potential Therapeutic Agents to Prevent Skeletal Muscle Atrophy. *Eur. J. Pharmacol.* 925, 174995. https://doi.org/10.1016/j.ejphar.2022.174995

Yadav, E., Yadav, P., Khan, M.M.U., Singh, H., Verma, A., 2022. Resveratrol: A Potential Therapeutic Natural Polyphenol for Neurodegenerative Diseases Associated with Mitochondrial Dysfunction. *Front. Pharmacol.* 13, 922232. https://doi.org/10.3389/fphar.2022.922232

Zhang, M., Niu, H., Li, Q., Jiao, L., Li, H., Wu, W., 2024. Active Compounds of *Panax ginseng* in the Improvement of Alzheimer's Disease and Application of Spatial Metabolomics. *Pharmaceuticals* 17, 38. https://doi.org/10.3390/ph17010038

Zhou, B., Chen, X., Zhang, C., Wang, Y., Ma, P., Hao, S., Hui, L., Bai, Y., 2024. Analysis of Spinal Muscular Atrophy Carrier Screening Results in 32,416 Pregnant Women and 7,231 Prepregnant Women. *Front. Neurol.* 15. https://doi.org/10.3389/fneur.2024.1357476

9 Synthetic Way of Management in General

Magham Sai Varshini,[1,2] *Madhu Tanya Singh,*[1]
Sai Kiran S.S. Pindiprolu,[3]
Praveen Thaggikuppe Krishnamurthy,[1]
and R. Aishwarya Reddy[4]

[1]Department of Pharmacology, JSS College of Pharmacy, JSS Academy of Higher Education & Research, Ooty, The Nilgiris, Tamil Nadu, India

[2]Department of Pharmacology, Vignan Pharmacy College, Vadlamudi, Guntur, Andhra Pradesh, India

[3]Department of Pharmacology, Aditya Pharmacy College, Surampalem, India

[4]Department of Pharmaceutics, JSS College of Pharmacy, JSS Academy of Higher Education & Research, Ooty, The Nilgiris, Tamil Nadu, India

9.1 INTRODUCTION

Neurodegenerative diseases (NDs), including Alzheimer's disease (AD), Parkinson's disease (PD), Huntington's disease (HD), amyotrophic lateral sclerosis (ALS), diabetic neuropathy (DN), and multiple sclerosis (MS), represent a diverse group of disorders marked by progressive deterioration of neuronal function and structure. These debilitating conditions lead to cognitive decline, dementia, and impaired motor functions, posing significant challenges due to their complex etiology and lack of curative therapies (1). According to the World Health Organization, NDs are the second leading cause of morbidity and mortality worldwide.

Current therapeutic approaches primarily focus on managing symptoms rather than halting or reversing disease progression. Marketed drugs include donepezil, rivastigmine, galantamine, and memantine for AD; levodopa, piribedil, pramipexole, apomorphine, selegiline, rasagiline, safinamide, and entacapone for PD; xenazine for HD; riluzole and edaravone for ALS; duloxetine and pregabalin for DN; and cladribine, teriflunomide, and dimethyl fumarate for MS. Despite their availability, these treatments have limitations, including insufficient disease-modifying effects and limited specificity, highlighting the urgent need for novel therapies involving innovative chemical entities.

The underlying mechanisms of neurodegeneration are multifactorial, encompassing oxidative stress, mitochondrial dysfunction, neuronal loss, viral inflammation, ischemia-reperfusion injury, and abnormal protein aggregation in neuronal tissues (2). For instance, AD is characterized by the accumulation of amyloid-beta (Aβ) plaques and hyperphosphorylated tau (p-tau) tangles, while PD involves α-synuclein (α-syn) aggregates in Lewy bodies. ALS features aggregates of mutant superoxide dismutase 1 (SOD1), TAR DNA binding protein 43 (TDP-43), and fused in sarcoma (FUS), as well as dipeptide repeats resulting from non-canonical translation of mutant C9ORF72. Additionally, disruptions in metal ion homeostasis further complicate the pathology of these disorders (3). The causes of NDs involve a complex interplay of genetic, environmental, and lifestyle factors. Although current treatments aim to alleviate symptoms and decelerate disease progression, a definitive cure remains elusive. Recent advances in synthetic compounds have shown promise in targeting the molecular pathways implicated in

DOI: 10.1201/9781003478652-10

neurodegeneration, providing hope for more effective therapies. These compounds offer potential through diverse mechanisms of action, including antioxidant, anti-inflammatory, and protein aggregation-inhibiting properties.

This chapter delves into recent developments in synthetic compounds for treating NDs, emphasizing their mechanisms of action, therapeutic potential, and challenges in clinical translation. With ongoing research focusing on understanding disease mechanisms and improving treatment efficacy, the exploration of novel chemical moieties may pave the way for groundbreaking interventions in neurodegenerative disorders.

9.2 ADAMANTINE DERIVATIVES

Numerous studies have reported the synthesis of adamantane derivatives exhibiting diverse biological activities. Several reviews highlight the development and applications of these compounds, which have been effectively utilized as pharmaceuticals with varying therapeutic properties (4). Among the most recognized adamantane derivatives are amantadine and memantine. Memantine, marketed as Ebixa in Canada, received approval for the symptomatic treatment of moderate to severe Alzheimer's dementia in 2004. It was also approved in the United States and Australia in 2003. Amantadine (1-amino-adamantane), known by trade names such as Gocovri, Symadine, and Symmetrel, was first synthesized in the late 1950s. Its antiviral activity against influenza was identified in the early 1960s, and, by October 1968, it received approval from the U.S. Food and Drug Administration (FDA) as a prophylactic treatment for Asian influenza. In 1976, it was authorized for the treatment of influenza A virus infections. Amantadine's therapeutic applications expanded to PD in 1969, and more recently, it has been utilized in managing HD chorea, Creutzfeldt–Jakob disease, and AD. Interestingly, synthetic derivatives of adamantane with enhanced pharmacological potential compared to amantadine and memantine are being explored for the treatment of neurodegenerative disorders such as dementia, AD, and PD. These include compounds based on 1-fluoro- and 1-phosphonic acid adamantane derivatives, which are regarded as promising candidates for further experimental research.

9.3 PYRIMIDINE DERIVATIVES

Pyrimidine represents a vital heterocyclic framework that has garnered significant attention for its therapeutic potential in addressing NDs. Studies have highlighted the role of pyrimidine-based compounds in enhancing cognitive functions in the brain. Moreover, numerous pyrimidine-containing compounds have demonstrated promise as anti-neurodegenerative agents in both clinical and preclinical evaluations (3). Herein, we discuss some of the pyrimidine derivatives and their pharmacological interventions in NDs.

Han et al. developed, synthesized, and assessed a new series of 2–(2-oxoethyl)pyrimidine-5-carboxamide compounds as inhibitors of acetylcholinesterase (AChE), aiming for AD therapy (5). Jain et al. designed and synthesized a series of 2,4,6-substituted pyrimidine derivatives as inhibitors of BACE-1 (β-secretase) for AD treatment. The synthesized compounds were found to demonstrate improved oral bioavailability and effective central nervous system (CNS) penetration (6). In a study by Fang et al., 28 pyrimidine-based compounds were designed, synthesized, and assessed for their anti-neuroinflammatory properties in BV2 cells as potential treatments for AD. Among these, compound 8 exhibited the strongest inhibitory effects on nitric oxide (NO), tumor necrosis factor-alpha (TNF-α), and interleukin-1β (IL-1β (7).

Qiu et al. introduced a novel deep-red fluorogenic probe based on a pyrimidine ring for detecting hydrogen peroxide (H_2O_2) levels in *Drosophila* brain tissue models of PD. The study's findings highlighted the potential of this pyrimidine-based probe as a promising tool for PD treatment (8). Robinson and colleagues synthesized and evaluated a new series of 2-amino-4,6-diphenylpyrimidines substituted with carbamates as antagonists of adenosine A1 and A2A receptors for managing PD (9). Toledo-Sherman et al. developed a series of pyrimidine carboxylic acids as potent inhibitors of kynurenine monooxygenase (KMO) (10). Wityak et al. explored the effectiveness of pyrimidine-based pan-JNK inhibitors for HD treatment through X-ray crystallography and computational modeling. Their results suggested that co-administering compound 23 with the pan-cytochrome P450 inhibitor, 1-aminobenzotriazole, improved brain exposure to the drug, supporting the modulation of a pharmacodynamic marker in mice (11). Xia et al. conducted an in-depth study of pyrimidine-2,4,6-triones (PYTs) and their analogs as inhibitors of superoxide dismutase (SOD1) for clinical trials in ALS (12).

The biological data of these compounds suggest their potential as multitarget enzyme inhibitors, demonstrating pharmacological activity through the inhibition of various targets, including ChE, Aβ, MAO B, KMO, SOD1, Kv, and CX3CR1, which are involved in different neurodegenerative disorders. Therefore, it is evident that pyrimidine-based compounds offer a wide range of applications in the development of drugs for treating various NDs.

9.4 BENZOTHIAZOLES

Heterocyclic compounds, like benzothiazole, play a crucial role in organic chemistry and drug discovery due to their broad spectrum of physiological effects. Extensive research on benzothiazole and its derivatives has revealed their potential for various biological activities. In the 1950s, benzothiazole derivatives, particularly 2-aminobenzothiazoles, were studied for their muscle relaxant effects. Since then, benzothiazole analogs have been shown to possess a wide array of pharmacological properties, including antibacterial, anti-inflammatory, analgesic, anticonvulsant, antiviral, anthelmintic, antioxidant, and anticancer activities (13). Additionally, numerous successful clinical drugs, including ethoxzolamide, frentizole, riluzole, zopolrestat, and the widely recognized amyloid imaging agent thioflavin T, feature the benzothiazole core structure.

Given the importance of multitarget medications for NDs, tacrine-benzothiazole hybrid compounds (14) were synthesized as disease-modifying drugs. These hybrids act as inhibitors of AChE and Aβ oligomerization, helping to prevent toxicity induced by Aβ42 peptides. The study highlighted the protective properties of this novel class of tacrine-benzothiazole hybrids in mitigating the pathogenesis of AD (15). Cui et al. reported that 2-arylbenzothiazole-based metal complexes exhibit a high affinity for Aβ plaques. In vitro competitive binding studies demonstrated that these compounds have moderate affinity for Aβ aggregates. Autoradiography experiments showed that these Tc-99m-labeled molecules can specifically bind to human cerebrovascular Aβ plaques with high affinity, making them potential imaging agents for diagnosing CNS diseases linked to Aβ deposition (16). In a separate study, Zhang et al. introduced benzodioxane–benzoxazole/benzothiazole piperidine derivatives (formula 57) that display strong affinity for 5-HT1A and 5-HT2A receptors. Animal studies indicated a significant reduction in immobility time in mice, suggesting their potential therapeutic effect on depression (17). Byeon et al. examined novel ferulic acid and benzothiazole dimers for their specific binding affinities to Aβ fibrils, with compound 403 showing excellent binding affinity to Aβ fibrils (18). Furthermore, some benzothiazole agents which were approved by the FDA for neurological conditions include pramipexole (anti-Parkinson's agent), perospirone (antipsychotic agent), revospirone (anxiolytic agent), riluzole (anti-AD agent), and thioflavin T (anti-AD agent).

Recently, Elbatrawy et al. developed benzothiazole- and indole-based compounds through a structural hybridization approach, combining the benzothiazole N744 cyanine dye with the diphenyl pyrazole Anle138b. These compounds demonstrated anti-aggregation activity against 2N4R tau and α-syn, respectively. The anti-aggregation effects of the synthesized compounds were evaluated using the thioflavin T (ThT) fluorescence assay, while transmission electron microscopy was utilized to detect fibrils after conducting a time-course study with the ThT assay (19).

9.5 INDOLE DERIVATIVES

Indole is a naturally occurring heterocyclic aromatic structure found in both plants and animals. It features a six-membered benzene ring fused with a five-membered pyrrole ring, possessing unique electronic and steric properties that enhance bioavailability and pharmacological activity. Known as a versatile pharmacophore, indole can interact with a variety of receptors, leading to diverse biological effects. This makes it one of the most valued frameworks in medicinal chemistry.

Studies on the central indole core indicated that modifications, such as oxidation or reduction, resulted in reduced inhibition of BChE. Additionally, substituting the core with indazole, quinoline, and 4-phenyltriazole also led to a decrease in potency. Building on previously established structure–activity relationships, further research focused on a new series of tryptophan-based selective hBChE inhibitors. These tryptophan-derived compounds exhibited strong selectivity for hBChE over hAChE, which was attributed to the indole group interacting with the hydrophobic acyl-binding pocket and forming π–π interactions with Phe329 and Trp231. In contrast, replacing the indole ring of tryptophan with the isopropyl group of leucine (structure not provided) drastically reduced inhibitory potency by three orders of magnitude. Substituting with a smaller phenyl ring also diminished potency, though to a lesser degree (20). Subsequently, indole-based sulfonamide derivatives and indole-based hydrazide-hydrazone derivatives were synthesized, and their specificity for hAChE and hBChE was evaluated. In 2019, a study investigated the potential of a small custom library of therapeutics to alleviate a tau-induced rough-eye phenotype in a model of frontotemporal dementia (21). The phosphorylation of tau in vivo was also evaluated, and among the promising candidates, the indole derivative Ro 31–8220, previously identified as a potent protein kinase C alpha (PKCα) inhibitor, was further studied. Ro 31–8220 significantly improved the rough-eye phenotype, decreased phosphorylated tau species both in vitro and in vivo, reversed tau-induced memory deficits, and enhanced motor functions in midges. Additionally, in a human neuroblastoma cell line, Ro 31–8220 lowered PKC activity and altered the tau phosphorylation pattern. In 2019, EI-Sayed et al. synthesized phosphazine and phosphazide derivatives, demonstrating their anti-AD effects by targeting acetylcholinesterase and Aβ aggregation (22).

In 2022, Knez et al. synthesized 23 indole analogs (series 14) to identify reversible MAO inhibitors. Among these compounds, two, 14a (IC50 = 12.63 μM) and 14b (IC50 = 8.65 μM), were found to be selective inhibitors of human MAO B at low micromolar concentrations (23). In fact, patients with HD and coexisting type 2 diabetes who were treated with metformin showed improved cognitive functions, indicating that the pharmacological activation of AMPK could be a promising strategy for managing HD (24). In 2022, a series of indole-derived compounds were synthesized and tested for their ability to activate AMPK at the cellular level using HEK293 cell lines (25). Among these, compounds 19–21 demonstrated a dose-dependent increase in AMPKα phosphorylation compared to the untreated control cells, confirming AMPK activation. Notably, compound 21 was identified as the most potent AMPK activator in this study. Recent research on NDs has focused on various molecular targets, with the indole-based derivatives discussed here primarily interacting with ChE, Aβ, MAO B, A2AAR, PERK signaling, AMPK, and 5-HT6R.

9.6 PYRAZOLINES

Pyrazolines are five-membered heterocyclic rings containing two nitrogen atoms in their structure. These compounds exist in three isomeric forms: 1-pyrazoline, 2-pyrazoline, and 3-pyrazoline. In 2-pyrazolines, medicinal chemists have introduced two or three substitutions to explore their biological potential (26). Furthermore, pyrazolines can serve as isosteres for various heterocyclic rings, such as imidazole, thiazole, and oxazole. Known for their high stability, pyrazolines have been modified structurally by chemists to create a wide range of compounds with diverse pharmacological activities. Extensive research in medicinal chemistry has focused on pyrazoline scaffolds for treating various conditions. These compounds exhibit a broad spectrum of biological activities, including anticancer, anti-AD, antitubercular, anticonvulsant, anti-inflammatory, antidepressant, antimicrobial, anti-HIV, antimalarial, MAO inhibition, anticholinesterase, cannabinoid antagonism, anti-EGFR, anti-Parkinsonian effects, and more (14, 27–32). As a result, pyrazolines are an important class of heterocyclic compounds found in the structures of numerous therapeutic agents.

In the context of AD, pyrazolines have been identified as effective scaffolds for inhibiting the degradation of ACh and Aβ plaques, potentially offering therapeutic benefits for treating AD. Yamali et al. synthesized eight new 4-(3-(difluorophenyl)-5-(dimethoxyphenyl)-4,5-dihydropyrazol-1-yl) benzenesulfonamides and assessed their AChE inhibitory activity using Ellman's method. These novel compounds demonstrated significant AChE inhibition at the nanomolar level (33). Singh and Bansal developed a new series of 16,17-N′-(alkyl/arylsulfonyl)pyrazolinyl steroids and evaluated their neuroprotective effects. One compound, 1′-(4-fluorobenzenesulfonyl)-5′-(pyridin-4-yl)-androst-5-en-[17,16-c]-1′H-pyrazoline-3β-ol (A02), showed promising results by suppressing AChE activity, reducing lipid peroxidation, and lowering TNF-α levels in the brain serum of lipopolysaccharide-treated rats (34). Tan et al. synthesized a series of 2-pyrazolines and reported their AChE inhibitory activity. One compound, 1–(5-(4–(4-methylpiperazin-1-yl)phenyl)-3–(3,4-dimethoxyphenyl)-4,5-dihydro-1H-pyrazol-1-yl)ethanone (A07), effectively inhibited AChE and significantly reduced Aβ1–42 aggregation by 72.6% (28).

Monoamine oxidase B (MAO B) and catechol-O-methyltransferase (COMT) are crucial neurotransmitters involved in the pathophysiology of PD. Singh and Bansal synthesized a series of 16,17-N′-(alkyl/arylsulfonyl)pyrazolinyl steroids to assess their neuroprotective properties. One compound, 1′-(4-fluorobenzenesulfonyl)-5′-(pyridin-4-yl)-androst-5-en-[17,16-c]-1′H-pyrazoline-3β-ol (A02), showed promising anti-Parkinsonian effects, improving ambulation, rearing, and reducing catatonia (34). Gokhan et al. developed 12 pyrazolines and studied their MAO inhibitory effects. Four compounds, P03, P04, P05, and P06, selectively and irreversibly inhibited MAO B in a classic noncompetitive manner (35). Hitge et al. synthesized a series of pyrazoline derivatives and evaluated the MAO and COMT inhibitory effects of nitro-catechol derivatives of chalcone and pyrazoline. The compound P07 was the most potent inhibitor of COMT and showed moderate inhibition of MAO B (36). Sahoo et al. synthesized 10 pyrazoline analogs and assessed their MAO inhibitory activity, finding that most of the compounds were selective and reversible inhibitors of MAO B (37).

9.7 CHALCONES

The chalcone structure is a widely studied scaffold for various medicinal purposes, including anticancer, antiulcer, antituberculosis, anti-inflammatory, and antibacterial activities. The presence of an α, β-unsaturated carbonyl group, which functions as a potential Michael acceptor, allows chalcone molecules to interact with the sulfhydryl group of cysteine residues or other thiol groups. This interaction is thought to contribute to their biological effects. Based on promising preclinical findings, some chalcone-based compounds, such as metochalcone (a choleretic drug), hesperidin methyl chalcone (vascular protective), and sofalcone (antiulcer and mucoprotective), have been approved for clinical use (38). Due to chalcone's low molecular weight, ease of synthesis, and the ability to optimize lipophilicity (log P) through the addition of specific substituents, this scaffold has garnered significant interest in the development of drugs for CNS diseases.

Chandrika et al. developed chalcone-donepezil hybrids with notable inhibition of cholinesterase and Aβ fibril assembly (39). They synthesized a series of 1,3- (compounds 1a–f) and 1,4- (compounds 2a–f) chalcone-donepezil hybrids. The 1,4-chalcone-donepezil hybrids (2a–2f) demonstrated superior AChE inhibition compared to the 1,3-chalcone-donepezil hybrids. A similar strategy of combining chalcone with the clinically used rivastigmine was also explored (40). In this study, the researchers modified the chalcone rings A and B with functional groups, including carbamates found in rivastigmine, and tested their AChE and BuChE inhibitory effects. Several compounds exhibited BuChE inhibitory activity at micromolar to submicromolar concentrations. Burmaoglu et al. synthesized tris-chalcones with fluorine groups positioned on various sites of the phenyl ring. Nine derivatives were prepared and tested for their cholinesterase inhibitory activity, with all compounds showing inhibition of both AChE and BuChE at nanomolar concentrations (41). Additionally, Shah et al. reported quinoline and piperidyl chalcone derivatives as cholinesterase inhibitors (42).

Youn et al. investigated the BACE-1 inhibitory activities of three flavonoids derived from *Boesenbergia rotunda*, including the chalcone cardamonin (compound 48). Among these flavonoids, compound 48 exhibited the strongest noncompetitive inhibition of BACE-1 (43). Building on the moderate BACE-1 inhibitory activity of the natural chalcone isoliquiritigenin, Ma et al. synthesized a series of 21 hydroxyl-substituted chalcone derivatives to evaluate their BACE-1 inhibition (44). Several of these compounds displayed enhanced BACE-1 inhibitory effects compared to isoliquiritigenin, with compound 53 being the most potent. Mphahlele et al. explored several halogenated chalcone derivatives for their BACE-1 and cholinesterase inhibitory properties (45). They found that compound 54 was the most potent BACE-1 inhibitor in their series and also demonstrated moderate BuChE inhibition. The natural product curcumin, closely related to chalcone, has been shown to prevent the oligomerization of Aβ and disaggregate preformed Aβ fibrils (46). Hirohata et al. studied the mechanisms behind the anti-amyloidogenic effects of five flavonoids (47), noting that myricetin bound reversibly to Aβ fibrils rather than the monomers. In a separate study, Cong et al. synthesized a series of hydroxylated chalcones to assess their dual effects on Aβ aggregation and ferroptosis (48). They found that the trihydroxy compounds 59a–c exhibited superior inhibition of Aβ aggregation compared to the other compounds.

9.8 CHALLENGES IN CLINICAL TRANSLATION

Despite promising preclinical results, the clinical development of synthetic compounds faces several challenges. Effective delivery to the CNS is hindered by the blood–brain barrier, requiring chemical modifications or novel delivery systems like nanoparticles to enhance bioavailability. Long-term safety profiles must be rigorously evaluated to minimize toxicity. Furthermore, many compounds fail to replicate preclinical efficacy in human trials due to differences in biological models and the multifactorial nature of NDs. Multitargeted approaches may offer better results by addressing the interconnected pathways involved in neurodegeneration. Drug metabolism and excretion also pose challenges, as compounds must be designed to achieve optimal half-lives and avoid rapid degradation. Additionally, side effects such as hepatotoxicity, immunosuppression, or off-target interactions need to be carefully monitored. The heterogeneity of NDs further complicates treatment strategies, necessitating personalized approaches to therapy development.

9.9 FUTURE DIRECTIONS

Advances in medicinal chemistry, nanotechnology, and drug delivery systems offer opportunities to overcome existing limitations. Nanocarriers, liposomes, and polymer-based systems can enhance drug stability and brain delivery. Additionally, combining synthetic compounds with gene therapy or stem cell approaches may amplify therapeutic outcomes. Future research should also focus on precision medicine approaches to tailor therapies based on individual genetic and biomarker profiles. Synthetic compounds designed to target multiple pathways simultaneously represent a promising direction. Developing hybrid molecules that combine antioxidant, anti-inflammatory, and anti-aggregation properties may yield more effective therapies. Integrating computational modeling and artificial intelligence into drug design could also accelerate the discovery of new compounds with improved efficacy and safety profiles.

9.10 CONCLUSION

Synthetic compounds provide a promising avenue for addressing the complex pathology of NDs. Classes such as adamantine derivatives, pyrimidines, benzothiazoles, indole derivatives, pyrazolines, and chalcones target key molecular pathways, including protein aggregation, oxidative stress, and inflammation. However, significant challenges remain in translating these findings into clinical practice. Continued research focusing on optimizing drug design, improving bioavailability, and integrating multimodal therapies is essential to harness their full potential.

REFERENCES

1. Magham SV, Thaggikuppe Krishnamurthy P, Shaji N, Mani L, Balasubramanian S. Cannabinoid receptor 2 selective agonists and Alzheimer's disease: An insight into the therapeutic potentials. *J Neurosci Res*. 2021 Nov;99(11):2888–905.
2. Barresi E, Baglini E, Poggetti V, Castagnoli J, Giorgini D, Salerno S, et al. Indole-based compounds in the development of anti-neurodegenerative agents. *Molecules*. 2024 Jan;29(9):2127.
3. Pant S, Kapri A, Nain S. Pyrimidine analogues for the management of neurodegenerative diseases. *Eur J Med Chem Rep*. 2022 Dec 1;6:100095.
4. Dembitsky VM, Gloriozova TA, Poroikov VV. Pharmacological profile of natural and synthetic compounds with rigid adamantane-based scaffolds as potential agents for the treatment of neurodegenerative diseases. *Biochem Biophys Res Commun*. 2020 Sep 3;529(4):1225–41.
5. Han C, Wei BB, Shang PP, Guo XY, Bai LG, Ma ZY. Design, synthesis and evaluation of 2–(2-oxoethyl) pyrimidine-5-carboxamide derivatives as acetylcholinesterase inhibitors. *Bioorg Med Chem Lett*. 2022 Sep;72:128873.
6. Jain P, Wadhwa PK, Jadhav HR. Design, synthesis and evaluation of 2,4,6-substituted pyrimidine derivatives as BACE-1 inhibitor: Plausible lead for Alzheimer's disease. *Med Chem*. 2021 Dec;17(10):1194–206.
7. Fang Y, Xia W, Cheng B, Hua P, Zhou H, Gu Q, et al. Design, synthesis, and biological evaluation of compounds with a new scaffold as anti-neuroinflammatory agents for the treatment of Alzheimer's disease. *Eur J Med Chem*. 2018 Apr;149:129–38.
8. Qiu X, Xin C, Qin W, Li Z, Zhang D, Zhang G, et al. A novel pyrimidine based deep-red fluorogenic probe for detecting hydrogen peroxide in Parkinson's disease models. *Talanta*. 2019 Jul;199:628–33.
9. Robinson SJ, Petzer JP, Rousseau AL, Terre'Blanche G, Petzer A, Lourens ACU. Carbamate substituted 2-amino-4,6-diphenylpyrimidines as adenosine receptor antagonists. *Bioorg Med Chem Lett*. 2016 Feb;26(3):734–8.
10. Toledo-Sherman LM, Prime ME, Mrzljak L, Beconi MG, Beresford A, Brookfield FA, et al. Development of a series of aryl pyrimidine kynurenine monooxygenase inhibitors as potential therapeutic agents for the treatment of Huntington's disease. *J Med Chem*. 2015 Feb 12;58(3):1159–83.
11. Wityak J, McGee KF, Conlon MP, Song RH, Duffy BC, Clayton B, et al. Lead optimization toward proof-of-concept tools for Huntington's disease within a 4-(1 *H*-pyrazol-4-yl)pyrimidine class of pan-JNK inhibitors. *J Med Chem*. 2015 Apr 9;58(7):2967–87.
12. Xia G, Benmohamed R, Kim J, Arvanites AC, Morimoto RI, Ferrante RJ, et al. Pyrimidine-2,4,6-trione derivatives and their inhibition of mutant SOD1-dependent protein aggregation. Toward a treatment for amyotrophic lateral sclerosis. *J Med Chem*. 2011 Apr 14;54(7):2409–21.
13. Yadav KP, Rahman MA, Nishad S, Maurya SK, Anas M, Mujahid M. Synthesis and biological activities of benzothiazole derivatives: A review. *Intell Pharm*. 2023 Oct 1;1(3):122–32.
14. Ahsan MJ, Samy JG, Khalilullah H, Bakht MA, Hassan Mohd Z. Synthesis and antimycobacterial evaluation of 3a,4-dihydro-3*H*-indeno [1,2-c] pyrazole-2-carboxamide analogues. *Eur J Med Chem*. 2011 Nov 1;46(11):5694–7.
15. DeKosky ST, Marek K. Looking backward to move forward: Early detection of neurodegenerative disorders. *Science*. 2003 Oct 31;302(5646):830–4.
16. 崔孟超, 刘伯里, 贾建华. 2-aryl benzothiazole compound with high affinity with A(beta) plaque and preparation method and application thereof [Internet]. CN103497217A, 2014 [cited 2025 Jan 4]. Available from: https://patents.google.com/patent/CN103497217A/en
17. Kamal A, Syed MAH, Mohammed SM. Therapeutic potential of benzothiazoles: A patent review (2010–2014). *Expert Opin Ther Pat*. 2015 Mar 4;25(3):335–49.
18. Byeon SR, Jin YJ, Lim SJ, Lee JH, Yoo KH, Shin KJ, et al. Ferulic acid and benzothiazole dimer derivatives with high binding affinity to β-amyloid fibrils. *Bioorg Med Chem Lett*. 2007 Jul 15;17(14):4022–5.

19. Elbatrawy AA, Ademoye TA, Alnakhala H, Tripathi A, Zami A, Ostafe R, et al. Discovery of small molecule benzothiazole and indole derivatives tackling tau 2N4R and α-synuclein fibrils. *Bioorg Med Chem*. 2024 Feb 15;100:117613.
20. Meden A, Knez D, Malikowska-Racia N, Brazzolotto X, Nachon F, Svete J, et al. Structure-activity relationship study of tryptophan-based butyrylcholinesterase inhibitors. *Eur J Med Chem*. 2020 Dec 15;208:112766.
21. Shim KH, Kim SH, Hur J, Kim DH, Demirev AV, Yoon SY. Small-molecule drug screening identifies drug Ro 31-220 that reduces toxic phosphorylated tau in *Drosophila melanogaster*. *Neurobiol Dis*. 2019 Oct 1;130:104519.
22. El-Sayed NF, El-Hussieny M, Ewies EF, Fouad MA, Boulos LS. New phosphazine and phosphazide derivatives as multifunctional ligands targeting acetylcholinesterase and β-Amyloid aggregation for treatment of Alzheimer's disease. *Bioorganic Chem*. 2020 Jan 1;95:103499.
23. Knez D, Hrast M, Frlan R, Pišlar A, Žakelj S, Kos J, et al. Indoles and 1-(3-(benzyloxy)benzyl)piperazines: Reversible and selective monoamine oxidase B inhibitors identified by screening an in-house compound library. *Bioorganic Chem*. 2022 Feb 1;119:105581.
24. Sanchis A, García-Gimeno MA, Cañada-Martínez AJ, Sequedo MD, Millán JM, Sanz P, et al. Metformin treatment reduces motor and neuropsychiatric phenotypes in the zQ175 mouse model of Huntington disease. *Exp Mol Med*. 2019 Jun;51(6):1–16.
25. Vela M, García-Gimeno MA, Sanchis A, Bono-Yagüe J, Cumella J, Lagartera L, et al. Neuroprotective effect of IND1316, an indole-based AMPK activator, in animal models of Huntington disease. *ACS Chem Neurosci*. 2022 Jan 19;13(2):275–87.
26. Ahsan MJ, Ali A, Ali A, Thiriveedhi A, Bakht MA, Yusuf M, et al. Pyrazoline containing compounds as therapeutic targets for neurodegenerative disorders. *ACS Omega*. 2022 Nov 1;7(43):38207–45.
27. Tok F, İrem Abas B, Çevik Ö, Koçyiğit-Kaymakçıoğlu B. Design, synthesis and biological evaluation of some new 2-Pyrazoline derivatives as potential anticancer agents. *Bioorganic Chem*. 2020 Sep 1;102:104063.
28. Unsal-Tan O, Küçükkılınç TT, Ayazgök B, Balkan A, Ozadali-Sari K. Synthesis, molecular docking, and biological evaluation of novel 2-pyrazoline derivatives as multifunctional agents for the treatment of Alzheimer's disease. *MedChemComm*. 2019 Jun 19;10(6):1018–26.
29. Ahsan MJ, Khalilullah H, Stables JP, Govindasamy J. Synthesis and anticonvulsant activity of 3a, 4-dihydro-3H-indeno[1,2-c]pyrazole-2-carboxamide/carbothioamide analogues. *J Enzyme Inhib Med Chem*. 2013 Jun 1;28(3):644–50.
30. Tripathi AC, Upadhyay S, Paliwal S, Saraf SK. N1-benzenesulfonyl-2-pyrazoline hybrids in neurological disorders: Syntheses, biological screening and computational studies. *EXCLI J*. 2018 Jan 19;17:126–48.
31. Badavath VN, Baysal İ, Ucar G, Sinha BN, Jayaprakash V. Monoamine oxidase inhibitory activity of novel pyrazoline analogues: Curcumin based design and synthesis. *ACS Med Chem Lett*. 2016 Jan 14;7(1):56–61.
32. Singh R, Thota S, Bansal R. Studies on 16,17-pyrazoline substituted heterosteroids as anti-Alzheimer and anti-Parkinsonian agents using LPS induced neuroinflammation models of mice and rats. *ACS Chem Neurosci*. 2018 Feb 21;9(2):272–83.
33. Yamali C, Gul HI, Kazaz C, Levent S, Gulcin I. Synthesis, structure elucidation, and *in vitro* pharmacological evaluation of novel polyfluoro substituted pyrazoline type sulfonamides as multi-target agents for inhibition of acetylcholinesterase and carbonic anhydrase I and II enzymes. *Bioorganic Chem*. 2020 Mar 1;96:103627.
34. Singh R, Bansal R. 16,17-*N'*-(alky/arylsulfonyl)pyrazoline substituted neuroprotective heterosteroids: Synthesis, molecular docking and preclinical efficacy/toxicity studies in rodents. *Steroids*. 2019 Aug 1;148:114–24.
35. Gökhan N, Yeşilada A, Uçar G, Erol K, Bilgin AA. 1-N-substituted thiocarbamoyl-3-phenyl-5-thienyl-2-pyrazolines: Synthesis and evaluation as MAO inhibitors. *Arch Pharm (Weinheim)*. 2003;336(8):362–71.
36. Hitge R, Smit S, Petzer A, Petzer JP. Evaluation of nitrocatechol chalcone and pyrazoline derivatives as inhibitors of catechol-O-methyltransferase and monoamine oxidase. *Bioorg Med Chem Lett*. 2020 Jun 15;30(12):127188.
37. Sahoo A, Yabanoglu S, Sinha BN, Ucar G, Basu A, Jayaprakash V. Towards development of selective and reversible pyrazoline based MAO-inhibitors: Synthesis, biological evaluation and docking studies. *Bioorg Med Chem Lett*. 2010 Jan 1;20(1):132–6.
38. Thapa P, Upadhyay SP, Suo WZ, Singh V, Gurung P, Lee ES, et al. Chalcone and its analogs: Therapeutic and diagnostic applications in Alzheimer's disease. *Bioorganic Chem*. 2021 Mar 1;108:104681.

39. Thamban Chandrika N, Fosso MY, Tsodikov OV, LeVine H, Garneau-Tsodikova S. Combining chalcones with donepezil to inhibit both cholinesterases and Aβ fibril assembly. *Molecules*. 2020 Jan;25(1):77.
40. Wang L, Wang Y, Tian Y, Shang J, Sun X, Chen H, et al. Design, synthesis, biological evaluation, and molecular modeling studies of chalcone-rivastigmine hybrids as cholinesterase inhibitors. Bioorg Med Chem. 2017 Jan 1;25(1):360–71.
41. Burmaoglu S, Yilmaz AO, Polat MF, Kaya R, Gulcin İ, Algul O. Synthesis and biological evaluation of novel tris-chalcones as potent carbonic anhydrase, acetylcholinesterase, butyrylcholinesterase and α-glycosidase inhibitors. *Bioorganic Chem*. 2019 Apr 1;85:191–7.
42. Shah MS, Najam-ul-Haq M, Shah HS, Farooq Rizvi SU, Iqbal J. Quinoline containing chalcone derivatives as cholinesterase inhibitors and their *in silico* modeling studies. *Comput Biol Chem*. 2018 Oct 1;76:310–7.
43. Youn K, Jun M. Biological evaluation and docking analysis of potent BACE1 inhibitors from *Boesenbergia rotunda*. *Nutrients*. 2019 Mar;11(3):662.
44. Ma L, Yang Z, Li C, Zhu Z, Shen X, Hu L. Design, synthesis and SAR study of hydroxychalcone inhibitors of human β-secretase (BACE1). *J Enzyme Inhib Med Chem*. 2011 Oct 1;26(5):643–8.
45. Mphahlele MJ, Agbo EN, Gildenhuys S. Synthesis and evaluation of the 4-substituted 2-hydroxy-5-iodochalcones and their 7-substituted 6-iodoflavonol derivatives for inhibitory effect on cholinesterases and β-secretase. *Int J Mol Sci*. 2018 Dec;19(12):4112.
46. Yang F, Lim GP, Begum AN, Ubeda OJ, Simmons MR, Ambegaokar SS, et al. Curcumin inhibits formation of amyloid β oligomers and fibrils, binds plaques, and reduces amyloid *in vivo**. *J Biol Chem*. 2005 Feb 18;280(7):5892–901.
47. Hirohata M, Hasegawa K, Tsutsumi-Yasuhara S, Ohhashi Y, Ookoshi T, Ono K, et al. The anti-amyloidogenic effect is exerted against Alzheimer's β-amyloid fibrils in vitro by preferential and reversible binding of flavonoids to the amyloid fibril structure. *Biochemistry*. 2007 Feb 1;46(7):1888–99.
48. Cong L, Dong X, Wang Y, Deng Y, Li B, Dai R. On the role of synthesized hydroxylated chalcones as dual functional amyloid-β aggregation and ferroptosis inhibitors for potential treatment of Alzheimer's disease. *Eur J Med Chem*. 2019 Mar 15;166:11–21.

10 Credentials of Natural Therapy over Synthetic Treatment

Vetriselvan Subramaniyan,[1] Muralikrishnan Dhanasekaran,[2] Mukesh Kumar,[3] Sivaraman Dhanasekaran,[4] Srikanth Jeyabalan,[5] Gowthamarajan Kuppusamy,[6] and Karri V. V. S. Narayana Reddy[6]

[1]Sir Jeffery Cheah Sunway Medical School, Faculty of Medical and Life Sciences, Sunway University, Malaysia

[2]Department of Drug Discovery & Development, Harrison College of Pharmacy, Auburn University, Auburn, AL, USA

[3]Botany and Microbiology, Gurukul Kangri University, Haridwar, Uttarakhand, India

[4]Department of Biotechnology, School of Energy Technology, Pandit Deendayal Energy University, Knowledge Corridor, Gandhinagar, Gujarat, India

[5]Department of Pharmacology, Sri Ramachandra Faculty of Pharmacy, SRIHER (DU) Porur, Chennai

[6]Department of Pharmaceutics, JSS College of Pharmacy, JSS Academy of Higher Education & Research, Ooty, Tamil Nadu, India

10.1 OVERVIEW

Natural therapy has been increasingly popular in contemporary healthcare in recent years. The rising incidence of lifestyle-related illnesses, increased worries about antibiotic resistance, and the negative side effects of synthetic medications are some of the causes of this change. Natural therapy has gained recognition for its holistic approach, which incorporates herbal medicine, dietary changes, and alternative healing techniques to boost general well-being, even if synthetic medicines have been crucial to medical progress. The move to preventative healthcare is one of the main causes of this change [1]. Instead of waiting for illnesses to strike to seek treatment, people are now more proactive in preserving their health. This paradigm change is also reflected in the growing popularity of integrative medicine, which blends traditional and alternative therapies. Furthermore, scientific developments have produced empirical data demonstrating the effectiveness of conventional therapeutic modalities like naturopathy, Ayurveda, and traditional Chinese medicine (TCM). These methods provide a more sustainable approach to long-term health because they place more emphasis on re-establishing body equilibrium than just treating symptoms. Furthermore, the pharmaceutical sector has acknowledged the promise of plant-based bioactive chemicals. Research and development expenditures have resulted in a better comprehension of natural substances and how they might be used to treat chronic diseases [2]. Herbal medications are now much safer and more effective thanks to greater regulation, standardisation, and control. By providing patients with safer and more complete treatment options than synthetic alternatives, natural therapy's acceptance into mainstream healthcare highlights its legitimacy [3].

DOI: 10.1201/9781003478652-11

10.1.1 Natural Therapy's Historical Context and Development

Natural treatments have been used for thousands of years, and several civilisations have developed their own medical systems based on natural resources. Ayurveda, which originated in India more than 5,000 years ago, uses yoga, dietary recommendations, and herbal compositions to treat and maintain health. Similarly, Qi Gong, herbal medicines, and acupuncture are all part of TCM, which has been practised for more than 2,500 years. Plant-based medicine served as the main form of treatment in many traditional civilisations. European folk cures, African traditional medicine, and Native American herbalism all made use of the local natural resources. These conventional methods were gradually superseded by synthetic medications that were developed with the development of contemporary pharmacology [4]. However, there is now more interest in natural medicines due to worries about side effects, high costs, and drug resistance. Integrative medicine is currently bridging the gap between alternative and traditional therapies. Traditional treatments are being more and more studied and validated by healthcare professionals and research institutes, which has led to their official inclusion in medical procedures. This historical continuity emphasises how credible and relevant natural medicines are today [5].

10.1.2 A Comparative Study of Synthetic and Natural Therapies

10.1.2.1 Action Mechanism

As shown in Figure 10.1, treat a disease's symptoms, synthetic medications frequently target particular metabolic processes. Although this method works well for acute situations, it doesn't always deal with the root causes. Conversely, natural therapies take a comprehensive approach, emphasising the

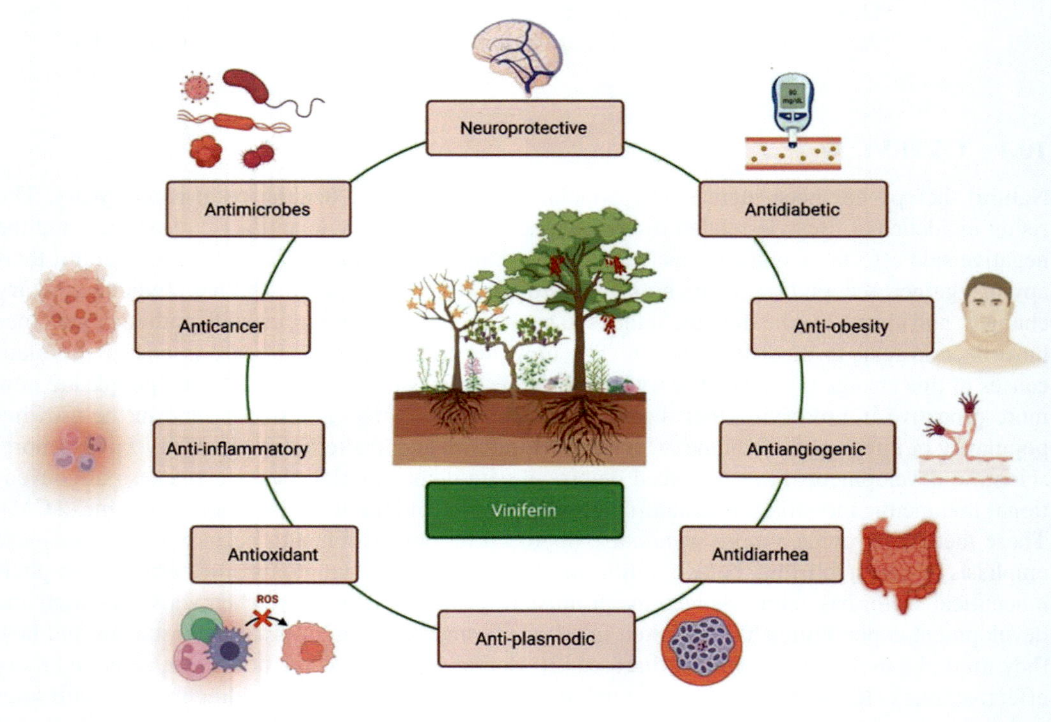

FIGURE 10.1 A drug's natural source for treating various medical diseases (copyright © 2022, Fuloria et al.). This open-access publication is distributed under the terms of the Creative Commons Attribution Licence (CC BY).

improvement of the body's self-healing systems. For example, herbal treatments are more suited to the physiological processes of the body since they contain several active ingredients that function in concert [6].

10.1.2.2 Safety Overview and Adverse Reactions

The adverse effects of synthetic drugs are one of their main disadvantages. A lot of prescription medications have side effects that might cause problems and necessitate further care. Because natural medicines come from biocompatible sources, they typically have a better safety profile. The adverse effects of herbal remedies are usually milder than those of synthetic medications, while some may be mild [7].

10.1.2.3 Management of Chronic Illnesses Effectiveness

Long-term medication is frequently necessary for chronic conditions like diabetes, hypertension, and arthritis. Long-term synthetic drug use can result in secondary health problems and dependency. To manage these diseases, natural therapies, on the other hand, emphasise dietary adjustments, lifestyle changes, and the use of herbal supplements. For instance, bitter melon is well-known for its hypoglycaemic benefits in the treatment of diabetes, while curcumin, which is present in turmeric, has shown anti-inflammatory qualities helpful in the treatment of arthritis [8].

10.1.2.4 Economical

Around the world, healthcare costs are a major problem. Because of the expenditures associated with research, production, and marketing, synthetic treatments – particularly copyrighted pharmaceutical drugs – can be costly. Natural treatments are typically less expensive, especially when they use easily accessible botanicals and dietary changes. Because of its accessibility, natural medicine is a desirable choice for people in lower economic groups and in developing nations [9].

10.1.3 Current Research and Scientific Validation

In the past, the absence of scientific support for natural medicine was one of the primary objections. Nonetheless, recent developments in pharmacognosy, biotechnology, and clinical trials have produced compelling proof of the effectiveness of some natural remedies. For example, the plant *Artemisia annua*, a crucial part of TCM, was used to make the anti-malarial medication artemisinin. Similarly, willow bark was the original source of aspirin, one of the most popular painkillers. The medicinal potential of plant-based chemicals is still being investigated in modern research, which has produced findings like berberine (from *Berberis* species) for metabolic problems and resveratrol (from grapes) for cardiovascular health. The significance of traditional medicine has also been acknowledged by institutions like the World Health Organization (WHO). By incorporating traditional healing systems within national healthcare frameworks, the WHO's traditional medicine strategy seeks to increase the legitimacy of natural medicine [5].

10.1.3.1 Quality Assurance and the Regulatory Framework

As natural therapy gains popularity around the world, regulatory bodies have set standards to guarantee the quality, safety, and effectiveness of herbal remedies. The European Medicines Agency controls herbal medical goods in the US, while the Food and Drug Administration (FDA) is in charge of nutritional supplements. Assuring standardised practices, the Ministry of AYUSH in India is committed to Ayurveda, yoga, naturopathy, Unani, siddha, and homoeopathy. The production of herbal medicines is now subject to quality assurance procedures like clinical testing, appropriate labelling, and good manufacturing practices. These initiatives address issues that previously prevented the broad use of natural medicines by reducing the hazards associated with contamination, adulteration, and improper dosages [10].

10.1.3.2 Natural Therapy's Prospects in Healthcare

It is anticipated that the use of natural therapy will increase as healthcare changes. Natural therapy is in line with personalised medicine, which adjusts therapies based on a patient's surroundings, lifestyle, and genetic composition. Furthermore, the expanding discipline of pharmacogenomics is investigating the ways in which natural substances interact with genetic circuits, opening the door to more accurate and potent therapies. Additionally, biopharmaceutical firms are funding phytomedicine research to investigate new plant-based medications that may compete with synthetic substitutes. The bioavailability of herbal substances is being improved by advances in nanotechnology, which also increases their therapeutic potential and efficacy. Additionally, the move towards plant-based therapy is supported by growing environmental and sustainability consciousness. In contrast to natural remedies, which frequently have a smaller ecological imprint, the overuse of synthetic drugs leads to pollution and antibiotic resistance. Natural therapy is becoming more and more popular than synthetic treatment due to reasons including scientific confirmation, cost-effectiveness, improved safety profiles, and holistic therapeutic concepts. Natural remedies provide a sustainable and complementary approach to long-term health management, even though synthetic pharmaceuticals are still essential in acute care and emergencies. Natural therapies will probably become even more integrated into mainstream medicine in the future as long as research on traditional remedies continues to be conducted and regulatory frameworks guarantee quality control [11].

10.2 A COMPARISON OF EFFICACY AND MECHANISM OF ACTION

10.2.1 A Comparison of Synthetic Drugs and Natural Therapies

Because natural remedies contain a wide range of bioactive substances that interact with multiple physiological pathways, they have made substantial contributions to both traditional and modern medicine. The natural substance usually has a polypharmacological activity with many interactions on synergistically related pathways, as opposed to manufactured medications that are often targeted towards a particular receptor or biochemical process. This characteristic lessens adverse effects and drug resistance while simultaneously boosting its therapeutic potency. In this way, contrasting the workings and effectiveness of synthetic medications with natural remedies provides clearer insights into their place in contemporary medicine [12].

10.2.1.1 Natural Compounds' Multi-Targeting and Synergistic Effects

The way that synthetic medications and natural remedies work is one of their primary distinctions. Synthetic medications act on a single enzyme, receptor, or signalling system, making them extremely specialised. Although this great specificity frequently yields strong therapeutic outcomes, it may also be a sign of medication resistance and adverse consequences. Natural substances, on the other hand, have various bioactive molecules and can interact with a variety of biological targets. Its effectiveness against a wide range of medical diseases is based on its polypharmacology, which also lowers the likelihood of resistance. For instance, the anti-inflammatory and antioxidant qualities of turmeric, or Curcuma longa, have been extensively researched and are typically ascribed to curcumin. Curcumin works via modulating NF-κB, TNF-α, interleukins, and other inflammatory mediators, in contrast to non-steroidal anti-inflammatory medicines (NSAIDs), which primarily work by inhibiting COX enzymes. Curcumin's broad-spectrum action allows it to reduce inflammation without causing the gastrointestinal harm that NSAIDs are known to cause. Curcumin reduces inflammatory biomarkers in rheumatoid arthritis patients, according to a 2023 meta-analysis. Its effectiveness is on par with that of traditional anti-inflammatory medications, but it has less side effects. By lowering cortisol levels, adaptogenic herbs like *Panax ginseng* and *Withania somnifera* (ashwagandha) have a stress-resistant effect that develops via the hypothalamic–pituitary–adrenal axis. Their adaptogenic qualities are credited with improving general health, which is enhanced by enhanced immune function, enhanced cognitive function, and improved metabolic balance.

In contrast to synthetic anxiolytics, which typically cause dependence and withdrawal symptoms, these phyto-analogues offer low side effects and long-lasting effects [13].

10.2.1.2 Use in Long-Term Conditions

These phytoconstituents significantly improve a number of chronic illness-related aspects, such as diabetes, heart disease, and neurological conditions. Green tea, berries, and grapes are a few foods high in polyphenolic chemicals, which lower oxidative stress, improve insulin sensitivity, and alter inflammatory pathways to control blood sugar levels. For instance, in patients with type 2 diabetes, the green tea polyphenol EGCG has been shown to improve glucose metabolism and lower insulin resistance. A potent antioxidant found in red wine and grapes, resveratrol has also been considered a potential treatment for neurological illnesses in recent years. Resveratrol does, in fact, support mitochondrial activity and reduce the development of beta-amyloid plaques, two characteristics that are common in Alzheimer's disease, according to a growing body of research. The chemical works by enhancing cellular energy metabolism and activating sirtuins, which is a benefit not typically shared by synthetic medications. It also has no negative side effects [14].

10.2.1.3 Use of Natural Substances in Cancer Treatment

Cytotoxic chemotherapy is the standard treatment for cancer, but because of its non-selective nature, it frequently causes major side effects. Nonetheless, a number of secondary metabolites originating

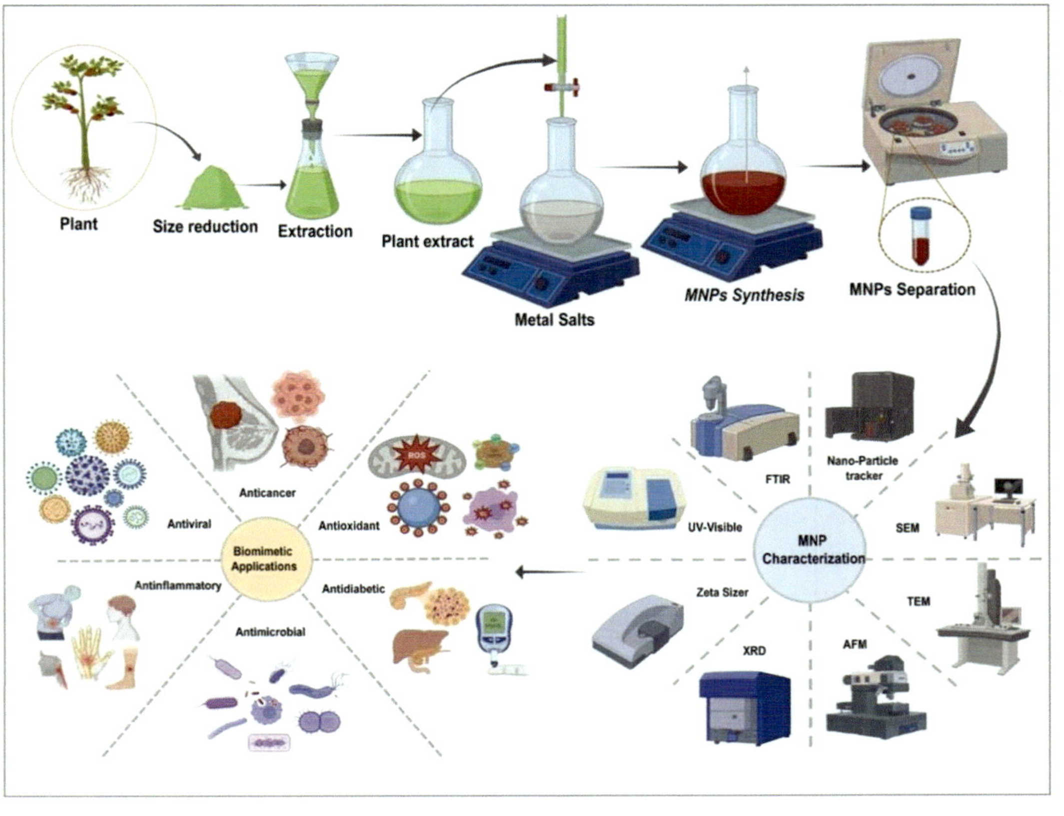

FIGURE 10.2 The multi-targeting and synergistic effects of a natural compound (copyright © 2024). Puri and associates, The Creative Commons Attribution Licence (CC BY), governs the distribution of this open-access article.

from plants have shown remarkable anticancer effects with comparatively low toxicity. For instance, by stabilising microtubules and stopping the division of cancer cells, paclitaxel, which is produced from *Taxus brevifolia*, has transformed chemotherapy. Likewise, vincristine, which is isolated from *Catharanthus roseus*, prevents tumour growth by interfering with the creation of mitotic spindles. Additionally, recent research has highlighted the function of polyphenolic chemicals in oncology. To stop tumour growth, they interfere with tumour suppressor genes, alter apoptotic pathways, and prevent angiogenesis. As illustrated in Figure 10.2, research highlights a viable strategy for using natural substances as supplemental or alternative cancer treatments, which are less harmful than traditional chemotherapy and more successful [15].

10.2.1.4 Obstacles and Opportunities

Despite their great therapeutic potential, natural treatments face challenges such as regulatory approval, bioavailability, and standardisation. Because natural chemicals are complex, determining a dosage schedule can be challenging. Low bioavailability can also result from inadequate absorption or fast metabolism. However, natural medicines are getting past these obstacles to increase their effectiveness thanks to developments in drug delivery methods and nanotechnology. Furthermore, because of their standardised formulations and well-understood actions, synthetic pharmaceuticals are always preferred by regulatory systems. To prove efficacy and incorporate natural substances into mainstream medicine, a significant increase in funding for clinical research and randomised controlled trials is required. Using multi-targeted mechanisms that maximise efficacy while reducing negative effects, natural treatments take a comprehensive approach to disease care. Natural substances are a more sustainable and integrative treatment choice than manufactured medications, which are typically linked to side effects and resistance. The effectiveness of bioactive substances like curcumin, resveratrol, and paclitaxel in treating everything from inflammation and chronic illnesses to cancer highlights their significance in contemporary medicine. Natural remedies have enormous potential to transform healthcare and close the gap between conventional and traditional medicine as research continues to elucidate their molecular underpinnings [16].

10.3 NATURAL THERAPY'S SAFETY PROFILE AND DECREASED SIDE EFFECTS

The higher safety profile of natural medicines over synthetic pharmaceuticals is one of the strongest arguments for their increasing popularity. Unquestionably, the contemporary pharmaceutical industry has made a substantial contribution to the management of sickness; nonetheless, synthetic drugs frequently have a number of negative side effects. Because many prescription medications have the potential to induce major side effects, such as organ toxicity, hormone imbalances, and reliance, they must be strictly monitored by pharmacovigilance. On the other hand, natural treatments, especially those based on Ayurveda and TCM, have been utilised for millennia with comparatively fewer adverse effects [17].

10.3.1 Examining Safety Profiles of Synthetic and Natural Drugs

Despite their effectiveness, synthetic medications can occasionally have serious side effects if taken over an extended period of time. For example, overuse of NSAIDs, which are frequently recommended to treat pain and inflammation, has been associated with higher cardiovascular risks, kidney damage, and gastrointestinal bleeding. Similarly, osteoporosis, hypertension, and weakened immunological function can result from long-term corticosteroid use. These drugs are closely monitored and regulated, which emphasises the dangers of using them. Natural remedies, on the other hand, have generally been linked to a better safety record. Green tea, ginger, and garlic are examples of medicinal plants that are well known for their health advantages and low risk of side effects. For instance, ginger is well-known for its anti-inflammatory qualities and capacity to reduce nausea without causing the gastrointestinal issues linked to synthetic antiemetics. Without the serious bleeding dangers associated with synthetic anticoagulants like warfarin, garlic functions as a natural anticoagulant, lowering the risk of blood clots. Green tea has shown promise as a cardiovascular supplement, supporting heart health without the negative effects of prescription lipid-lowering medications. These illustrations highlight how natural medicines can offer therapeutic advantages while being quite safe [18].

10.3.1.1 Traditional Medicine's Contribution to Safety

Finding harmony and balance inside the body is emphasised by both TCM and Ayurveda, which adds to their long-term safety. By emphasising holistic therapy, traditional medical practices make sure that therapies target the underlying causes of illnesses rather than just their symptoms. For example, in Ayurveda, people are categorised according to their doshas, or biological energies, and herbal formulations are prescribed according to their distinct requirements. This individualised method improves general well-being and reduces the possibility of negative reactions. The qualities, doses, and combinations of therapeutic herbs are documented in centuries-old pharmacopoeias, which are also used in TCM. To lessen the possibility of toxicity and negative reactions, practitioners take into account how various herbs interact with one another as well as how they could affect the body's energy systems. These old systems' legitimacy and ongoing usage in contemporary healthcare settings can be attributed to the long-standing empirical knowledge they contain [19].

10.3.1.2 Problems and Fixes for Guaranteeing the Security of Natural Medicines

Natural remedies are not risk-free, even if they are inherently safe. Concerns about herb–drug interactions, contamination, and uneven potency have arisen as a result of some regions' lack of standardisation, quality control, and regulation. It has been discovered that certain herbal items include pesticides, heavy metals, or adulterants that are extremely harmful to human health. Additionally, self-medication without a doctor's supervision or incorrect dosage can result in unexpected adverse effects. Regulatory bodies throughout the world have been attempting to create thorough standards for the manufacture, testing, and distribution of herbal medications in order to allay these worries. The U.S. FDA released new recommendations in 2023 that were centred on clinical validation and quality assurance for herbal products. These recommendations place a strong emphasis on:

- Standardisation of Herbal Medicines: determining consistent active component concentrations to guarantee reliable safety and efficacy [20].
- Quality Control Measures: Putting in place strict testing for heavy metals, pollutants, and adulterants.

Clinical validation is the process of carrying out thorough scientific research to verify the effectiveness and safety of herbal remedies.

- Control of Herb–Drug Interactions: To avoid negative effects, possible interactions between synthetic and herbal treatments are investigated.

10.3.1.3 The Future of Natural Therapies and Scientific Validation

Scientific interest in confirming the effectiveness and safety of natural medicines is growing as pharmacognosy and biotechnology improve. Integrating herbal medicine into traditional healthcare requires clinical trials and evidence-based research. Pharmacological characteristics of standardised extracts of therapeutic plants are being examined to make sure they adhere to contemporary medical standards. For example, the anti-inflammatory and antioxidant qualities of curcumin, the active ingredient in turmeric, have been thoroughly investigated. Its ability to treat diseases like cancer, metabolic problems, and arthritis has been confirmed by scientific studies. In a similar vein, adaptogenic herbs with few adverse effects, such as ginseng and ashwagandha, have demonstrated encouraging outcomes in lowering stress and improving cognitive performance. Because of their superior safety record and all-encompassing approach to treatment, natural therapies present a strong substitute for synthetic medications. Although synthetic drugs are still necessary for treating severe and acute diseases, the dangers of long-term use underscore the need for safer alternatives. Because of their long history and profound understanding, TCM and Ayurveda offer important insights into attaining optimal health without suffering from serious side effects. Continuous efforts in standardisation, quality control, and scientific validation are necessary to guarantee the efficacy and safety of natural medicines. The integrity of herbal medicine is mostly preserved by regulatory frameworks, such as the most recent FDA guidelines. Natural remedies will probably become more widely accepted as a common, secure, and efficient alternative in healthcare as long as research keeps bridging the gap between conventional wisdom and contemporary science. We can progress towards a more sustainable and comprehensive healthcare system that puts patient safety and well-being first by encouraging a balanced strategy that incorporates the best aspects of both worlds [21].

10.4 ENVIRONMENTAL ASPECTS AND SUSTAINABILITY IN THE PRODUCTION OF PHARMACEUTICALS

Despite the fact that the pharmaceutical industry has greatly improved healthcare around the world, the use of synthetic pharmaceuticals has raised severe environmental problems. Extraction, chemical synthesis, purification, and the disposal of leftover byproducts are all steps in the production of synthetic medications, and they all contribute to environmental contamination. The majority of the waste produced by pharmaceuticals contaminates water sources, resulting in ecological imbalance and endangering biodiversity. The inappropriate disposal of pharmaceutical residues, the majority of which have been shown to wind up in water bodies, is one of the main issues associated with the production of synthetic pharmaceuticals. The active pharmaceutical compounds found in wastewater from pharmaceutical firms are not usually fully broken down, leaving behind persistent contaminants in the environment. These have been linked to ecological imbalance, antibiotic resistance, and endocrine disturbance in aquatic life. Such an excessive reliance on non-renewable resources, such as the production of pharmaceuticals solely from petrochemicals, is very detrimental to the environment and greatly increases carbon emissions and resource depletion [22].

However, the procedures used in natural treatments derived from plants are more sustainable and kind to the environment. Medicinal plants have long been used in traditional medical systems, and their extraction typically involves far less complex chemical processing than that of modern medications. The majority of ethnobotanical studies emphasise that plant-based medications often have less of an adverse effect on the environment. The potential for producing plant-based medications sustainably using ethical farming methods that do not deplete resources for ongoing, long-term availability exists, in contrast to synthetic drugs. Due to the sustainability of their cultivation and gathering, medicinal plants would have a less negative impact on the environment. However, the methods of fermentation and plant tissue culture have dissected biotechnological approaches to environmentally friendly remedies for certain health issues. Plant tissue culture preserves biodiversity for ongoing sourcing of bioactive chemicals by enabling the mass production of medicinal plants without necessary clearing forests. Additionally, fermentation techniques allow for the in vitro synthesis of natural substances, which lessens the strain on wild plants and the loss of their habitat. Organic farming and permaculture are two practices that support the objective of sustainability as stated by the therapies themselves. In order to cultivate medicinal plants in an environment that will restore the ecosystem's equilibrium, the permaculture strategy involves cultivation in harmony with nature, which includes biodiversity, careful water management, and soil health maintenance. Synthetic insecticides and fertilisers are not used in organic farming. This promotes the establishment of naturally occurring medicinal plants with high bioactive potential and lessens contamination of the soil and water [23].

With international organisations starting to acknowledge the need for ecologically conscious medical practices, the worldwide push to transition to green healthcare has reached a crescendo. In 2023, a United Nations report on sustainable healthcare emphasised the necessity of integrating environmentally friendly medical supplies into contemporary healthcare systems. In order to find sustainable healthcare models with little ecological impact, it was necessary to combine traditional knowledge with recent scientific discoveries. The pharmaceutical sector can ensure effective healthcare solutions for future generations while moving towards a greener future by adopting sustainable medicinal practices. This study comes to the conclusion that in order to address some urgent environmental issues, the production of synthetic pharmaceuticals must give way to more sustainable and natural solutions. By utilising biotechnological advancements, sustainable farming, and the ethical production of medicinal plants, we may promote a more environmentally friendly approach to healthcare that promotes both environmental preservation and human well-being [24].

10.5 INCLUDING NATURAL THERAPY IN TRADITIONAL MEDICAL PRACTICES

One of the trickiest, yet most promising, aspects of contemporary healthcare is the incorporation of natural therapy with conventional medicine. Both patients and professionals now accept natural therapy, which includes acupuncture, herbal medicine, and dietary changes. There are some obstacles in the way of their full incorporation into standard medical practice, though. To create a framework that allows traditional and modern medicine to coexist successfully, integration entails cooperation between researchers, medical personnel, and policy makers [25].

10.5.1 Integrative Medicine's Promise

Integrative medicine is a method of treating patients that takes into account all of their requirements, combining evidence-based complementary therapies and conventional medical treatments with therapies including dietary and lifestyle modifications. Numerous chronic diseases, such as diabetes, heart disease, and autoimmune diseases, could be effectively managed with this strategy. In addition to standard medical care, CAM services are now offered by a number of hospitals and healthcare facilities worldwide. This change reflects a greater understanding of the impact natural treatments have on patient outcomes. Integrative medicine's primary focus on a patient-centred approach is one of its most notable advantages. This type of treatment takes into account a person's lifestyle, mental health, and eating habits, whereas the majority of traditional treatments have solely addressed the disease's symptoms. In addition to natural therapy like yoga, meditation, and herbal supplements, pharmaceutical intervention can be used to control conditions, including metabolic syndrome and hypertension [26].

10.5.1.1 The Function of WHO in Herbal Medicine Standardisation

The role of traditional medicine in global healthcare has been acknowledged by the World Health Organisation. However, there are still many concerns about making sure natural therapies are safe, effective, and subject to quality control. WHO started a significant initiative in 2024 to standardise herbal medicines by conducting clinical studies with a number of top research institutes worldwide. This programme's goal is to ensure that herbal products are produced in accordance with strict guidelines for dose, production, and therapeutic claims so that their application in clinical practice may be trusted. It will be extremely important because, in the majority of civilisations, herbal medicine is used on a regular basis. Traditional treatments are a component of primary healthcare in the majority of Asia and Africa. Herbal formulations cannot be standardised without regulatory supervision or standardisation, leaving them vulnerable to a number of contamination hazards. A connection between natural and conventional medicine can be established by adhering to well-recognised WHO rules for clinical trials and manufacturing procedures [27].

10.5.1.2 Natural Therapy's Prospects in Customised Medicine

The legitimacy of natural therapy is being strengthened by developments in pharmacognosy, biotechnology, and systems biology. Numerous novel bioactive compounds with possible therapeutic uses are being discovered as our understanding of active plant chemical bases grows (Figure 10.3).

These will help develop even more specialised therapies that may use both synthetic and natural substances. Another fascinating field where integrative medicine will play a significant role is personalised medicine, which treats patients according to their genetic makeup and metabolic profile. AI-driven research has fundamentally altered the process of optimising herbal formulations, leading to more targeted and efficient treatment plans. Large datasets of patient responses to herbal medicines can be analysed by artificial intelligence, which can also be used to improve dose recommendations and anticipate possible interactions with prescription medications [28].

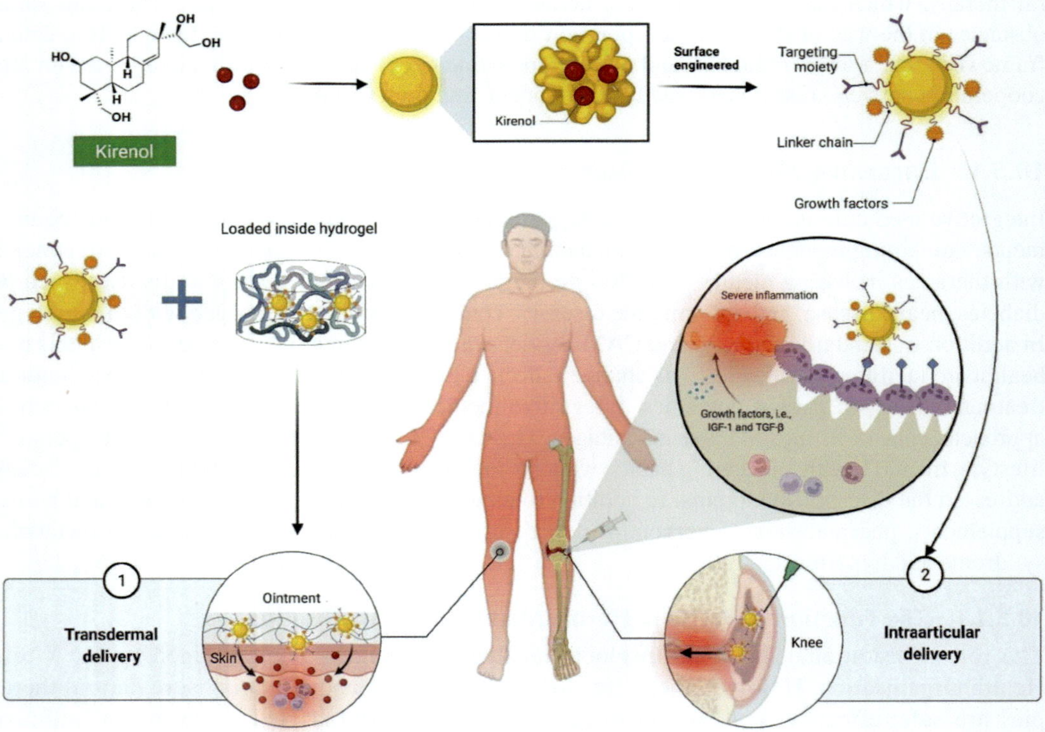

FIGURE 10.3 The potential of natural therapy in customised medicine (copyright © 2022; Nasir et al.). This open-access publication is distributed under the terms of the Creative Commons Attribution Licence (CC BY).

10.6 CONCLUSION

Natural therapy, supported by scientific progress and regulatory standards, offers safer and more comprehensive options in contemporary healthcare. Despite existing hurdles in validation and regulation, the integration of natural and conventional treatments may facilitate customised medicine strategies that enhance patient outcomes and world health.

REFERENCES

1. Nanda S. Integrating traditional and contemporary systems for health and well-being. *Ann Neurosci.* 2023;30(2):77–78. https://doi.org/10.1177/09727531231185648.
2. Nasim N, Sandeep IS, Mohanty S. Plant-derived natural products for drug discovery: Current approaches and prospects. *Nucleus (Calcutta).* 2022;65(3):399–411. https://doi.org/10.1007/s13237-022-00405-3.
3. Aizuddin AN, Zamzuri M'IA, Mansor J, Nurumal SR, Yunus SZSA, Razak MAA, Jamhari MN, Fah TS, Miskam HM, Hod R, Yusoff HM. Perception of integrating complementary and alternative medicine practice in patient's treatment among the healthcare practitioners: A systematic review. *Pan Afr Med J.* 2022;43:19. https://doi.org/10.11604/pamj.2022.43.19.31133.
4. Matos LC, Machado JP, Monteiro FJ, Greten HJ. Understanding traditional chinese medicine therapeutics: An overview of the basics and clinical applications. *Healthcare (Basel).* 2021;9(3):257. https://doi.org/10.3390/healthcare9030257.
5. Wang H, Chen Y, Wang L, Liu Q, Yang S, Wang C. Advancing herbal medicine: Enhancing product quality and safety through robust quality control practices. *Front Pharmacol.* 2023;14:1265178. https://doi.org/10.3389/fphar.2023.1265178.
6. Czigle S, Nagy M, Mladěnka P, Tóth J; OEMONOM. Pharmacokinetic and pharmacodynamic herb-drug interactions-part I. Herbal medicines of the central nervous system. *PeerJ.* 2023;11:e16149. https://doi.org/10.7717/peerj.16149.
7. Kenda M, Kočevar Glavač N, Nagy M, Sollner Dolenc M. Medicinal plants used for anxiety, depression, or stress treatment: An update. *Molecules.* 2022;27(18):6021. https://doi.org/10.3390/molecules27186021.
8. Kou H, Huang L, Jin M, He Q, Zhang R, Ma J. Effect of curcumin on rheumatoid arthritis: A systematic review and meta-analysis. *Front Immunol.* 2023;14:1121655. https://doi.org/10.3389/fimmu.2023.1121655.
9. Ekor M. The growing use of herbal medicines: Issues relating to adverse reactions and challenges in monitoring safety. *Front Pharmacol.* 2014;4:177. https://doi.org/10.3389/fphar.2013.00177.
10. Hoenders R, Ghelman R, Portella C, Simmons S, Locke A, Cramer H, Gallego-Perez D, Jong M. A review of the WHO strategy on traditional, complementary, and integrative medicine from the perspective of academic consortia for integrative medicine and health. *Front Med (Lausanne).* 2024;11:1395698. https://doi.org/10.3389/fmed.2024.1395698.
11. Park JE, Yi J, Kwon O. Twenty years of traditional and complementary medicine regulation and its impact in Malaysia: Achievements and policy lessons. *BMC Health Serv Res.* 2022;22(1):102. https://doi.org/10.1186/s12913-022-07497-2.
12. Karimi A, Majlesi M, Rafieian-Kopaei M. Herbal versus synthetic drugs; beliefs and facts. *J Nephropharmacol.* 2015;4(1):27–30.
13. Garakani A, Murrough JW, Freire RC, Thom RP, Larkin K, Buono FD, Iosifescu DV. Pharmacotherapy of anxiety disorders: Current and emerging treatment options. *Front Psychiatry.* 2020;11:595584. https://doi.org/10.3389/fpsyt.2020.595584.
14. Stewart SA. The effects of benzodiazepines on cognition. *J Clin Psychiatry.* 2005;66 Suppl 2:9–13.
15. Anand U, Dey A, Chandel AKS, Sanyal R, Mishra A, Pandey DK, De Falco V, Upadhyay A, Kandimalla R, Chaudhary A, Dhanjal JK, Dewanjee S, Vallamkondu J, Pérez de la Lastra JM. Cancer chemotherapy and beyond: Current status, drug candidates, associated risks and progress in targeted therapeutics. *Genes Dis.* 2022 Mar 18;10(4):1367–1401. https://doi.org/10.1016/j.gendis.2022.02.007. Erratum in: *Genes Dis.* 2024;11(4):101211. https://doi.org/10.1016/j.gendis.2024.101211.
16. Yuan H, Ma Q, Ye L, Piao G. The traditional medicine and modern medicine from natural products. *Molecules.* 2016;21(5):559. https://doi.org/10.3390/molecules21050559.
17. Marshall AC. Traditional Chinese medicine and clinical pharmacology. *Drug Discovery and Evaluation: Methods in Clinical Pharmacology.* 2020:455–482. https://doi.org/10.1007/978-3-319-68864-0_60.
18. Rizvi SAA, Einstein GP, Tulp OL, Sainvil F, Branly R. Introduction to traditional medicine and their role in prevention and treatment of emerging and re-emerging diseases. *Biomolecules.* 2022;12(10):1442. https://doi.org/10.3390/biom12101442.

19. McKeown A. Ethical challenges and principles in integrated care. *Br Med Bull*. 2023;146(1):4–18. https://doi.org/10.1093/bmb/ldac030.
20. Hassen G, Belete G, Carrera KG, Iriowen RO, Araya H, Alemu T, Solomon N, Bam DS, Nicola SM, Araya ME, Debele T, Zouetr M, Jain N. Clinical implications of herbal supplements in conventional medical practice: A US perspective. *Cureus*. 2022;14(7):e26893. https://doi.org/10.7759/cureus.26893.
21. Hibbert PD, Stewart S, Wiles LK, Braithwaite J, Runciman WB, Thomas MJW. Improving patient safety governance and systems through learning from successes and failures: Qualitative surveys and interviews with international experts. *Int J Qual Health Care*. 2023;35(4):0. https://doi.org/10.1093/intqhc/mzad088.
22. Moon SEJ, Hogden A, Eljiz K. Sustaining improvement of hospital-wide initiative for patient safety and quality: A systematic scoping review. *BMJ Open Qual*. 2022;11(4):e002057. https://doi.org/10.1136/bmjoq-2022-002057.
23. Abdallah EM, Alhatlani BY, de Paula Menezes R, Martins CHG. Back to nature: Medicinal plants as promising sources for antibacterial drugs in the post-antibiotic era. *Plants (Basel)*. 2023;12(17):3077. https://doi.org/10.3390/plants12173077.
24. Marcelino S, Hamdane S, Gaspar PD, Paço A. Sustainable agricultural practices for the production of medicinal and aromatic plants: Evidence and recommendations. *Sustainability*. 2023;15(19):14095. https://doi.org/10.3390/su151914095
25. Ikhoyameh M, Okete WE, Ogboye RM, Owoyemi OK, Gbadebo OS. Integrating traditional medicine into the African healthcare system post-traditional medicine global summit: Challenges and recommendations. *Pan Afr Med J*. 2024;47:146. https://doi.org/10.11604/pamj.2024.47.146.43011.
26. Jang S, Jang BH, Ko Y, Sasaki Y, Park JS, Hwang EH, Song YK, Shin YC, Ko SG. Herbal medicines for treating metabolic syndrome: A systematic review of randomized controlled trials. *Evid Based Complement Alternat Med*. 2016;2016:5936402. https://doi.org/10.1155/2016/5936402. Epub 2016 Jun 19.
27. Kalariya Y, Kumar A, Ullah A, Umair A, Neha F, Madhurita F, Varagantiwar V, Ibne Ali Jaffari SM, Ahmad A, Aman M, Sapna F, Varrassi G, Kumar S, Khatri M. Integrative medicine approaches: Bridging the gap between conventional and renal complementary therapies. *Cureus*. 2023;15(9):e46033. https://doi.org/10.7759/cureus.46033.
28. Nasir NN, Sekar M, Fuloria S, Gan SH, Rani NNIM, Ravi S, Begum MY, Chidambaram K, Sathasivam KV, Jeyabalan S, Dhiravidamani A, Thangavelu L, Lum PT, Subramaniyan V, Wu YS, Azad AK, Fuloria NK. Kirenol: A potential natural lead molecule for a new drug design, development, and therapy for inflammation. *Molecules*. 2022;27(3):734.

11 Nano-Delivery System for the Management of Neurodegenerative Disorder

Sai Kiran S.S. Pindiprolu, Ramakrishna Gummadi, Nagasen Dasari, and Chirravuri Satyanarayana Phani Kumar
School of Pharmacy, Aditya University, Surampalem, Andhra Pradesh, India

11.1 INTRODUCTION

Neurodegenerative disorders, such as Alzheimer's disease (AD), Parkinson's disease (PD), and Huntington's disease (HD), pose substantial challenges to healthcare systems due to their chronic and progressive nature, which results in the gradual degradation of neuronal cells. These disorders are characterized by cognitive decline, motor impairments, and loss of neurological functions, significantly affecting patients, caregivers, and healthcare resources. The pathology behind these conditions is highly complex, involving [1, 2] mechanisms like protein misfolding, oxidative stress, neuroinflammation, and mitochondrial dysfunction. Effectively treating neurodegenerative disorders is particularly challenging, with drug delivery to the brain being one of the primary obstacles. The blood–brain barrier (BBB) restricts the movement of most therapeutic agents into the central nervous system (CNS), limiting their bioavailability [3]. Additionally, many available treatments are associated with systemic toxicity, which complicates long-term use. Consequently, there is a pressing need for innovative drug delivery systems capable of enhancing CNS access, reducing adverse effects, and ultimately improving treatment outcomes for patients [4].

11.2 CURRENT THERAPEUTIC APPROACHES AND LIMITATIONS

Traditional therapies for neurodegenerative diseases primarily utilize small molecules, biologic agents, and symptomatic interventions, each aiming to manage symptoms or slow the progression of the disease. Small-molecule drugs, which are chemically synthesized, are widely used in neurodegenerative conditions for their capacity to interact with specific molecular targets [5]. For instance, acetyl cholinesterase inhibitors in AD and dopaminergic therapies in PD have shown potential in alleviating symptoms. Increasingly, biologics, such as monoclonal antibodies and peptides, are investigated for their potential to target specific disease-associated proteins and pathways. Additionally, symptomatic therapies are employed to manage particular symptoms, such as motor or cognitive impairments, to enhance patients' quality of life. Despite their common use, conventional drug treatments face substantial challenges in effectively reaching the CNS [6]. One of the primary obstacles is the BBB, a selective membrane that restricts the entry of most therapeutic agents into the brain. The chemical characteristics of small molecules can limit their BBB permeability, while larger biologic molecules typically struggle even more to penetrate this barrier, reducing their efficacy in CNS disorders [7]. Traditional drugs for neurodegenerative diseases often exhibit poor solubility, which reduces their bioavailability, as listed in Table 11.1. Coupled with a short half-life, these drugs are frequently metabolized and cleared from the body before they can achieve sustained therapeutic levels in the CNS. This often necessitates repeated dosing to maintain

DOI: 10.1201/9781003478652-12

effective drug concentrations, raising the risk of systemic toxicity and causing adverse effects that may impact patient safety and adherence to treatment. To address these limitations, advanced drug delivery systems are essential for improving therapeutic outcomes in neurodegenerative disorders. Emerging delivery platforms, including nano-delivery systems, show promise in enhancing drug solubility, extending circulation time, and enabling targeted delivery across the BBB. By increasing drug bioavailability within the CNS and reducing systemic exposure, these systems have the potential to minimize adverse effects and improve the overall therapeutic efficacy of treatments [8].

TABLE 11.1
List of Traditional Therapies

Therapeutic Technique	Description	Advantages	Disadvantages	References
Pharmacotherapy	Use of medications like cholinesterase inhibitors, dopamine agonists, or NMDA antagonists	Provides symptomatic relief and improves quality of life temporarily	Limited efficacy in halting disease progression Potential for severe side effects High cost	[9]
Physical therapy	Exercises to improve mobility, balance, and strength	Enhances physical function Reduces risk of falls Non-invasive	Requires consistency and commitment Limited impact on disease progression May be costly long-term	[10]
Occupational therapy	Techniques to help patients perform daily activities	Improves independence Customized to individual needs	Requires ongoing professional input Limited effect on disease pathology	[11]
Speech therapy	Exercises to improve speech and swallowing difficulties	Improves communication Reduces the risk of aspiration	Time-intensive Effects may diminish if therapy is stopped	[10]
Dietary interventions	Tailored diets to manage symptoms or improve overall health	May reduce symptoms like constipation or malnutrition Generally safe and non-invasive	Limited evidence for disease-modifying effects Requires patient compliance	[11]
Deep brain stimulation (DBS)	Surgical implantation of electrodes to stimulate specific brain areas (e.g., for Parkinson's disease)	Effective for motor symptoms Reduces medication dependence	Invasive procedure with surgical risks Not suitable for all patients High cost	[9, 11]
Cognitive therapy	Mental exercises to slow cognitive decline	Helps maintain cognitive function Improves quality of life	Effects may be temporary Requires patient motivation and engagement	[9–11]
Psychological counseling	Support for emotional and mental health challenges associated with neurodegenerative diseases	Improves coping mechanisms Enhances mental health and quality of life	Does not address physical symptoms Requires long-term commitment	[9–11]
Palliative care	Holistic care focusing on comfort and quality of life	Provides symptom management Supports emotional and spiritual well-being	Does not modify disease progression Often initiated in later stages of disease	[9–11]

11.3 NANO-DELIVERY SYSTEMS IN CNS THERAPEUTICS

Nanotechnology is rapidly advancing in the medical field, providing innovative approaches to treat and manage a variety of diseases. Through manipulating materials on the nanoscale, typically ranging from 1 to 100 nanometers, nanotechnology allows for the creation of highly specialized delivery systems with unique properties that enhance therapeutic efficacy [12]. In neurotherapeutics, where traditional methods often encounter challenges, nanotechnology has shown promising potential in improving the delivery of therapeutic agents to the CNS. Nano-delivery systems are specifically engineered to transport drugs across the BBB, a major obstacle in CNS therapy [13]. These nanoscale carriers, with their small size and adaptable surface properties, can either bypass or penetrate barriers that usually prevent larger molecules from entering the brain. Utilizing various nanoscale carriers, such as nanoparticles, liposomes, and dendrimers, these delivery systems enable the direct administration of medications to specific targets within the CNS, potentially enhancing drug bioavailability and reducing systemic side effects. In CNS applications, nanotechnology further supports targeted drug delivery, directing medications to specific brain regions or cell types. This precision not only increases therapeutic effectiveness but also minimizes the impact on non-target tissues, which helps reduce toxicity. Additionally, nano-delivery systems can be designed to allow for controlled drug release over time, offering prolonged therapeutic effects that may improve patient adherence and enhance outcomes in managing chronic neurodegenerative diseases [14]. By addressing many of the limitations inherent to conventional CNS therapies, nanotechnology offers a promising step forward in neurotherapeutics. This chapter examines various nano-delivery platforms under investigation or in use for CNS disorders, exploring their mechanisms, benefits, and potential for clinical application. A notable challenge encountered with traditional medications is their poor solubility in biological fluids. Numerous therapeutic agents aimed at CNS applications are characterized by hydrophobic properties or limited solubility, significantly hindering their absorption and bioavailability. When these drugs do not dissolve adequately in the bloodstream, their ability to reach specific target sites within the CNS is adversely affected, resulting in reduced therapeutic efficacy and frequently necessitating increased dosages. The BBB also represents a significant impediment to effective drug delivery [15]. This selectively permeable barrier is designed to safeguard the brain from potentially harmful substances but simultaneously complicates the entry of many conventional drugs, particularly larger molecules and biologics. Consequently, the concentration of these medications in the CNS may remain below optimal levels, undermining the overall effectiveness of therapies for various neurodegenerative diseases, such as AD and PD. The short half-life of many conventional drugs presents another considerable challenge. This characteristic can lead to rapid metabolism and elimination from the body, hindering the achievement of desired therapeutic effects. As a result, patients often find themselves needing to take medications more frequently to maintain sufficient drug levels in their bloodstream, which can be burdensome and may complicate adherence to treatment plans. Conventional therapies carry the potential risk of systemic toxicity and adverse effects. Many traditional drugs lack specificity in targeting, which can result in interactions with unintended sites and negative repercussions in other organ systems. These side effects not only pose safety concerns for patients but may also discourage adherence to prescribed treatment regimens, further complicating the management of chronic CNS disorders. Nanocarriers exhibit a remarkable ability to traverse or circumvent the BBB, significantly increasing drug concentrations within the CNS. This barrier serves as a crucial protective mechanism, selectively allowing certain substances to enter the brain while obstructing potentially harmful agents. However, this selectivity presents considerable challenges for drug delivery, particularly for medications aimed at treating CNS disorders [15]. The nanoscale size of these carriers enables them to navigate through the tight junctions formed by the endothelial cells that constitute the BBB. Their effectiveness is further improved by modifying their surface properties, which allows them to interact with specific receptors or leverage various transport mechanisms that facilitate their entry into the brain. Methods such as receptor-mediated endocytosis and the utilization of transport proteins enable nanocarriers to deliver therapeutic agents directly to their intended sites of action. The increased permeability of nanocarriers across the BBB results in higher drug concentrations in the CNS, which is essential for achieving the desired therapeutic effects. This enhanced drug delivery can lead to improved treatment outcomes for a variety of neurological conditions, including neurodegenerative diseases, brain tumors, and psychiatric disorders. Moreover, the targeted delivery capabilities of nanocarriers help reduce the risk of systemic side effects, ensuring that a larger proportion of the administered drug reaches its intended destination without affecting non-targeted tissues [16, 17]. The capacity of nanocarriers to effectively cross or bypass the BBB represents a significant advancement in drug delivery systems. By facilitating increased drug concentrations within the CNS, these innovative platforms hold the potential to transform the treatment landscape for numerous neurological disorders, ultimately improving patient outcomes and enhancing quality of life [18, 19].

Types of Nano-Delivery Systems for Neurodegenerative Disorders. The pursuit of effective treatments for neurodegenerative disorders has increasingly centered on the utilization of liposomes and nanoparticles as advanced nano-delivery systems. These cutting-edge strategies are designed to improve the delivery of therapeutic agents to the CNS. This chapter examines the structural characteristics, manufacturing techniques, functionalization processes, and clinical applications related to these nanocarriers [19–21].

Liposomes are spherical vesicles constructed from phospholipid bilayers, which allow them to encapsulate both hydrophilic and hydrophobic drugs. This distinctive architecture enables liposomes to act as versatile carriers for a wide range of therapeutic agents, as shown in Figure 11.1. Each liposome features an aqueous core enveloped by a lipid membrane, with variations in both composition and thickness customized for specific applications. Common manufacturing methods for liposomes include thin-film hydration, where a lipid film is hydrated to form vesicles, and reverse-phase evaporation, which provides precise control over liposome size and drug encapsulation efficiency [22]. To enhance the targeting efficacy and overall performance of liposomal formulations for brain delivery, functionalization techniques are implemented. Researchers attach ligands – such as antibodies, peptides, or small molecules – to the surface of liposomes, which improves the delivery of drugs to specific receptors or transport mechanisms within the

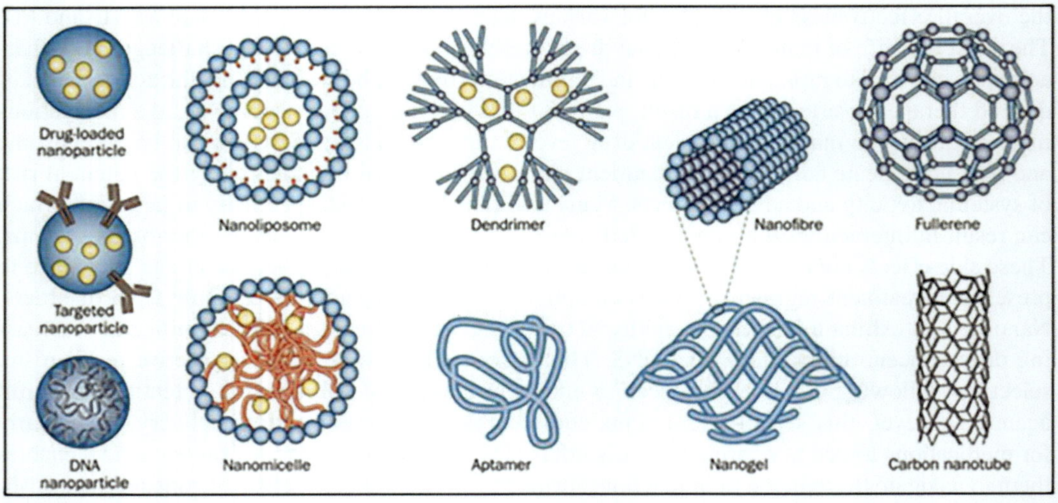

FIGURE 11.1 Representation of different types of nanoparticles used for CNS therapeutics.

https://www.cd-bioparticles.net/cns-targeted-drug-delivery-systems

CNS. This tailored approach aims to facilitate the transportation of therapeutic agents across the BBB, ensuring that they reach their intended sites of action [23]. Numerous case studies illustrate the effectiveness of liposomal formulations in managing neurodegenerative diseases like AD and PD. For instance, liposomes loaded with anti-amyloid compounds have demonstrated the capability to improve the bioavailability of these drugs while reducing systemic side effects. Similarly, in the case of PD, liposomal formulations containing dopamine agonists have shown promise in enhancing drug delivery to the brain, resulting in improved patient outcomes. Nanoparticles represent another category of nano-delivery systems, encompassing various materials and designs, including polymeric, metallic, and lipid-based nanoparticles. Each of these types offers distinct advantages for drug delivery to the CNS [24]. Polymeric nanoparticles, made from biodegradable polymers, are particularly recognized for their ability to provide controlled drug release and enhanced stability in circulation. Their size and surface properties can be tailored to improve BBB permeability and optimize therapeutic efficacy. Metallic nanoparticles, such as those composed of gold or silver, have attracted attention not only for their drug delivery potential but also for their applications in imaging and diagnostics. These nanoparticles can be functionalized with targeting agents to facilitate specific delivery to brain tissues, enhancing treatment effectiveness for neurological conditions. Lipid-based nanoparticles, including solid lipid nanoparticles and nanostructured lipid carriers, combine the advantages of liposomes with traditional lipid formulations. They enhance drug solubility and stability while enabling sustained release of encapsulated therapeutic agents [25, 26]. The versatility of these lipid-based carriers makes them well-suited for administering a broad range of drugs, particularly those that have low solubility in biological fluids. The integration of liposomes and nanoparticles into therapies for neurodegenerative disorders marks significant progress in drug delivery systems. By enhancing the ability to traverse the BBB and improving the localization of therapeutic agents within the CNS, these innovative nanocarriers hold great promise for transforming treatment strategies for various neurological diseases, ultimately leading to improved patient outcomes and enhanced quality of life. To enhance the delivery of therapeutic agents to the CNS, it is crucial to establish effective design strategies that enable the penetration of the BBB and ensure precise targeting within the brain. Among the advanced nanocarrier options, dendrimers and nanogels are particularly significant due to their unique structural characteristics and functionalities. Dendrimers are highly branched macromolecules that facilitate targeted drug loading and controlled release. Their complex architecture allows for the customization of surface properties, which enables the attachment of targeting ligands [27]. These modifications enhance the ability of dendrimers to navigate through the BBB and deliver therapeutic agents to specific brain regions. Ongoing clinical investigations are examining the effectiveness of dendrimers in treating various neurological conditions, highlighting their potential in managing diseases such as brain tumors, AD, and other neurodegenerative disorders. Conversely, nanogels are three-dimensional hydrophilic networks capable of encapsulating drugs and providing a sustained release profile. Their specialized structure allows them to respond to physiological changes, such as variations in pH or temperature, thus enabling controlled drug delivery [28]. This responsive behavior is particularly advantageous for neuroprotection and neuro-regeneration, as nanogels can be engineered to release neurotherapeutics in a regulated manner over extended periods, thereby enhancing therapeutic efficacy while minimizing side effects. Nanogels employ various mechanisms for drug delivery, including passive diffusion and active transport, underscoring their versatility in treating neurological disorders. This adaptability positions nanogels as valuable assets for targeted therapies, potentially leading to significant improvements in treatment outcomes. Dendrimers and nanogels represent promising strategies for advancing drug delivery to the brain. By focusing on improving BBB permeability and ensuring specific targeting within the CNS, these innovative nanocarriers have the potential to revolutionize treatment approaches for a range of neurological conditions, ultimately contributing to better patient care and improved health outcomes [29, 30].

11.4 DESIGN AND FABRICATION OF NANO-DELIVERY SYSTEMS

In the field of nanomedicine, the creation of nanoparticles tailored for neurodegenerative disorders requires an in-depth comprehension of various physical and chemical properties. These characteristics are essential in determining the effectiveness of drug delivery systems targeted at the CNS [31]. Among the critical design considerations are particle size, surface charge, and functionalization. Particle Size plays a crucial role in the ability of nanoparticles to cross biological barriers, particularly the BBB. For applications in the CNS, nanoparticles with diameters ranging from 1 to 100 nanometers are typically favored, as this size range facilitates optimal cellular uptake and enhances BBB permeability. Smaller nanoparticles can navigate biological membranes more easily, while larger particles may encounter obstacles that hinder their therapeutic effectiveness. Surface Charge is another important factor influencing how nanoparticles behave in biological environments. The electrical charge present on the nanoparticle surface can significantly impact their interactions with cell membranes and plasma proteins. Generally, positively charged nanoparticles exhibit enhanced cellular uptake due to electrostatic attraction with the negatively charged membranes of cells. However, excessive positive charge can lead to cytotoxicity. Conversely, negatively charged nanoparticles may experience lower cellular uptake but can demonstrate improved stability during circulation in the bloodstream. Therefore, determining the optimal surface charge is crucial for maximizing therapeutic delivery and minimizing adverse reactions. Functionalization refers to the modification of nanoparticle surfaces to enhance their targeting abilities and therapeutic effectiveness [32, 33]. By incorporating specific ligands – such as antibodies, peptides, or small molecules – researchers can improve the ability of nanoparticles to bind to particular receptors or transport mechanisms within the CNS. This tailored strategy not only enhances targeting accuracy but also increases the retention time of therapeutic agents at their desired sites of action. Additionally, successful functionalization can aid in evading the immune system, thereby prolonging the circulation time of nanoparticles. The design of nanoparticles for neurodegenerative applications is a multifaceted process that requires a comprehensive understanding of particle size, surface charge, and functionalization methods. By optimizing these parameters, researchers can develop effective nano-delivery systems that enhance drug transport to the CNS, ultimately leading to improved treatment outcomes for a range of neurodegenerative diseases. The advancement of nanotechnology in drug delivery systems has significantly contributed to addressing neurological disorders by overcoming the challenges associated with the BBB and improving the bioavailability of therapeutic agents [34]. A range of techniques is utilized for the fabrication and functionalization of nanocarriers, which are specifically designed to achieve effective penetration through the BBB and enhance the availability of drugs. One notable fabrication technique is solvent evaporation, in which selected polymers are dissolved in a volatile organic solvent. As the solvent gradually evaporates, nanoparticles are formed, allowing for precise control over their size and the amount of drug they can load. This technique is advantageous for creating nanocarriers that possess optimized characteristics tailored for crossing the BBB. Another widely used method is nanoprecipitation, where a polymer solution is rapidly mixed with a nonsolvent, leading to the precipitation of nanoparticles. This approach is appreciated for its simplicity and its ability to produce uniform nanoparticles with high drug encapsulation efficiency. Researchers can effectively manipulate the size and distribution of the nanoparticles by adjusting the mixing conditions and concentrations employed [35]. Emulsion polymerization is also a key technique, involving the creation of an emulsion system with water as the continuous phase and organic solvents containing the polymer as the dispersed phase. Through polymerization, nanoparticles capable of encapsulating therapeutic agents are formed. This method is particularly useful for generating nanoparticles with specific surface properties that can be customized for targeted BBB delivery. Electrospinning represents another versatile approach for creating nanofibers from polymer solutions through the application of high voltage. This process results in fibrous scaffolds that can encapsulate drugs, forming nanocarriers that facilitate sustained drug release and improve retention within target tissues. Top-down approaches, such as milling and laser ablation, involve breaking down larger particles into nanoscale materials. While these methods can produce nanoparticles with defined sizes and shapes, achieving uniformity may be more complex compared to bottom-up techniques. When it comes to functionalization, one strategy involves surface coating, in which nanocarriers are covered with specific materials like polyethylene glycol (PEG) [36]. This coating improves the stability of the nanocarriers and prolongs their circulation time in the bloodstream. PEGylation minimizes protein adsorption and immune recognition, which increases the likelihood of successful BBB penetration. Ligand attachment represents another functionalization strategy, where targeting ligands – such as antibodies, peptides, or small molecules – are conjugated to the surfaces of nanocarriers. This enhances the selectivity of the nanocarriers for particular receptors on the BBB, promoting their uptake and transcytosis into the brain. For example, RGD peptides targeting integrins can significantly enhance brain delivery. The development of pH-responsive nanocarriers allows for the selective release of therapeutic agents in response to specific pH changes within the body. By incorporating pH-sensitive materials into the nanocarrier design, drugs can be released in the acidic microenvironments typical of tumors or other pathological sites. Temperature-sensitive materials can also be integrated into nanocarriers to facilitate drug release under defined thermal conditions. This strategy is especially beneficial for localized therapies, where localized hyperthermia can trigger the release of the drug. Additionally, magnetic targeting can be applied using magnetic nanoparticles that respond to external magnetic fields, enabling targeted delivery to specific regions of the brain [37]. By applying a magnetic field, researchers can concentrate nanoparticles at desired sites, thereby enhancing therapeutic effectiveness. The fabrication and functionalization of nanocarriers aimed at penetrating the BBB and improving bioavailability are essential for the effective delivery of therapeutic agents to the CNS. Employing a diverse range of advanced techniques allows researchers to optimize the design of nanocarriers, ensuring successful drug delivery while minimizing systemic side effects. These innovative strategies have the potential to transform treatment modalities for neurological disorders, ultimately leading to improved patient outcomes. Targeting strategies in drug delivery systems that utilize nanocarriers are vital for enhancing therapeutic effectiveness and minimizing side effects. Various methods, including ligand-based approaches, receptor-mediated transport, and stimuli-responsive designs, are utilized to improve the precision of drug delivery, particularly for challenging conditions such as neurological disorders [38]. Ligand-based approaches emphasize the attachment of specific ligands to the surfaces of nanocarriers. These ligands may consist of antibodies, peptides, or small molecules that selectively bind to target receptors on cells or tissues. By choosing ligands with a high affinity for specific receptors, researchers can enhance the selectivity of drug delivery, ensuring that therapeutic agents reach their designated sites of action. This strategy not only increases the treatment's efficacy but also reduces the potential for unintended side effects. Receptor-mediated transport represents a more refined strategy that takes advantage of the natural mechanisms through which cells internalize substances using specific receptors. Nanocarriers can be designed to mimic natural ligands that interact with these receptors, facilitating their uptake into target cells [39]. This method is particularly beneficial for traversing biological barriers such as the BBB, allowing for more effective delivery of therapeutic agents directly into the brain. By harnessing receptor-mediated endocytosis, nanocarriers can significantly improve bioavailability and therapeutic impact. Stimuli-responsive designs offer a novel approach within drug delivery systems, wherein nanocarriers are engineered to release their payloads in response to specific external stimuli. These stimuli can include variations in pH, temperature changes, light exposure, or the presence of specific enzymes. For instance, a nanocarrier designed to release its drug in acidic environments may be particularly effective for targeting tumor tissues, which generally exhibit lower pH levels than healthy tissues. Similarly, temperature-sensitive nanocarriers can be activated to release their contents upon localized heating, making them suitable for hyperthermia treatments [40]. This responsive behavior enables controlled and precise drug delivery, enhancing therapeutic outcomes while minimizing systemic exposure. Targeting strategies such as ligand-based approaches, receptor-mediated transport, and stimuli-responsive designs are essential for optimizing drug delivery systems. By applying these techniques, researchers can enhance the specificity and effectiveness of therapeutic agents, ultimately improving treatment outcomes for a variety of medical conditions, particularly complex disorders like neurological diseases [41].

11.5 MECHANISMS OF BBB PENETRATION AND TARGETED DRUG DELIVERY

The BBB serves as a crucial protective layer for the CNS, effectively regulating what substances can enter the brain while safeguarding it from potentially harmful compounds present in the bloodstream. Composed of tightly knit endothelial cells that form the capillary walls in the brain, the BBB is further supported by astrocytic end-feet and pericytes. This intricate structure poses significant obstacles for drug delivery, as it prevents many drugs and larger molecules from diffusing passively into the CNS. Consequently, therapeutic agents often face difficulties in reaching their intended brain targets, complicating treatment strategies for various neurological conditions, including AD, PD, and multiple sclerosis [42]. To overcome these barriers, nanocarriers utilize several mechanisms to facilitate their passage through the BBB. One prominent method is receptor-mediated transcytosis. In this process, nanocarriers are engineered to mimic natural ligands that specifically bind to receptors on the endothelial cells of the BBB. Once binding occurs, it triggers endocytosis, allowing the nanocarrier to be taken up and transported across the endothelial barrier. For instance, nanoparticles can be modified with ligands like transferrin or insulin, enabling them to interact with their respective receptors and promote the direct delivery of therapeutic agents into the brain. Another vital mechanism is adsorptive-mediated transcytosis, which depends on the electrostatic interactions between positively charged nanocarriers and the negatively charged elements of the endothelial cell membrane. This interaction enhances the internalization of nanocarriers, allowing them to cross the BBB effectively. Researchers can exploit this mechanism to design nanoparticles that efficiently penetrate the barrier while delivering their therapeutic cargo [43].

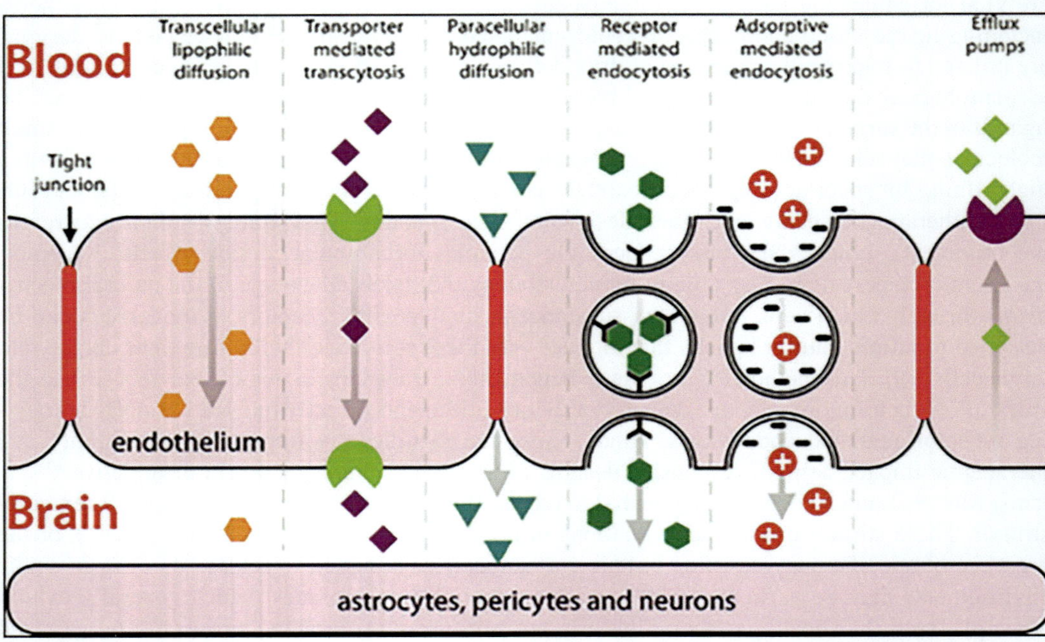

FIGURE 11.2 Representation of different mechanisms for transport of nano-delivery systems across blood–brain barrier (BBB).

https://link.springer.com/article/10.1007/s11095-010-0291-7

Targeted drug delivery and localized release of medications within the brain are essential for maximizing therapeutic efficacy while minimizing off-target effects. By employing targeted strategies, such as ligand-based techniques or antibody conjugation, nanocarriers can be specifically directed to certain cell types or regions in the brain. This targeted approach is particularly advantageous for addressing conditions like tumors, neurodegenerative diseases, or brain injury, as shown in Figure 11.2. The development of stimuli-responsive nanocarriers represents an innovative strategy for achieving controlled drug release within the brain. These nanocarriers can be designed to respond to specific physiological conditions, such as pH changes, temperature variations, or the presence of specific enzymes. For example, nanocarriers that release their therapeutic payloads in the acidic environments characteristic of tumors can improve localized treatment effectiveness while minimizing systemic exposure. The BBB poses significant challenges for effective CNS drug delivery, highlighting the need for innovative strategies to facilitate the transport of therapeutic agents into the brain. By utilizing mechanisms like receptor-mediated transcytosis and adsorptive-mediated transcytosis, in conjunction with targeted delivery methods and stimuli-responsive designs, researchers are developing advanced nanocarrier systems. These innovations hold tremendous potential for enhancing drug delivery to the CNS, ultimately leading to improved therapeutic outcomes for individuals suffering from neurological disorders [44, 45].

11.6 APPLICATIONS OF NANO-DELIVERY SYSTEMS IN SPECIFIC NEURODEGENERATIVE DISORDERS

AD is a complex neurodegenerative disorder marked by progressive cognitive decline, largely driven by the buildup of amyloid-beta plaques and tau protein tangles in the brain. Current treatment strategies struggle to effectively reduce these pathological markers due to difficulties in drug delivery across the BBB. However, emerging nano-delivery systems hold great promise in overcoming these challenges. Recent case studies have shown the potential of various nanoparticles in targeting amyloid-beta and tau proteins. For example, lipid-based nanoparticles functionalized with specific antibodies against amyloid-beta have been created to facilitate the targeted delivery of therapeutic agents. These nanoparticles can selectively bind to amyloid plaques, improving the clearance of amyloid-beta from the brain. Additionally, nanoparticles loaded with small interfering RNA (siRNA) that target tau protein expression have demonstrated effectiveness in lowering tau levels, potentially preventing the formation of neurofibrillary tangles. Such innovations in nano-delivery systems not only enhance treatment specificity but also help reduce systemic side effects, making them a promising avenue for managing AD [46]. PD is characterized by the degeneration of dopaminergic neurons in the substantia nigra, resulting in motor dysfunction and other debilitating symptoms. A primary therapeutic goal in PD management is to restore dopamine levels in the brain, and nanocarriers have emerged as valuable tools in enhancing dopamine delivery and protecting vulnerable neurons. Recent studies emphasize the use of nanocarriers, such as polymeric nanoparticles and liposomes, to deliver dopamine precursors like L-DOPA directly to the brain. These nanocarriers can effectively bypass the BBB, ensuring that higher concentrations of therapeutic agents reach the target sites and improving treatment efficacy. Additionally, some nanocarrier systems are engineered to release their cargo in a controlled manner, helping maintain stable dopamine levels over extended periods. Moreover, nanoparticles containing neuroprotective agents can shield dopaminergic neurons from oxidative stress and apoptosis, further enhancing their survival and function. HD is a genetic neurodegenerative disorder characterized by motor dysfunction, cognitive decline, and neuroinflammation [47]. A key area of research focuses on modulating gene expression and reducing neuroinflammatory processes associated with HD. Nanoparticles present a unique platform for delivering therapeutic agents that can address these challenges. Research indicates that nanoparticles can be engineered to deliver gene-editing technologies, such as CRISPR–Cas9 systems, aimed at silencing the mutant huntingtin gene responsible for HD. By employing targeted delivery mechanisms, these nanoparticles can effectively introduce gene-editing components into affected neurons, resulting in reduced expression of the toxic huntingtin protein. Furthermore, nanoparticles loaded with anti-inflammatory drugs or RNA molecules designed to modulate inflammatory pathways have shown potential in alleviating neuroinflammation linked to HD. By addressing both genetic and inflammatory aspects of the disease, these nanocarrier systems could provide a comprehensive approach to treating HD. The use of nano-delivery systems in managing neurodegenerative diseases like AD, PD, and HD holds considerable promise for transforming therapeutic strategies. Through targeted delivery mechanisms aimed at reducing amyloid-beta and tau protein accumulation, enhancing dopamine levels, and modulating gene expression, these innovative systems have the potential to change treatment paradigms for these complex disorders. Ongoing research and development in nanotechnology will be crucial for overcoming existing challenges in drug delivery and improving the quality of life for patients affected by these devastating conditions [48].

11.7 SAFETY AND TOXICITY CONSIDERATIONS IN NANOMEDICINE FOR CNS

The incorporation of nano-delivery systems into therapeutic approaches, particularly for CNS applications, presents significant issues related to systemic toxicity and biocompatibility. As these technologies gain traction in clinical settings, it becomes essential to confront these challenges to ensure patient safety and effective treatment outcomes. Systemic toxicity refers to the adverse effects that therapeutic agents may have on the body beyond their intended targets. Due to their unique physicochemical properties, nanoparticles can exhibit behaviors that differ from those of traditional drug formulations. Their small size and large surface area can enhance bioavailability but may also lead to unintended interactions within biological systems. Such interactions can trigger inflammatory responses, cause organ toxicity, or alter cellular functions. Thus, understanding biocompatibility in the context of nano-delivery systems is crucial [49]. Biocompatibility denotes a material's ability to elicit an appropriate biological response when introduced into the body, ensuring that it does not provoke adverse reactions. To address these concerns, existing safety protocols incorporate comprehensive preclinical evaluations prior to the initiation of clinical trials. These evaluations typically include a range of in vitro and in vivo studies aimed at assessing the safety profile of nano-delivery systems. Initial assessments often focus on cytotoxicity to examine how these systems interact with various cell types. Subsequently, animal models are used to evaluate systemic effects, biodistribution, and potential accumulation of nanoparticles in critical organs. These studies provide vital insights into the pharmacokinetics and pharmacodynamics of nanoparticle formulations, allowing researchers to refine their designs for enhanced safety and efficacy. Ethical considerations also hold significant importance in the development and implementation of nano-delivery systems for CNS therapies [50]. The use of nanotechnology in medicine introduces specific ethical challenges, particularly concerning informed consent, as patients may not fully understand the implications of using novel delivery methods. Additionally, the potential for long-term effects of nanoparticles – often unknown at the time of clinical application – raises ethical questions regarding patient safety and the responsibilities of researchers and healthcare providers. Regulatory authorities play a crucial role in establishing guidelines for the development and use of nano-delivery systems. These guidelines aim to ensure that safety and efficacy are thoroughly evaluated before products are made available on the market. As the field of nanomedicine evolves, collaboration among researchers, clinicians, and regulatory bodies will be essential in creating effective protocols that address safety concerns while fostering innovation [30]. Effectively addressing systemic toxicity and biocompatibility issues associated with nano-delivery systems is critical for their successful application in CNS therapies. By implementing rigorous safety protocols, conducting comprehensive preclinical assessments, and emphasizing ethical considerations, potential risks can be effectively managed. By prioritizing patient safety and adhering to responsible research practices, the promise of nano-delivery systems can be fully realized, leading to improved therapeutic outcomes for various CNS disorders [16, 51].

11.8 CHALLENGES IN CLINICAL TRANSLATION OF NANO-DELIVERY SYSTEMS

The transition of innovative nano-delivery systems from the laboratory to clinical applications faces several significant challenges, including issues related to manufacturing scalability, regulatory compliance, and long-term safety considerations, as stated in Table 11.2. Addressing these challenges effectively is essential for the successful adoption of these technologies in clinical practice. A primary challenge lies in manufacturing scalability. While the production of nano-delivery systems at the laboratory scale allows for detailed formulation adjustments, scaling up to meet the demands of clinical and commercial markets can be intricate. This process requires that these systems maintain their efficacy, safety, and quality as they are produced in larger quantities [52]. Key factors such as raw material procurement, equipment capabilities, and consistency in manufacturing processes need careful management to ensure that the final product complies with the stringent standards required for clinical use. Variations in manufacturing techniques can also affect the physicochemical properties of nanoparticles, potentially influencing their therapeutic effectiveness and safety profiles. Regulatory challenges further complicate the path to clinical application for nano-delivery systems. Regulatory agencies necessitate thorough documentation and substantial evidence to demonstrate the safety and efficacy of new therapies before they can enter clinical trials [53, 54]. The unique characteristics of nanoparticles can complicate this evaluation, as conventional pharmacological guidelines may not adequately reflect their behavior. Consequently, developers must navigate a complex regulatory landscape that entails demonstrating biocompatibility, long-term safety, and overall effectiveness through rigorous preclinical and clinical studies. Long-term safety concerns are another critical issue. As nano-delivery systems are relatively new in the therapeutic landscape, their long-term impacts on human health remain largely unknown. Potential risks associated with chronic exposure to nanoparticles, such as tissue accumulation or unexpected interactions with biological systems, necessitate comprehensive safety assessments [6, 55]. Researchers must design studies that not only evaluate immediate outcomes but also monitor potential long-term effects over extended periods. To address these challenges, various strategies have been proposed. For manufacturing scalability, researchers are exploring advanced production techniques, such as continuous flow synthesis and automated manufacturing processes, which can enhance both consistency and efficiency in nanoparticle production. Collaborations between academic institutions and industry can facilitate knowledge transfer, offering insights into best practices for scaling up manufacturing operations [56, 57].

To navigate regulatory challenges, developers are increasingly seeking early engagement with regulatory agencies. This proactive approach clarifies regulatory expectations and enables developers to address potential issues before formal applications are submitted. Additionally, the development of standardized testing methods and specific guidelines for nanomedicine can streamline the regulatory process. Recent advancements in clinical trial designs also offer promising solutions to these challenges [63, 64]. Adaptive trial designs allow for modifications to ongoing studies based on interim results, enabling researchers to respond more effectively to emerging data. Furthermore, incorporating biomarker-based endpoints can expedite the assessment of treatment effects, facilitating faster approval processes. While there are numerous challenges in transitioning nano-delivery systems from laboratory research to clinical applications – such as manufacturing scalability, regulatory hurdles, and long-term safety concerns – various strategies are being developed to address these issues [21, 65]. By leveraging advancements in manufacturing techniques, fostering collaboration with regulatory bodies, and employing innovative clinical trial methodologies, researchers can enhance the feasibility of translating promising nano-delivery technologies into effective therapeutic options. Continued progress in these areas is crucial for unlocking the full potential of

TABLE 11.2
Challenges of Clinical Transitions and Their Description

Challenge	Description	Details	Reference
Biocompatibility and Safety	Ensuring that nanomaterials are non-toxic and do not elicit adverse immune responses.	Nanoparticles may interact unpredictably with biological systems, leading to potential toxicity or unintended side effects.	[58]
Manufacturing and scalability	Producing nano-delivery systems consistently on a large scale is challenging.	Achieving uniformity in size, shape, and functionalization of nanoparticles is essential for reproducibility and regulatory approval.	[59]
Regulatory hurdles	Lack of standardized regulatory frameworks for nanomedicines complicates approval processes.	Regulatory bodies require comprehensive data on pharmacokinetics, toxicity, and efficacy, which can be complex for nano-delivery systems.	[60]
Stability and shelf-life	Maintaining the stability of nano-formulations during storage and transportation is difficult.	Nanoparticles may aggregate or degrade over time, affecting their efficacy and safety.	[61]
Targeting and delivery efficiency	Achieving precise targeting to specific tissues or cells remains a significant obstacle.	Non-specific distribution can lead to reduced therapeutic efficacy and increased side effects.	[62]
Intellectual property and commercialization	Navigating patent landscapes and commercial viability poses challenges.	Protecting intellectual property while ensuring freedom to operate requires strategic planning.	[58–62]

nano-delivery systems in clinical applications, ultimately improving patient outcomes and advancing medical science [51, 66].

11.9 FUTURE DIRECTIONS AND PROSPECTS FOR NANO-DELIVERY IN NEUROTHERAPEUTICS

The domain of nanomedicine is experiencing a swift evolution, opening new pathways for the treatment of neurodegenerative diseases, including AD, PD, and HD. Central to these advancements are gene therapy and CRISPR-based nano-systems, which present innovative strategies aimed at targeting the underlying causes of these complex conditions. Furthermore, the incorporation of artificial intelligence (AI) into nano-drug design is anticipated to enhance personalized treatment strategies, while the creation of multi-functional nanocarriers holds the potential to improve therapeutic efficacy [31]. Gene therapy is emerging as a transformative technique for addressing neurodegenerative diseases by targeting genetic mutations and restoring normal cellular functions. By utilizing nanocarriers to deliver therapeutic genes directly to affected neurons, researchers can significantly enhance the effectiveness and precision of treatments. These nanoscale systems provide protection to therapeutic agents from degradation, improve their uptake by cells, and facilitate targeted delivery to specific brain regions. The introduction of CRISPR–Cas9 technology further amplifies the potential of gene therapy in managing neurodegenerative disorders. Through CRISPR-based nano-systems, scientists can perform precise gene edits that are critical for combating neurodegeneration, potentially arresting or even reversing disease progression. For example, in HD, where a particular genetic mutation is implicated in neurodegeneration, CRISPR technology can be leveraged to disrupt the problematic gene, thereby alleviating its harmful effects [67]. The delivery of CRISPR components via nanoscale carriers improves the accuracy of gene editing, allowing for targeted interventions with minimized off-target impacts. AI is making substantial strides across various fields, including nanomedicine. In the context of neurodegenerative diseases, AI can optimize nano-drug design by analyzing large datasets to identify the most effective formulations, predict biological interactions, and enhance delivery methods. Machine learning algorithms can process complex biological data to uncover patterns that inform the development of nanoparticles tailored to specific therapeutic goals. Additionally, AI can assist in personalizing treatments by forecasting individual patient responses to nano-delivery systems. By incorporating personalized patient data – such as genetic profiles and specific disease characteristics – AI can facilitate the creation of custom nano-drugs that maximize therapeutic efficacy while minimizing adverse reactions [36]. This approach aligns with the broader shift toward precision medicine, ensuring that treatments are specifically crafted to address the unique needs of each patient. The innovation of multi-functional nanocarriers presents another exciting avenue in the treatment of neurodegenerative diseases. These advanced carriers can be designed to deliver a combination of therapeutic agents, including drugs, genes, and imaging agents, all at once. This multi-pronged approach not only enhances the effectiveness of treatments but also enables real-time monitoring of therapeutic progress. For instance, a multi-functional nanocarrier could be engineered to deliver a neuroprotective drug in conjunction with a gene therapy agent while also featuring imaging capabilities [68]. Such design would allow healthcare professionals to track the delivery and effectiveness of treatments in real-time, facilitating prompt adjustments to therapeutic strategies based on patient feedback. The adaptability of multi-functional nanocarriers positions them as invaluable tools in the fight against neurodegenerative diseases, with the potential to significantly improve patient outcomes. The emergence of new technologies in nanomedicine, particularly gene therapy and CRISPR-based nano-systems, offers renewed hope for the treatment of neurodegenerative diseases by addressing their fundamental causes. The integration of AI into the design of nano-drugs fosters personalized treatment strategies, ensuring that interventions are tailored to meet the specific profiles of individual patients. Furthermore, advancements in multi-functional nanocarriers promise to enhance therapeutic outcomes through the simultaneous delivery of combined therapies and real-time monitoring capabilities. As these innovative technologies continue to advance, they hold great potential to transform the management of neurodegenerative disorders, paving the way for more effective and targeted treatment options [69].

11.10 CONCLUSION

Nano-delivery systems represent a significant advancement in the management of neurodegenerative disorders, offering a range of benefits that address the specific challenges associated with treating conditions such as AD, PD, and HD. These advantages stem from the unique properties of nanoparticles, which can be tailored to enhance drug delivery, improve therapeutic efficacy, and minimize adverse effects. One of the primary benefits of nano-delivery systems is their ability to improve the bioavailability of therapeutic agents. Treating neurodegenerative diseases often requires delivering medications across the BBB, a highly selective barrier that protects the CNS from harmful substances while also hindering the entry of beneficial drugs. By engineering nanoparticles to facilitate drug transport across the BBB, more effective treatment methods can be developed [46]. Researchers can modify the surface characteristics of nanoparticles to enhance their capacity to penetrate this barrier, thereby increasing the concentration of drugs at targeted sites within the brain. Nano-delivery systems allow for controlled and sustained release of therapeutic agents. This controlled release can lead to prolonged therapeutic effects and help mitigate the peaks and troughs in drug concentration that often contribute to side effects. By designing nanoparticles that respond to specific stimuli, such as changes in pH or temperature, researchers can create systems that release their therapeutic payloads in a targeted manner, further improving treatment efficacy while reducing the risk of adverse reactions [70]. The versatility of nanocarriers also enables the co-delivery of multiple therapeutic agents, including drugs, genes, and imaging agents. This multi-functional approach can enhance treatment outcomes by integrating various therapeutic modalities that may work synergistically to combat neurodegeneration. For instance, delivering neuroprotective agents alongside gene therapies can maximize therapeutic effectiveness and provide a comprehensive treatment strategy that addresses multiple dimensions of the disease. Nano-delivery systems can be designed to be biocompatible and biodegradable, minimizing the risk of toxicity and ensuring safer interactions within the body. The incorporation of natural or synthetic biodegradable materials in the design of nanoparticles aligns with green chemistry principles and enhances the feasibility of translating these systems from the laboratory to clinical applications. The field of nanomedicine holds immense potential to revolutionize CNS therapies [70, 71]. As ongoing research continues to develop and refine nano-delivery systems, the ability to overcome existing treatment limitations becomes increasingly attainable. Current therapies for neurodegenerative disorders often face challenges related to poor drug delivery, inadequate bioavailability, and significant side effects. Nanomedicine presents innovative solutions that tackle these critical issues through advanced delivery mechanisms, tailored formulations, and the capacity to specifically target cellular pathways. As the understanding of neurodegenerative diseases deepens, the role of nanomedicine in developing targeted therapies is likely to expand. The integration of nanotechnology with fields such as genetics, molecular biology, and AI may lead to personalized treatment regimens that cater to the unique genetic and phenotypic profiles of individual patients. This trend toward personalization aligns with the broader shift toward precision medicine, promising more effective and safer treatment options. Several challenges must be addressed before the widespread implementation of nano-delivery systems in clinical practice. These challenges include regulatory hurdles, the need for extensive safety and efficacy testing, and the development of robust manufacturing processes. Ongoing research and collaboration among academic institutions, industry, and regulatory bodies will be essential to navigate these challenges and facilitate the translation of nanomedicine from bench to bedside. The application of nano-delivery systems in managing neurodegenerative disorders represents a promising frontier in medical science. With the potential to enhance drug delivery, improve therapeutic outcomes, and reduce side effects, nanomedicine is poised to play a pivotal role in reshaping CNS therapies. As advancements in this field continue to unfold, the outlook for overcoming current treatment limitations becomes increasingly optimistic, paving the way for more effective interventions in the battle against neurodegenerative diseases [72, 73].

REFERENCES

1. Zhao, N., Francis, N.L., Calvelli, H.R. and Moghe, P.V., 2020. Microglia-targeting nanotherapeutics for neurodegenerative diseases. *APL Bioengineering*, *4*(3).
2. Spandana, K.A., Bhaskaran, M., Karri, V.R. and Natarajan, J., 2020. A comprehensive review of nano drug delivery system in the treatment of CNS disorders. *Journal of Drug Delivery Science and Technology*, *57*, p. 101628.
3. Shukla, R., Singh, A., Handa, M., Flora, S.J.S. and Kesharwani, P., 2021. Nanotechnological approaches for targeting amyloid-β aggregation with potential for neurodegenerative disease therapy and diagnosis. *Drug Discovery Today*, *26*(8), pp. 1972–1979.
4. Shen, J., Lu, Z., Wang, J., Hao, Q., Ji, W., Wu, Y., Peng, H., Zhao, R., Yang, J., Li, Y. and Shi, Z., 2021. Traceable nano-biohybrid complexes by one-step synthesis as CRISPR-chem vectors for neurodegenerative diseases synergistic treatment. *Advanced Materials*, *33*(27), p. 2101993.
5. Seo, M.W. and Park, T.E., 2021. Recent advances with liposomes as drug carriers for treatment of neurodegenerative diseases. *Biomedical Engineering Letters*, *11*(3), pp. 211–216.
6. Rahmani, M., Álvarez, S.E.N. and Hernández, E.B., 2022. The potential use of tetracyclines in neurodegenerative diseases and the role of nano-based drug delivery systems. *European Journal of Pharmaceutical Sciences*, *175*, p. 106237.
7. Rahman, M.M., Islam, M.R., Akash, S., Harun-Or-Rashid, M., Ray, T.K., Rahaman, M.S., Islam, M., Anika, F., Hosain, M.K., Aovi, F.I. and Hemeg, H.A., 2022. Recent advancements of nanoparticles application in cancer and neurodegenerative disorders: At a glance. *Biomedicine & Pharmacotherapy*, *153*, p. 113305.
8. Poudel, P. and Park, S., 2022. Recent advances in the treatment of Alzheimer's disease using nanoparticle-based drug delivery systems. *Pharmaceutics*, *14*(4), p. 835.

9. Palanisamy, C.P., Pei, J., Alugoju, P., Anthikapalli, N.V.A., Jayaraman, S., Veeraraghavan, V.P., Gopathy, S., Roy, J.R., Janaki, C.S., Thalamati, D. and Mironescu, M., 2023. New strategies of neurodegenerative disease treatment with extracellular vesicles (EVs) derived from mesenchymal stem cells (MSCs). *Theranostics*, *13*(12), p. 4138.
10. Lamptey, R.N., Chaulagain, B., Trivedi, R., Gothwal, A., Layek, B. and Singh, J., 2022. A review of the common neurodegenerative disorders: Current therapeutic approaches and the potential role of nanotherapeutics. *International Journal of Molecular Sciences*, *23*(3), p. 1851.
11. Cai, C., Fang, J. and Ke, H., 2023. Traditional medicine and phytochemicals for neurodegenerative diseases treatment: Application of interdisciplinary technologies in novel therapeutic target and drug discovery. *Frontiers in Neuroscience*, *17*, p. 1268710.
12. Pinto, M., Silva, V., Barreiro, S., Silva, R., Remião, F., Borges, F. and Fernandes, C., 2022. Brain drug delivery and neurodegenerative diseases: Polymeric PLGA-based nanoparticles as a forefront platform. *Ageing Research Reviews*, *79*, p. 101658.
13. Paramanick, D., Singh, V.D. and Singh, V.K., 2022. Neuroprotective effect of phytoconstituents via nanotechnology for treatment of Alzheimer diseases. *Journal of Controlled Release*, *351*, pp. 638–655.
14. Panzarini, E., Mariano, S., Tacconi, S., Carata, E., Tata, A.M. and Dini, L., 2020. Novel therapeutic delivery of nanocurcumin in central nervous system related disorders. *Nanomaterials*, *11*(1), p. 2.
15. Padmanabhan, P., Palanivel, M., Kumar, A., Máthé, D., Radda, G.K., Lim, K.L. and Gulyás, B., 2020. Nanotheranostic agents for neurodegenerative diseases. *Emerging Topics in Life Sciences*, *4*(6), pp. 645–675.
16. Ozceylan, O. and Sezgin-Bayindir, Z., 2024. Current overview on the use of nanosized drug delivery systems in the treatment of neurodegenerative diseases. *ACS Omega*, *9*(33), pp. 35223–35242.
17. Ouyang, Q., Meng, Y., Zhou, W., Tong, J., Cheng, Z. and Zhu, Q., 2022. New advances in brain-targeting nano-drug delivery systems for Alzheimer's disease. *Journal of Drug Targeting*, *30*(1), pp. 61–81.
18. Neganova, M.E., Aleksandrova, Y.R., Sukocheva, O.A. and Klochkov, S.G., 2022, November. Benefits and limitations of nanomedicine treatment of brain cancers and age-dependent neurodegenerative disorders, in *Seminars in Cancer Biology* (Vol. 86, pp. 805–833). Academic Press.
19. Naz, F. and Siddique, Y.H., 2021. Nanotechnology: Its application in treating neurodegenerative diseases. *CNS Neurol Disord Drug Targets, 20*(1), pp. 34–53. https://doi.org/10.2174/1871527319666200916121515
20. Mohi-Ud-Din, R., Mir, R.H., Wani, T.U., Shah, A.J., Mohi-Ud-Din, I., Dar, M.A. and Pottoo, F.H., 2022. Novel drug delivery system for curcumin: Implementation to improve therapeutic efficacy against neurological disorders. *Combinatorial Chemistry & High Throughput Screening*, *25*(4), pp. 607–615.
21. Mohi-Ud-Din, R., Mir, R.H., Wani, T.U., Shah, A.J., Banday, N. and Pottoo, F.H., 2022. Berberine in the treatment of neurodegenerative diseases and nanotechnology enabled targeted delivery. *Combinatorial Chemistry & High Throughput Screening*, *25*(4), pp. 616–633.
22. Mishra, N., Ashique, S., Garg, A., Rai, V.K., Dua, K., Goyal, A. and Bhatt, S., 2022. Role of siRNA-based nanocarriers for the treatment of neurodegenerative diseases. *Drug Discovery Today*, *27*(5), pp. 1431–1440.
23. Marino, A., Battaglini, M., Moles, N. and Ciofani, G., 2022. Natural antioxidant compounds as potential pharmaceutical tools against neurodegenerative diseases. *ACS Omega*, *7*(30), pp. 25974–25990.
24. Maan, G., Sikdar, B., Kumar, A., Shukla, R. and Mishra, A., 2020. Role of flavonoids in neurodegenerative diseases: Limitations and future perspectives. *Current Topics in Medicinal Chemistry*, *20*(13), pp. 1169–1194.
25. Lopes, D., Lopes, J., Pereira-Silva, M., Peixoto, D., Rabiee, N., Veiga, F., Moradi, O., Guo, Z.H., Wang, X.D, Conde, J., Makvandi, P., Paiva-Santos, A.C., 2023. Bioengineered exosomal-membrane-camouflaged abiotic nanocarriers: Neurodegenerative diseases, tissue engineering and regenerative medicine. *Mil Med Research, 10*(1), p. 19. https://doi.org/10.1186/s40779-023-00453-z
26. Kumari, N., Daram, N., Alam, M.S. and Verma, A.K., 2022. Rationalizing the use of polyphenol nano-formulations in the therapy of neurodegenerative diseases. *CNS & Neurological Disorders-Drug Targets-CNS & Neurological Disorders)*, 21(10), pp. 966–976.
27. Kumar, R., Aadil, K.R., Mondal, K., Mishra, Y.K., Oupicky, D., Ramakrishna, S. and Kaushik, A., 2022. Neurodegenerative disorders management: State-of-art and prospects of nano-biotechnology. *Critical Reviews in Biotechnology*, *42*(8), pp. 1180–1212.
28. Keshavarz Shahbaz, S., Koushki, K., Keshavarz Hedayati, S., McCloskey, A.P., Kesharwani, P., Naderi, Y. and Sahebkar, A., 2024. Polymer nanotherapeutics: A promising approach toward microglial inhibition in neurodegenerative diseases. *Medicinal Research Reviews*, *44*(6), pp. 2793–2824.

29. Hernandez C, Shukla S. Liposome based drug delivery as a potential treatment option for Alzheimer's disease. *Neural Regeneration Research, 17*(6), pp. 1190–1198. https://doi.org/10.4103/1673-5374.327328
30. Goyal, R., Mittal, G., Khurana, S., Malik, N., Kumar, V., Soni, A., Chopra, H. and Kamal, M.A., 2024. Insights on quercetin therapeutic potential for neurodegenerative diseases and its nano-technological perspectives. *Current Pharmaceutical Biotechnology, 25*(9), pp. 1132–1141.
31. González, L.F., Bevilacqua, L.E., Naves, R., 2021. Nanotechnology-based drug delivery strategies to repair the mitochondrial function in neuroinflammatory and neurodegenerative diseases. *Pharmaceutics, 13*(12), p. 2055. https://doi.org/10.3390/pharmaceutics13122055
32. Fonseca-Santos, B., Chorilli, M., 2020. The uses of resveratrol for neurological diseases treatment and insights for nanotechnology based-drug delivery systems. The *International Journal of Pharmaceutics, 589*, p. 119832. https://doi.org/10.1016/j.ijpharm.2020.119832
33. Eleftheriadou, D., Kesidou, D., Moura, F., Felli, E., and Song, W., 2020. Redox-responsive nanobiomaterials-based therapeutics for neurodegenerative diseases. *Small, 16*(43), p. e1907308. https://doi.org/10.1002/smll.201907308
34. Duan, L., Li, X., Ji, R., Hao, Z., Kong, M., Wen, X., Guan, F., Ma, S., 2023. Nanoparticle-based drug delivery systems: An inspiring therapeutic strategy for neurodegenerative diseases. *Polymers (Basel), 15*(9), p. 2196. https://doi.org/10.3390/polym15092196
35. Silveira, G.B., Muller, A.P., Machado-de-Ávila, R.A., Silveira, P.C.L., 2021. Advance in the use of gold nanoparticles in the treatment of neurodegenerative diseases: New perspectives. *Neural Regeneration Research, 16*(12), pp. 2425–2426. https://doi.org/10.4103/1673-5374.313040
36. Chaudhary, V., 2022. Prospects of green nanotechnology for efficient management of neurodegenerative diseases. *Frontiers in Nanotechnology, 4*, p. 1055708.
37. Cano, A., Sanchez-Lopez, E., Ettcheto, M., Lopez-Machado, A., Espina, M., Souto, E.B., Galindo, R., Camins, A., García, M.L. and Turowski, P., 2020. Current advances in the development of novel polymeric nanoparticles for the treatment of neurodegenerative diseases. *Nanomedicine, 15*(12), pp. 1239–1261.
38. Bilal, M., Barani, M., Sabir, F., Rahdar, A. and Kyzas, G.Z., 2020. Nanomaterials for the treatment and diagnosis of Alzheimer's disease: An overview. *NanoImpact, 20*, p. 100251.
39. Bhattacharya, T., Soares, G.A.B.E., Chopra, H., Rahman, M.M., Hasan, Z., Swain, S.S. and Cavalu, S., 2022. Applications of phyto-nanotechnology for the treatment of neurodegenerative disorders. *Materials, 15*(3), p. 804.
40. Bandiwadekar, A., Jose, J., Khayatkashani, M., Habtemariam, S., Khayat Kashani, H.R. and Nabavi, S.M., 2022. Emerging novel approaches for the enhanced delivery of natural products for the management of neurodegenerative diseases. Journal of Molecular Neuroscience, 72(3), pp. 653–676.
41. Awad, R., Avital, A. and Sosnik, A., 2023. Polymeric nanocarriers for nose-to-brain drug delivery in neurodegenerative diseases and neurodevelopmental disorders. *Acta Pharmaceutica Sinica B, 13*(5), pp. 1866–1886.
42. Asil, S.M., Ahlawat, J., Barroso, G.G. and Narayan, M., 2020. Nanomaterial based drug delivery systems for the treatment of neurodegenerative diseases. Biomaterials science, 8(15), pp. 4109–4128.
43. Ashok, A., Andrabi, S.S., Mansoor, S., Kuang, Y., Kwon, B.K. and Labhasetwar, V., 2022. Antioxidant therapy in oxidative stress-induced neurodegenerative diseases: Role of nanoparticle-based drug delivery systems in clinical translation. *Antioxidants, 11*(2), p. 408.
44. Asefy, Z., Hoseinnejhad, S. and Ceferov, Z., 2021. Nanoparticles approaches in neurodegenerative diseases diagnosis and treatment. *Neurological Sciences, 42*(7), pp. 2653–2660.
45. Altinoglu, G. and Adali, T., 2020. Alzheimer's disease targeted nano-based drug delivery systems. *Current Drug Targets, 21*(7), pp. 628–646.
46. Akhtar, A., Andleeb, A., Waris, T.S., Bazzar, M., Moradi, A.R., Awan, N.R. and Yar, M., 2021. Neurodegenerative diseases and effective drug delivery: A review of challenges and novel therapeutics. Journal of controlled release, 330, pp. 1152–1167.
47. Agraharam, G., Saravanan, N., Girigoswami, A. and Girigoswami, K., 2022. Future of Alzheimer's disease: Nanotechnology-based diagnostics and therapeutic approach. *BioNanoScience, 12*(3), pp. 1002–1017.
48. Akhtar, A., Andleeb, A., Waris, T.S., Bazzar, M., Moradi, A.R., Awan, N.R. and Yar, M., 2021. Neurodegenerative diseases and effective drug delivery: A review of challenges and novel therapeutics. *Journal of Controlled Release, 330*, pp. 1152–1167.
49. Panghal, A. and Flora, S.J.S., 2024. Nanotechnology in the diagnostic and therapy for Alzheimer's disease. *Biochimica et Biophysica Acta (BBA)-General Subjects, 1868*(3), p. 130559.

50. Panda, P. and Mohapatra, R., 2024. Herbal Nanoparticles: A targeted approach for Neurodegenerative disorder Treatment. *Journal of Drug Targeting*, *32*(10), pp. 1233–1246.
51. Hussain, M.S., Chaturvedi, V., Goyal, S., Singh, S. and Mir, R.H., 2024. An update on the application of nano phytomedicine as an emerging therapeutic tool for neurodegenerative diseases. *Current Bioactive Compounds*, *20*(5), pp. 78–91.
52. Wen, M.M., El-Salamouni, N.S., El-Refaie, W.M., Hazzah, H.A., Ali, M.M., Tosi, G., Farid, R.M., Blanco-Prieto, M.J., Billa, N. and Hanafy, A.S., 2017. Nanotechnology-based drug delivery systems for Alzheimer's disease management: Technical, industrial, and clinical challenges. *Journal of Controlled Release*, *245*, pp. 95–107.
53. Siafaka, P.I., Okur, M.E., Erim, P.D., Çağlar, E.Ş., Özgenç, E., Gündoğdu, E., Köprülü, R.E.P., Karantas, I.D. and Üstündağ Okur, N., 2022. Protein and gene delivery systems for neurodegenerative disorders: Where do we stand today?. *Pharmaceutics*, *14*(11), p. 2425.
54. Riccardi, C., Napolitano, F., Montesarchio, D., Sampaolo, S. and Melone, M.A.B., 2021. Nanoparticle-guided brain drug delivery: Expanding the therapeutic approach to neurodegenerative diseases. *Pharmaceutics*, *13*(11), p. 1897.
55. López-Espinosa, J., Park, P., Holcomb, M., Godin, B. and Villapol, S., 2024. Nanotechnology-driven therapies for neurodegenerative diseases: A comprehensive review. *Therapeutic Delivery*, *15*(12), pp. 997–1024.
56. Poovaiah, N., Davoudi, Z., Peng, H., Schlichtmann, B., Mallapragada, S., Narasimhan, B. and Wang, Q., 2018. Treatment of neurodegenerative disorders through the blood–brain barrier using nanocarriers. *Nanoscale*, *10*(36), pp. 16962–16983.
57. Palaniyappan, R., Arthanari, S., Subramanian, M., Periyasamy, P., Pethappachetty, P., Ganesan, R., Prakash, M. and Govindaraj, A., 2024. Nanotechnology-driven approaches in overcoming drug delivery challenges for neurodegenerative diseases. *Indian Journal of Pharmaceutical Education and Research*, *58*(4s), pp. s1185–s1200.
58. Younis, M.A., Tawfeek, H.M., Abdellatif, A.A., Abdel-Aleem, J.A. and Harashima, H., 2022. Clinical translation of nanomedicines: Challenges, opportunities, and keys. *Advanced Drug Delivery Reviews*, *181*, p. 114083.
59. Min, Y., Caster, J.M., Eblan, M.J. and Wang, A.Z., 2015. Clinical translation of nanomedicine. *Chemical Reviews*, *115*(19), pp. 11147–11190.
60. Metselaar, J.M. and Lammers, T., 2020. Challenges in nanomedicine clinical translation. *Drug Delivery and Translational Research*, *10*(3), pp. 721–725.
61. Đorđević, S., Gonzalez, M.M., Conejos-Sánchez, I., Carreira, B., Pozzi, S., Acúrcio, R.C., Satchi-Fainaro, R., Florindo, H.F. and Vicent, M.J., 2022. Current hurdles to the translation of nanomedicines from bench to the clinic. *Drug Delivery and Translational Research*, *12*(3), pp. 500–525.
62. Agrahari, V. and Hiremath, P., 2017. Challenges associated and approaches for successful translation of nanomedicines into commercial products. *Nanomedicine*, 12(8), pp. 819–823.
63. Nirale, P., Paul, A. and Yadav, K.S., 2020. Nanoemulsions for targeting the neurodegenerative diseases: Alzheimer's, Parkinson's and Prion's. *Life sciences*, *245*, p. 117394.
64. Nayab, D.E., Din, F.U., Ali, H., Kausar, W.A., Urooj, S., Zafar, M., Khan, I., Shabbir, K. and Khan, G.M., 2023. Nano biomaterials based strategies for enhanced brain targeting in the treatment of neurodegenerative diseases: An up-to-date perspective. *Journal of Nanobiotechnology*, *21*(1), p. 477.
65. Mittal, K.R., Pharasi, N., Sarna, B., Singh, M., Rachana, Haider, S., Singh, S.K., Dua, K., Jha, S.K., Dey, A. and Ojha, S., 2022. Nanotechnology-based drug delivery for the treatment of CNS disorders. *Translational Neuroscience*, *13*(1), pp. 527–546.
66. Goyal, K., Koul, V., Singh, Y. and Anand, A., 2014. Targeted drug delivery to central nervous system (CNS) for the treatment of neurodegenerative disorders: Trends and advances. *Central Nervous System Agents in Medicinal ChemistryChemistry-Central Nervous System Agents)*, *14*(1), pp. 43–59.
67. Di Martino, P., Censi, R., Rosa Gigliobianco, M., Zerrillo, L., Magnoni, F., Agas, D., Quaglia, W. and Lupidi, G., 2017. Nano-medicine improving the bioavailability of small molecules for the prevention of neurodegenerative diseases. *Current Pharmaceutical Design*, *23*(13), pp. 1897–1908.
68. Babazadeh, A., Vahed, F.M. and Jafari, S.M., 2020. Nanocarrier-mediated brain delivery of bioactives for treatment/prevention of neurodegenerative diseases. *Journal of Controlled Release*, *321*, pp. 211–221.
69. Ashique, S., Afzal, O., Yasmin, S., Hussain, A., Altamimi, M.A., Webster, T.J. and Altamimi, A.S., 2023. Strategic nanocarriers to control neurodegenerative disorders: Concept, challenges, and future perspective. *International Journal of Pharmaceutics*, *633*, p. 122614.

70. Nagasen, D., et al., 2025. Molecular warriors: Proteases as sentinels in the war against cancer diagnosis and prognosis, in *Pathophysiological Aspects of Proteases in Cancer* (pp. 327–343). Elsevier.
71. Nagasen, D., Vemana, H.R.S.L. and Pindiprolu, S.K., 2025. Breaking barriers: Proteases as key players in cancer chemoresistance, in *Pathophysiological Aspects of Proteases in Cancer* (pp. 483–498). Elsevier.
72. Gummadi, R., Dasari, N., Kumar, D.S. and Pindiprolu, S.K.S., 2024. Evaluating the accuracy of large language model (ChatGPT) in providing information on metastatic breast cancer. *Advanced Pharmaceutical Bulletin*, *14*(3), p. 499.
73. Dasari, N., Guntuku, G.S. and Pindiprolu, S.K.S., 2024. Targeting triple negative breast cancer stem cells using nanocarriers. *Discover Nano*, *19*(1), p. 41.

Part II

Neurodegenerative Disease and Its Therapeutic Approaches

12 Synergistic Approach and the Treatment of Neurodegenerative Disorder

Nusrat K. Shaikh
Department of Quality Assurance, Smt. N. M. Padalia Pharmacy College, Ahmedabad, Gujarat, India

12.1 INTRODUCTION

Neurologists are motivated to study neurodegenerative diseases like Huntington's disease, multiple sclerosis (MS), Parkinson's disease (PD), spinal cord injury, and Alzheimer's disease (AD) because of their rising prevalence in the ageing population.[1] By impacting the nervous system in the central region, acute traumas also worsen chronic neurological diseases and raise systemic neuroinflammatory responses.[2] Within the next 20 years, neurological illnesses are predicted by the World Health Organisation (WHO) to overtake all other causes of death in humans.[3] Innovative methods of therapy must be created to address neurodegenerative illnesses to overcome the shortcomings of the available treatments.[4] Numerous sophisticated materials cannot be fully fabricated due to the absence of appropriate carriers for targeted medications.[5]

Cognitive and behavioural deficits are caused by a cognitive illness, which causes progressive neuronal loss and functionality reduction.[6] The central nervous system's (CNS's) cells without neurones are called lymphocytes or neuroglia. Neuroglia's four forms, like NG2a-glia, microglia, oligodendrocytes, and astrocytes, have a role in maintaining homeostasis, protecting neurones, and regulating brain plasticity.[7] Interactions between neurones, astrocytes, microglia, and oligodendrocytes cause additional harm to the CNS, foremost to myelin loss and apoptosis of neurons.[8]

A variety of disorders known as neurodegenerative illnesses occur when neurones in the brain progressively degenerate.

There is no known cure for most brain diseases, such AD, PD, and HD, although there are a few available treatments.[9] Currently used techniques comprise prescription drugs, work standards, rehabilitation, and therapy for stuttering. Modern techniques that can help control symptoms and reduce the progression of the disease include drugs, occupational and physical therapies, linguistic adaptive technology, and changes in habits. It is important to realise that these therapies do not stop or reverse the progression of the disease, even while they do alleviate symptoms and improve quality of life. The goal of progressing research is to create treatments that can slow the evolution of these illnesses by altering the root cause of destructive mechanisms.[10]

12.1.1 A Synergistic Approach for Neurodegenerative Conditions: Novel Beneficial Strategies

Even over a century after its emergence, the progression of memory loss poses a serious challenge in neurodegenerative illnesses since its causes are so obscure. The creation of medications aimed at several facets of the illness has become a viable approach to dealing with the complexity of AD and associated disorders. The function of the immune system, especially in AD, has drawn a lot of attention, and small particles offer a fresh approach to healthcare studies. Antibody production and

DOI: 10.1201/9781003478652-14

the creation of medicines have been transformed by developments in addressing weapons against amyloid-β (Aβ) and employing messenger RNA for genetic translation, which have opened new therapy options. Notwithstanding these advances, the long-term efficacy of traditional AD treatments like Cognex, Exelon, Razadyne, and Aricept is frequently low, highlighting the need for creative approaches. Because of this need, cutting-edge technologies like machine learning and artificial intelligence are now being used in the medication finding procedure to combat brain diseases. By assisting in the identification of treatment targets and the optimisation of lead compounds, these approaches provide a more efficient means of tackling the difficulties associated with AD and related disorders.[11]

12.1.2 International Cooperation–Synergistic Actions to Enhance Brain Function for Human Growth

The achievement of UN SDG 3 on health, and, in fact, all SDGs, depends on a synergistic, collaborative, and comprehensive approach to neurological conditions and brain health. All people and economies around the world will continue to suffer greatly from a lack of attention to brain health in terms of both economic and developmental outcomes. According to this chapter, to maximise the benefits for those immediately impacted and to minimise the cost on the community at large, neurological illnesses must be given top priority by all parties involved at the local, regional, and international levels. The four cornerstones of the brain quadrangle – surveillance; essential, primary, and secondary aversion; aid; and rehabilitation – urgently need to undergo synergistic novel endeavours throughout their lifespan course.

Through the harmonisation and consolidation of worldwide brain initiatives and political campaigns into a strong, cohesive force, the WHO IGAP,[12-13] and WHO Brain Health[14] programmes are advancing brain functioning to the pinnacle of the advancement and health of the world agenda. To utilise resources and promote interdisciplinary cooperation throughout each of the four elements and the key of the cerebral quadrangle, sustainable ecosystems are required in every nation. When paired with the cutting-edge ideas of comprehensive, mental, and planetary health, this new paradigm – which supports neurological and brain health – has the potential to enhance human welfare and ecological sustainability. This groundbreaking call to action has the potential to significantly improve human welfare, prosperity, and satisfaction worldwide by promoting brain health.[15]

12.2 THERAPIES FOR NEURODEGENERATIVE DISEASES

12.2.1 Traditional Therapies

The main goals of conventional treatments for neurodegenerative illnesses are symptom management and slowing the progression. Conventional treatments for neurodegenerative illnesses include the following:

Pharmacological Interventions: To control the manifestations of neurodegenerative illnesses, medications are frequently administered. Levodopa, for instance, is frequently used to restore dopamine levels and lessen the illness's physical signs. In AD, cholinesterase inhibitors, like donepezil, are used to improve cognitive performance. Additionally, there are drugs to treat symptoms including tremors, muscle stiffness, and irregular sleep patterns.[16]

Occupational and Physical Therapy: In neurodegenerative illnesses, both physical and occupational therapy can help with strength, balance, coordination, and mobility. To preserve autonomy and quality of life, these therapies may include flexing locomotor orientation, workouts, and assistive gadgets.[17]

Expression and ingesting Therapy: Speech and swallowing abilities may be impacted by neurodegenerative illnesses. By imparting skills that reinforce vocal muscles and enhance utterance, talking therapy can assist those who struggle with speaking.[18]

Being Supportive Care: Assistance with care addresses the additional needs of people with neurodegenerative diseases. Swallowing treatment, however, emphasises methods to reduce reflux and encourage secure and efficient choking. It covers therapy, academic achievement, counselling, and help with everyday tasks for clients and their careers.[19]

Auxiliary Technologies: People with neurodegenerative disorders can benefit from a variety of assistive devices that improve their abilities and standard of living. Encouragement organisations and local amenities can also offer an assortment of information and support. Assistance tools like scooters, walking sticks, and wooden sticks are illustrative, as are gadgets that facilitate interpersonal interaction, like visual tracking or speech-generating technologies.[20]

Changes in Lifestyle: Maintaining good health can help people with neurodegenerative disorders feel better overall. This could involve getting enough sleep, managing stress, keeping up a nutritious diet and doing regular workouts. Additionally, partaking in intellectually challenging activities and maintaining ties might be beneficial.

It is crucial to remember that standard medicines do not provide a remedy for neurodegenerative illnesses, even though medications can improve quality of life and assist in managing issues. To address the fundamental causes of these disorders, research is being conducted to provide disease-modifying treatments and investigate novel strategies, including gene and stem cell therapies.[21]

12.2.2 Drawbacks in Existing Treatments for Neurodegenerative Diseases

There are restrictions and disadvantages to traditional treatments for neurodegenerative illnesses. These encompass their failure to impede the development of the illness and offer longevity, the severe negative consequences that come with pharmaceutical remedies, inability to alter the course of illness, non-individualised tactic that ignores a person's differences, inability to cure, and the insufficient control of symptoms. Research is still being done to develop new medicines, including genetic treatments, vaccines, tailored therapies, and regenerative cell-based interventions, to get around these restrictions. The goal of these developments is to treat neurodegenerative disorders with more potent and disease-modifying therapies.[22–25]

12.2.3 A Handful of Novel Therapies for Brain Diseases

It is anticipated that developments in health sciences will aid research on neuroscience, particularly in the creation of more modern and focused therapy strategies. Innovative methods of disease treatment have gained popularity in medicine and therapies in recent years.[26] Demands about the creation, use, and oversight of novel technologies are growing as they are being developed, used in therapy, reviewed by regulators, and investigated after authorisation.[27]

Neurodegenerative disease treatments are now on the market (Figure 12.1).

Most neurodegenerative illnesses have no known cure, but there are several remedies and actions that may boost existence of individuals affected, lessen the severity of the condition, and alleviate symptoms.

12.2.3.1 Gene Therapy

Among the many effective treatments for neurological illnesses may be gene therapy.[28] To create successful treatments against neurological diseases, it is essential to comprehend the disease's root aetiology via the control of gene activity at the chronological and spatial level.[29] Fortunately, one of the difficult parts of introducing genes by propagation is avoiding escape into nearby areas or perivascular channels. Delivering biologically beneficial genes to specific organs or tissues with the goal of treating genetic abnormalities, producing therapeutic effects, or altering biological functioning is known as gene therapy. The method of transferring the curative genes into the intended target cells is known as transduction.[30] The benchmark for MRI-guided convection-enhanced release

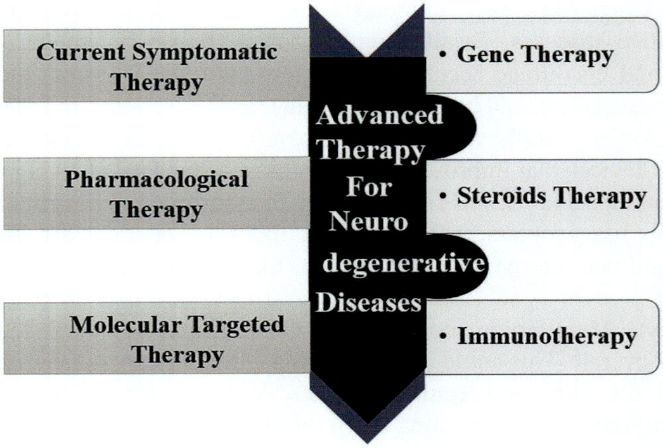

FIGURE 12.1 Advanced treatments for neurodegenerative diseases.

DNA therapy[31] is real-time vector delivery monitoring. If a new cutting-edge neurosurgical method is effectively used, promising preliminary treatments for neurodegenerative diseases could be translated into clinical trials.[32]

12.2.3.2 The Use of Steroids

One of the preventive measures to combat neurodegenerative illnesses may be improving neural activity.[33] By inhibiting neuronal death, DNA damage, mitochondrial malfunction, and triggering microglial,[34] neuroactive and sex steroids may boost neuronal health by promoting their existence, brain development, and memory retention. Sexual orientation and the brain injury phase of a patient[35] determine the positive effects of these corticosteroids. Furthermore, there is currently little study on the use of steroids to treat neurodegenerative illnesses.[36] Several criteria, including formulation, dosage, route of administration, and bioavailability, must be considered before using steroids to treat neurodegenerative illnesses.[37]

12.2.3.3 Immunological Treatments

Since the worsening neurological ailments are linked to the buildup of misfolded and compiled proteins, immunotherapy using monoclonal antibodies and certain antigens may help manage these conditions.[38] Utilising immunity may be a viable remedy, according to mounting data. Despite being a major healthcare concern, degenerative disorders linked to ageing and non-autoimmune conditions have received little attention when it comes to immune therapy. Immunotherapies have demonstrated promise in autoimmune disorders, but there is still much to learn about the possibilities in these illnesses. The development of immunotherapies is made more difficult by the complicated nature of these disorders, which include several pathogenic processes and complex relationships between ageing, neural inflammation, and cognitive confusion.[39] We now know more about how they function thanks to recent developments in immunity modulation in the brain and spinal cord. Investigating new targets for immunotherapy approaches is a crucial line of inquiry in the medical industry. Even if traditional illness biomarkers have proven useful for both diagnosis and treatment, finding and focusing on novel immune-related targets may present more chances to create cutting-edge immunotherapies.[40] Immunotherapy that targets aberrant accumulations of proteins or chemical messengers seems to have the opportunity to be a more efficient remedy.[41]

12.2.3.4 Molecularly Tailored Treatments

In advanced neurobiology,[42] molecular targeted therapies have been created for neurodegenerative illnesses. Several drugs have demonstrated efficacy in animal research; however, human trials have not confirmed their efficacy. It is crucial to investigate the effectiveness of possible medications that could alter the course of disease through scientific and clinical studies.[43] Positive outcomes from animal trials must be reproduced[44] to increase the validity of preclinical research. The low efficacy of genetically tailored medicines in humans.[45] This kind of experiment for prevention necessitates the development of a medical outcome measure that is resistant to individualism, fluctuations, and the effects of a placebo. Additionally, an increasing number of therapies are being evaluated before symptoms appear. Naturally occurring histories of living things and neurophysiological variables can be useful in this kind of precautionary study. Since traditional methods are not always effective for molecular targeted medicines, philosophical novelty is required to carry out basic as well as clinical studies.[46]

12.2.3.5 Pharmacological Therapy

Particularly intriguing are brain development variables and antibacterial and anti-beta-amyloid qualities of medications being researched or awaiting clearance for AD interventions. In addition to blocking acetylcholinesterase,[47] some medications also block beta-amyloid development by altering secretase activity and increasing the synthesis of beta-amyloid precursor proteins. Apart from immunotherapy, metal ions can also regulate metabolic and hormonal processes, as well as their interactions with beta-amyloid and oxidative reactions. A few glutamate AMPA receptor antagonists and dopamine uptake inhibitors are anticipated to be approved for the treatment of PD symptoms.[48] Drugs that block dopamine absorption include ion channel modulators, monoamine oxidase B inhibitors, and amino acid receptor blockers. Moreover, 5-HT analogues and medications that modulate astrocytes and alpha(2)-adrenergic receptor antagonists are utilised to treat PD-related or L-dopa[49] related dyskinesias. As part of the treatment for HD, LAX-101 has antiapoptotic qualities by preserving mitochondria and stabilising the membrane. Neurodegenerative illnesses with less prevalent symptoms are being treated with medications that have antiapoptotic, beneficial, and protective antioxidant properties.[50]

12.2.3.6 Current Symptomatic Treatments

For the treatment of AD, the FDA has authorised one *N*-methyl-D-aspartate (NMDA) receptor antagonist and four cholinesterase inhibitors (ChE-Is).[51] There are now three ChE-Is on the market: galantamine, rivastigmine, and donepezil. Memantine is antagonistic to the NMDA receptor. These medications improve mental function. Cholinesterase inhibitors and memantine show comparable effects on daily living activities, global function, and cognition.[52] In certain instances, symptomatic medications can be used to prevent new symptoms and alleviate neuropsychiatric symptoms.[53] In contrast to other neurodegenerative illnesses, AD causes post-synaptic cholinergic neurones to survive in the basalis nucleus of Meynert but presynaptic neurones to disappear. Neuronal expansion can be employed for the generation of post-synaptic active inducement in addition to cognitive enhancements.[54] The centre of the basalis is the primary source of acetylcholine for the cerebral cortex and amygdala. Brain impulses in the core of basalis are tightly linked to neuronal involvement in the motor, along with paralimbic tendons. The reason other transporter networks fail to boost mental ability is that pre- or post-synaptic detachment was not discovered in different transmission pathways that are critical to brain function. Unbroken links are necessary for brain activity, which includes both cholinergic output and the involvement of other neurotransmitters. Cortical circuit activity is typically elevated in fluorodeoxyglucose through positron emission tomography or fMRI during ChE-Is therapy.[55] New approaches to imaging could help improve brain function to build more beneficial treatments.[52, 56]

12.3 MSC TREATMENTS FOR NEUROLOGICAL CONDITIONS

Mesenchymal stem cell (MSC) treatments have shown promise in the management of brain disorders. Multipotent stem cells, or MSCs, could differentiate into a variety of cell types, such as neurones and glial cells. Additionally, they have nutritive and immunomodulatory qualities that can support tissue regeneration and repair. The following are a few MSC-based treatments that have been studied in relation to cognitive diseases:

PD: MSCs are currently being investigated for potential treatment for PD. They can release neurotrophic factors that help existing neurones survive or be implanted into the brain to replenish deceased dopaminergic neurones. MSC transplantation in PD patients has been shown to be safe and potentially effective in clinical trials.[57]

AD: MSCs have been investigated as an AD treatment option. They can release anti-inflammatory and neurotrophic substances, which may lessen neuroinflammation and increase neuronal survival. Although clinical trials for MSC-based therapeutics are still in their early phases, they have demonstrated promising results in experimental models of AD.

MSCs have been studied as a possible treatment for amyotrophic lateral sclerosis (ALS), a neurodegenerative illness that affects motor neurones and progresses over time. MSCs can be administered via IV or by direct injection into the spinal cord. In preclinical research, they have demonstrated positive outcomes, such as enhanced motor function and longer survival in ALS experimental animals.

MSCs have been investigated as a possible treatment for HD, an inherited neurological illness.

HD: MSCs have been explored as a potential therapy for HD, an inherited neurodegenerative disorder. In laboratory models of HD, research has demonstrated that MSCs can release growth factors and provide brain protection. To ascertain the safety and effectiveness of MSC-based treatments in humans, further investigation is necessary.

MSCs have been studied as a potential treatment for MS, an inflammatory illness that causes inflammation and demyelination in the CNS. MSCs could encourage tissue healing and regulate the immune system. MSC transplantation has been demonstrated in clinical trials to alleviate symptoms and lower activity levels in MS patients.

Although MSC-based treatments have potential, further study is required to completely comprehend their modes of action, improve delivery systems, and guarantee their foreseeable safety and effectiveness in the treatment of neurodegenerative illnesses. Prior to these treatments being made generally accessible, regulatory permission is needed, and clinical investigations are still active.[58]

12.4 TREATMENTS FOR NEURODEGENERATIVE DISEASES BASED ON TARGETS AND MECHANISMS

The development of small molecules with diverse modes of action and the synergistic application of two or more drugs targeted at these sites should be investigated in the quest for more potent medicinal therapies for neurodisorders. At each stage of the process, NMDA receptors, voltage-gated calcium channels, neuronal nitric oxide synthase, protein aggregation and oxidative stress from reactive oxygen species have been targeted by small-molecule therapies. Small compounds that target proteins implicated in neurodegenerative processes have the promise to be treatments as well as to shed light on the relationships and functions of these proteins. Finding out how targets link to one another may help identify routes that were initially undiscovered. Despite this, despite over a century of intensive research, the goal of all contemporary neurodegenerative therapeutics is still symptom alleviation rather than preventive outcomes.[59]

A multidisciplinary therapeutic session tactics, like those used to cure disease, attack germs, and treat excessive blood pressure, should be taken into consideration while creating tiny molecules to fight diseases of the brain. By attacking the disease from multiple angles, a multi-target therapy approach will significantly increase the likelihood that the condition will be modified. Inhibiting nNOS from generating ROS-producing species and preventing protein aggregation, for instance, offers a dual mechanism of action by lowering the root cause of protein folding defects and accumulation (oxidative stress damage) and preventing agglomeration brought on by ROS that can damage proteins.

Second, the potential synergy that could arise from a combination of drugs may enable the use of therapeutics at lower levels, facilitating the deployment of therapeutics with low potency, high toxicity, off-target effects, or small therapeutic windows. This would increase the number of drugs available to combat these debilitating, deadly diseases.

Although successful dual-acting compounds are starting to be described, they are still rare in the literature and make up only a small percentage of the total effort put into finding new drugs to treat neurodegeneration.[60] A dual NMDA and LTCC inhibitor is one example of a drug that targets two of the pathways. Three relevant targets for AD are activated by benzofuran-based compounds and their analogues: they suppress acetylcholine esterase, Aβ aggregation, and Aβ peptide-induced ROS formation. Instead of merely killing specific species, these chemicals have an advantage in disrupting the illness by fighting one of the fundamental sources of ROS in AD.[61] Out of an assortment of counterparts evaluated for nNOS inhibitory activity, imidazole is the most active molecule. The compounds were created as dual-target compounds, which scavenger free radicals and/or lower the peroxidation of lipids, along with functioning as nNOS inhibitors.[62] These substances seem to be perfect for screening in the drug development process for neurodegenerative diseases, even though they were created as a potential treatment for cerebral ischaemia.

The integration of two therapeutic structures via a linker chain serves as a last illustration. Huprine-tacrine heterodimers exhibit high potency (about 1 nM) by blocking both the active and outside locations of acetylcholinesterase at the same time. Furthermore, the heterodimers interact with two targets in the pathway, functioning as BACE-1 inhibitors (5–7 μM) and Aβ self-aggregation inhibitors (varying from 39% to 67% inhibition). It's interesting to note that although both targets are specific to AD, the drugs prevent a prion peptide that is essential to the assemblage of prions from aggregating.[63] To better probe and clarify the specific molecular networks and signalling cascades that underpin the biology of neurodegeneration, a multi-target strategy would offer therapeutic synergies and first-in-class molecular scaffolds. Recognising the underlying causes of neuronal degeneration is important because, as of right now, no one treatment, much less a combination of treatments, can appreciably slow the progression of the disease. Druggable domains that work and substances created to interact with them are described in this perspective, which is based on the similarities seen in neurodegeneration. As a result, the chapter has addressed numerous disease-specific targets (e.g., BACE1 inhibitors for AD). Our intention is to alert the reader to potential target different combinations and biochemical frameworks that, when paired, could help uncover the disease-modifying therapy that is the aim of neuroscience drug discovery.

12.5 NOVEL EXTRACELLULAR VEHICLES DERIVED FROM MSCS FOR NEUROLOGICAL DISORDERS

Several mechanisms of swelling, impaired growth of tissues, and neuronal harm are characteristics of neurological illnesses, including MS, AD, PD, stroke, and traumatic brain injury. Because MSC–extracellular vehicles (EVs) may transport therapeutic cargo, adjust inflammation, enhance neuronal survival and function, and promote tissue repair, they have shown promise in various situations. MSC–EVs have been studied for neurological disorders in the following ways: Extracellular vesicles made from MSCs have been demonstrated to help animals imitate neurological conditions like stroke and hypoxic-ischaemic brain injury.[64] The potential benefits of MSC-derived EVs for animals with AD were investigated in a study by Wang et al. (2018). MSC was administered to APP/PS1 animals and respective non-transgenic littermates (WT) after a two-week interval. A new water maze task and an item recognition task were then used to measure cognitive behaviours. Electrophysiological studies were performed to evaluate the plasticity of hippocampus synaptic connections. qRT-PCR and western blotting were used to evaluate the mRNA and protein levels of iNOS in primary cultured neurones that had been treated or prepared from APP/PS1 mice. INOS mRNA and protein expression are decreased by treatment with MSC-derived EVs. MSC-derived

EVs dramatically decreased the amounts of iNOS mRNA and protein in cells cultivated from APP/PS1 pups. They restored synaptic transmission deficits and enhanced cognitive behaviour in APP/PS1 mice. These findings imply that in a mouse model of AD, MSC-derived EVs may inhibit iNOS transcription.[65]

12.6 THERAPEUTIC USES AS CLINICAL APPLICATIONS OF MSC–EVS IN NEUROLOGICAL CONDITIONS

Proteins, nucleic acids, and metabolites are only a few of the diverse cargo types that make up EV composition. Initially, the regulatory and therapeutic effects of native EVs and their cargoes were investigated. As crucial mediators of intercellular communication, EVs are crucial for preserving and regaining neuronal function in the nervous system.[66] Improved therapeutic capabilities, like as tailored medicine delivery, have been developed for EVs due to the rapid advancement of nanotechnology. Some well-known artificial nanoparticles, such as liposomes, the first nano-drug delivery system, have received regulatory approval and marketing authorisation. Their therapeutic potential is limited by the innate immune system's activation and the quick removal of these substances from the circulation.

EVs show less immune response than liposomes and are more intricate and biocompatible than manufactured drug delivery platforms.[67] The enormous volume of dissemination and greater stability in the circulation of EVs are attributed to functionally linked proteins with pharmacokinetic properties. The chemicals on the surface of EVs allow them to pass through the wall between the blood and the brain, carry cargo, and cause cytoreactions that receive it. These qualities give EVs significant therapeutic potential for the treatment of CNS diseases.[68] Researchers and several pharmaceutical businesses have shown interest in EV-based treatments. Human clinical studies for EV-based treatments are growing quickly, and more are being carried out now.[69] NCT05490173 (neurodevelopmental disorders), NCT04388982 (AD), and NCT03384433 (cerebrovascular diseases) are the clinical studies that were recently filed in the area of neurological conditions. MSCs produced EVs in each of the three studies, albeit from different tissue sources. Allogenic bone marrow MSCs were used in NCT03384433 (cerebrovascular dysfunction), while allogenic adipose MSCs were used in NCT04388982 (AD). It should be mentioned that allogenic MSCs infected with miR-124 were employed in NCT03384433 (cerebrovascular diseases), which could offer proof of the effectiveness of EV-based gene treatments in people. EVs were given intravenously in NCT03384433 (cerebrovascular diseases) and intranasally in NCT05490173 (neurological issues) and NCT04388982 (AD). These example research studies for EV-based therapies in human neurodegenerative disorders will yield fascinating outcomes.[66, 70]

12.7 COMBINATION DRUG- AND CELL-BASED THERAPY'S SYNERGISTIC EFFECT AS A DIFFERENT STRATEGY FOR NEURODEGENERATIVE ILLNESSES

The benefits of the current therapeutic treatments for neurodegenerative illnesses are not very great. The *one gene, one drug, one disease* theory forms the basis of most of them. The beneficial impact of combination therapy is clearly understood. The PCD rodent, a vertebrate model of cerebellar degeneration, was used in this pilot investigation to evaluate the neuroprotective impact of the combination of medication and cell therapy. On postnatal day 12 (P12), PCD rats were given oleoylethanolamide (OEA), an endocannabinoid, intraperitoneally (10 mg/kg). They were then given bone marrow cells from healthy GFP mice via the retro-orbital plexus for bone marrow ablation (P19) and transplantation (P20). A superficial blood specimen was taken, the creatures were preserved, and the brains underwent dissection 15 days following transplantation (P35). Using flow cytometry, the proportion of GFP-positive blood cells was ascertained. Otherwise, immunohistochemistry was used to visualise the transplanted cells and assess any alterations in Purkinje cell density and morphology to assess the neuroprotective effect of this combined therapy. Our initial findings demonstrated that, in PCD rats, the combination of OEA and cartilage implantation provided neuroprotection in terms of Purkinje cell density and form. Indeed, it appeared that the PCD mice's cerebellum received more GFP-positive cells from transplants when they were treated with OEA. Unfortunately, at the present longevity, we have not been able to locate GFP-positive Purkinje cells, while most GFP-positive cells found were microglia.

12.8 CHALLENGES WITH BRAIN DRUG DELIVERY

Instead of eradicating the underlying causes of NDs, current medication has helped moderate the disease's progression. Most of these formulations fall short because neurodegeneration is the issue underlying the blood–brain barrier (BBB). The effectiveness of NDs as an intervention is limited by their incapacity to deliver adequate dosages to the brain. The paucity of effective treatment choices for NDs is explained by the mature nature of the BBB and the low permeability of many medications, if not all of them.

Drug transport to the brain is still difficult, though, because of several important characteristics, such as the BBB, the drug's molecular weight, and lipophilicity.[5] These variables make treating NDs more difficult and restrict the therapeutic effectiveness of medications. As a result, targeted drug delivery to the brain via nanoparticles has been investigated recently for neurogenesis and offers a viable platform for enhancing therapeutic approaches. Despite these prospective benefits, nanocarrier-mediated medication delivery has certain problematic issues, including safety, production, and legislation.

The main factors influencing a nanoparticle's toxicity are its size, shape, ionised dissolution, and charge on its surface. According to the official nanotoxicity guidelines,[71] these characteristics should be considered when creating drug delivery systems based on nanoparticles. The implications on the environment and human health should also be carefully considered before approving these nanocarriers.[72] A large body of research recommends a few changes to reduce the toxicity related to extent and cost, comprising its outer appearance using intrinsically derived cells or sustainable agents.[73,74]

Regarding the manufacturing of nanocarriers, their size and composition should remain consistent from batch to batch. There are several known processes for producing nanoparticles, including extrusion, microemulsion, and high-pressure homogenisation. Furthermore, the systemic features of the nanotechnology considerably determine the effectiveness and toxicity. Thus, it is crucial to examine the pharmacokinetic properties of nanotechnology in a suitable biological system. The behaviour of nanotherapeutics within cells can be predicted in this way using pharmacokinetic and more sophisticated biologically driven pharmacokinetic algorithms.[75] Furthermore, it is important to thoroughly evaluate the legal criteria for the clinical utilisation of nanotherapeutics.[76]

For NDs, nanomedicine offers a glimmer of hope and can be a powerful and potent weapon to overcome the constraints and limitations of conventional and existing treatment.[77] We emphasised papers that used nanoparticles to combat a variety of NDs, which could pave the way for nanomedicine. It is understandable that the creation of a cure takes time, but early studies in the area could pave the way for the eventual eradication of NDs. Still, further in vitro and in vivo data must be produced in order to demonstrate the effectiveness against NDs. To evaluate the effectiveness of nanoparticles, a comprehensive evaluation through in vitro and in vivo studies and confirmation of effective linkage are also necessary. This could benefit the scientific community in expanding or identifying the most efficacious nanomaterials for use in medicinal or diagnostic purposes.

12.9 BENEFITS OF SYNERGISTIC APPROACHES

In the discipline of regenerative therapy, there has been a lot of interest in using extracellular vesicles composed of MSCs. MSC–EVs are tiny, membrane-bound vesicles that are secreted by MSCs and contain a range of bioactive substances, such as lipids, amino acids, peptides, and nucleic acid molecules. The following are potential prospects for MSC–EV therapy in the future:

- Focused Delivery: MSC–EVs might be transported medicinal materials, like growth hormones, microRNA, or small interfering RNA (siRNA). These MSC–EVs with cargo can be engineered to target neurodegenerative disease-affected cells or tissues. Therapeutic compounds may be more stable, resistant to deterioration, and more effectively delivered to the site of action if they are encapsulated within EVs.
- Immunomodulation: Like their parent cells, MSC–EVs have immunological characteristics. They can control inflammation, influence immunological responses, and encourage tissue healing. MSC–EVs may be employed as a non-cellular alternative to MSC transplant, offering immunological advantages despite risks associated with cell therapy.
- Combination Remedies: To increase the efficacy of MSC–EVs, they might be used in combination with various kinds of treatment. MSC–EVs, for instance, can be utilised in conjunction with gene therapy techniques or in combination with other neuroprotective medications. Treatment results may be improved by the cumulative and beneficial impacts of this combo.

  Biomarkers and Diagnostics: Certain chemicals that represent the biological and ailments of their parent cells are carried by MSC–EVs. As possible biomarkers for the diagnosis and tracking of neurodegenerative illnesses, analysis of the cargo carried by MSC–EVs can yield significant insight into their root cause events. This can result in the creation of individualised treatment plans and non-invasive testing instruments.
- Approval via Regulation and Clinical Translation: Even while MSC–EV treatments have a lot of potential, obstacles must be removed before they can be used extensively in clinical settings. For MSC–EVs to receive regulatory approval, characterisation strategies, isolation procedures, and quality control procedures must be standardised. Large-scale production techniques must also be developed in order to satisfy the requirements of clinical applications.
- All things considered, MSC–EV treatments hold great promise for the treatment of neurodegenerative illnesses and regenerative treatments. To maximise the curative effects and prove its security and effectiveness, additional research and innovation are required.

12.10 FUTURE PERSPECTIVES

EVs made from MSCs have been used in recent years to treat neurological disorders. MSCs release these tiny, membrane-bound EVs, which are packed with different bioactive substances like lipids, proteins, and nucleic acids. Scientists have found that these EVs can have positive impacts on the neurological system and have therapeutic qualities. The capacity of EVs to penetrate the BBB and reach the afflicted regions of the brain is one of the main benefits of employing them to treat neurodegenerative diseases. It has been demonstrated that EVs made from MSCs improve the persistence of neurones, encourage the growth of new neurons, and control the fibrosis of neurones, all of which slow the evolution of neurological illnesses. Furthermore, certain molecules, such as microRNAs, which are essential for controlling gene expression, can be transferred by these EVs. Through transporting specific microRNAs to particular brain regions, EVs may alter sequences of transcription and promote cognitive protection. Moreover, therapeutic medications or compounds can be delivered to specific brain regions using EVs made by MSCs as carriers. This focused strategy increases the therapeutic effectiveness of the medicines administered while reducing the possibility of adverse effects. The potential of synergistic treatments for several neurodegenerative illnesses, such as ALS, PD, and AD, is being intensively investigated by researchers. Preclinical research has shown encouraging outcomes, including improvements in histological alterations and symptoms associated with the condition. In conclusion, using EVs derived from mast cells is a novel and exciting approach to the treatment of brain diseases. Such EVs are capable of transferring bioactive compounds, have therapeutic qualities, and function as specifically designed drug carriers. They can also pass through the BBB. To completely comprehend and utilise this approach's effectiveness in therapeutic settings, more investigation and clinical studies are required.

12.11 CONCLUSION

Since EVs are known to eliminate waste products and facilitate interfacial exchanges among cells, there is mounting evidence that they play a part in the nervous system of the body (both in wholesome and sick states). The potential for EVs to spread misfolded proteins linked to neurodegenerative diseases (like prion disease) suggests a shared pathogenic mechanism. Formal experiments, the majority of which have been conducted in cultured cells and recombinant mouse models, have not yet shown it to be associated with human disease. By quantifying the amount of polypeptide or gene data contained outside of cell sacs (such as miRNA expression) near the brain, neurological disorders can be identified. Methodologies still require standardisation. To properly utilise EVs' tremendous promise for brain disorder diagnostics, methods and testing across various cohorts still need to be standardised. Determining whether active EVs engage in intraneuronal transmission will also be made easier by the ability to bundle them or target cargo precisely. Furthermore, the BBB makes it easier for neuronal EVs to travel throughout the peripheral nervous system. Rather than connecting these characteristics with other ones, when EVs are contributing factors or an intentional result of biological systems, a computational analysis of these facets would be necessary.

REFERENCES

1. Holbrook JA, Jarosz-Griffiths HH, Caseley E, Lara-Reyna S, Poulter JA, Williams-Gray CH, Peckham D, McDermott MF. Neurodegenerative disease and the NLRP3 inflammasome. *Front Pharmacol.* 2021;12:643254. https://doi.org/10.3389/fphar.2021.643254
2. Faden AI, Barrett JP, Stoica BA, Henry RJ. Bidirectional brain-systemic interactions and outcomes after TBI. *Trends Neurosci.* 2021;44:406–18. https://doi.org/10.1016/j.tins.2020.12.004
3. Feigin VL, Vos T, Nichols E, Owolabi MO, Carroll WM, Dichgans M, Deuschl G, Parmar P, Brainin M, Murray C. The global burden of neurological disorders: Translating evidence into policy. *Lancet Neurol.* 2020 Mar;19(3):255–265. https://doi.org/10.1016/S1474-4422(19)30411-9
4. Khachaturian A, Dengel A, Dočkal V, Hrobon P, Tolar M. Accelerating innovations for enhanced brain health. Can artificial intelligence advance new pathways for drug discovery for Alzheimer's and other neurodegenerative disorders? *J Prev Alzheimers Dis.* 2023. pp. 1–4.
5. Lamptey RN, Chaulagain B, Trivedi R, Gothwal A, Layek B, Singh J. A review of the common neurodegenerative disorders: Current therapeutic approaches and the potential role of nanotherapeutics. *Int J Mol Sci.* 2022;23:1851. https://doi.org/10.3390/ijms23031851
6. Ullah MF, Ahmad A, Bhat SH, Abu-Duhier FM, Barreto GE, Ashraf GM. Impact of sex differences and gender specificity on behavioral characteristics and pathophysiology of neurodegenerative disorders. *Neurosci Biobehav Rev.* 2019;102:95–105. https://doi.org/10.1016/j.neubiorev.2019.04.003
7. Gazerani P. Contribution of central and peripheral glial cells in the development and persistence of itch: Therapeutic implication of glial modulation. *Neuroglia.* 2023;4:15–27.
8. Weigel M, Wang L, Fu Mm. Microtubule organization and dynamics in oligodendrocytes, astrocytes, and microglia. *Dev Neurobiol.* 2021;81:310–20. https://doi.org/10.1002/dneu.22753
9. Baloni P, Funk CC, Readhead B, Price ND. Systems modeling of metabolic dysregulation in neurodegenerative diseases. *Curr Opin Pharmacol.* 2021;60:59–65. https://doi.org/10.1016/j.coph.2021.06.012
10. Armstrong MJ, Okun MS. Diagnosis and treatment of Parkinson disease: A review. *JAMA.* 2020;323:548–60. https://doi.org/10.1001/jama.2019.22360

11. Niazi SK, Magoola M, Mariam Z. Innovative therapeutic strategies in Alzheimer's disease: A synergistic approach to neurodegenerative disorders. *Pharmaceuticals*. 2024;17(6):741. https://doi.org/10.3390/ph17060741
12. World Health Organization. *Sustainable development goals*. 2015. https://www.who.int/data/gho/data/themes/theme-details/GHO/sustainable-development-goals
13. World Health Organization. *Intersectoral global action plan on epilepsy and other neurological disorders 2022–2031*. 2022. https://www.who.int/publications/m/item/intersectoral-global-action-plan-on-eepilepsy-and-other-neurological-disorders-2022-2031
14. Hachinski V, Avan A, Gilliland J, Oveisgharan S. A new definition of brain health. *Lancet Neurol*. 2021;20:335–6.
15. Owolabi MO, Leonardi M, Bassetti C, Jaarsma J, Hawrot T, Makanjuola AI, Dhamija RK, Feng W, Straub V, Camaradou J, Dodick DW, Sunna R, Menon B, Wright C, Lynch C, Chadha AS, Ferretti MT, De A, Catsman-Berrevoets CE, Gichu M, Tassorelli C, Oliver D, Paulus W, Mohammed RK, Charway-Felli A, Rostasy K, Feigin V, Craven A, Cunningham E, Galvin O, Perry AH, Fink EL, Baneke P, Helme A, Laurson-Doube J, Medina MT, Roa JD, Hogl B, O'Bryan A, Trenkwalder C, Wilmshurst J, Akinyemi RO, Yaria JO, Good DC, Hoemberg V, Boon P, Wiebe S, Cross JH, Haas M, Jabalpurwala I, Mojasevic M, DiLuca M, Barbarino P, Clarke S, Zuberi SM, Olowoyo P, Owolabi A, Oyesiku N, Maly-Sundgren PC, Norrving B, Soekadar SR, van Doorn PA, Lewis R, Solomon T, Servadei F. Global synergistic actions to improve brain health for human development. *Nat Rev Neurol*. 2023; 19(6):371–383. https://doi.org/10.1038/s41582-023-00808-z
16. Drijgers RL, Aalten P, Winogrodzka A, Verhey FR, Leentjens AF. Pharmacological treatment of apathy in neurodegenerative diseases: A systematic review. *Dement Geriatr Cogn Disord*. 2009;28:13–22. https://doi.org/10.1159/000228840
17. Arbesman M, Lieberman D, Berlanstein DR. Method for the systematic reviews on occupational therapy and neurodegenerative diseases. *Am J Occup Ther*. 2014;68:15–9. https://doi.org/10.5014/ajot.2014.009308
18. Tye CB, Gardner PA, Dion GR, Simpson CB, Dominguez LM. Impact of fiberoptic endoscopic evaluation of swallowing outcomes and dysphagia management in neurodegenerative diseases. *Laryngoscope*. 2021;131:726–30. https://doi.org/10.1002/lary.28791
19. Armitage A, Fonkem E. Supportive care of neurodegenerative patients. *Front Oncol*. 2023;13:1029938. https://doi.org/10.3389/fonc.2023.1029938
20. Chandrasekaran S, Veeran R. Book Chapter-Assistive device for neurodegenerative disease patients using IoT. In book: Innovative Data Communication Technologies and Application, Springer International Publishing. 2020; Edition: 1, 447–452. https://doi.org/10.1007/978-3-030-38040-3_51
21. Santiago JA, Potashkin JA. Physical activity and lifestyle modifications in the treatment of neurodegenerative diseases. *Front Aging Neurosci*. 2023;15:1185671. https://doi.org/10.3389/fnagi.2023.1185671
22. Lee S, Bang SM, Lee JW, Cho KS. Evaluation of traditional medicines for neurodegenerative diseases using Drosophila models. *Evid Based Complementary Altern Med*. 2014; 2014:967462. doi: 10.1155/2014/967462
23. Ramanathan S, Archunan G, Sivakumar M, Tamil Selvan S, Fred AL, Kumar S, Gulyás B, Padmanabhan P. Theragnostic applications of nanoparticles in neurodegenerative disorders. *Int J Nanomedicine*. 2018;13:5561–5576. https://doi.org/10.2147/IJN.S149022
24. Vissers MF, Heuberger JA, Groeneveld GJ. Targeting for success: Demonstrating proof-of-concept with mechanistic early phase clinical pharmacology studies for disease-modification in neurodegenerative disorders. *Int J Mol Sci*. 2021;22:1615. https://doi.org/10.3390/ijms22041615.
25. Speed LJ, van Dam WO, Hirath P, Vigliocco G, Desai RH. Impaired comprehension of speed verbs in Parkinson's disease. *J Int Neuropsychol Soc*. 2017;23:412–20. https://doi.org/10.1017/S1355617717000248.
26. Modi G, Pillay V, Choonara YE. Advances in the treatment of neurodegenerative disorders employing nanotechnology. *Ann N Y Acad Sci*. 2010;1184:154–72. https://doi.org/10.1111/j.1749-6632.2009.05108.x.
27. Naziris N, Demetzos C. Advanced Health Technologies and Nanotechnologies in Neurodegenerative Diseases. *In: Vlamos, P. (eds) GeNeDis 2020. Advances in Experimental Medicine and Biology, Springer, Cham*. 2021, vol 1339. p. 317. https://doi.org/10.1007/978-3-030-78787-5_37.
28. Sudhakar V, Richardson RM. Gene therapy for neurodegenerative diseases. *Neurotherapeutics*. 2019;16:166–75. https://doi.org/10.1007/s13311-018-00694-0.
29. D'Souza GX, Rose SE, Knupp A, Nicholson DA, Keene CD, Young JE. The application of in vitro-derived human neurons in neurodegenerative disease modeling. *J Neurosci Res*. 2021;99:124–40. https://doi.org/10.1002/jnr.24615.

30. Piguet F, de Saint Denis T, Audouard E, Beccaria K, André A, Wurtz G, Schatz R, Alves S, Sevin C, Zerah M, Cartier N. The challenge of gene therapy for neurological diseases: Strategies and tools to achieve efficient delivery to the central nervous system. *Hum Gene Ther.* 2021;32:349–374. https://doi.org/10.1089/hum.2020.105
31. Elder JB, Lonser RR. Direct Convective Delivery for Nervous System Gene Therapy. *Neurosurg Clin N Am.* 2025 Jan; 36(1): 101–111. https://doi.org/10.1016/j.nec.2024.09.002
32. Rosser AE, Busse ME, Gray WP, Badin RA, Perrier AL, Wheelock V, Cozzi E, Martin UP, Salado-Manzano C, Mills LJ, Drew C, Goldman SA, Canals JM, Thompson LM. Translating cell therapies for neurodegenerative diseases: Huntington's disease as a model disorder. *Brain.* 2022; 145(5):1584–1597. https://doi.org/10.1093/brain/awac086.
33. Bonfili L, Cecarini V, Gogoi O, Gong C, Cuccioloni M, Angeletti M, Rossi G, Eleuteri AM. Microbiota modulation as preventative and therapeutic approach in Alzheimer's disease. *FEBS J.* 2021; 288(9):2836–2855. https://doi.org/10.1111/febs.15571.
34. Akwa Y. Steroids and Alzheimer's disease: Changes associated with pathology and therapeutic potential. *Int J Mol Sci.* 2020;21:4812. https://doi.org/10.3390/ijms21134812.
35. Ouanes S, Clark C, Richiardi J, Maréchal B, Lewczuk P, Kornhuber J, Kirschbaum C, Popp J. Cerebrospinal Fluid Cortisol and Dehydroepiandrosterone Sulfate, Alzheimer's Disease Pathology, and Cognitive Decline. *Front Aging Neurosci.* 2022 Jul 7;14:892754. https://doi.org/10.3389/fnagi.2022.892754.
36. Deeb O, Nabulsi M. Exploring Multiple Sclerosis (MS) and Amyotrophic Lateral Scler osis (ALS) as neurodegenerative diseases and their treatments: A review study. *Curr Top Med Chem.* 2020;20:2391–403. https://doi.org/10.2174/1568026620666200924114827.
37. Renaud J, Martinoli M-G. Considerations for the use of polyphenols as therapies in neurodegenerative diseases. *Int J Mol Sci.* 2019;20:1883. https://doi.org/10.3390/ijms20081883.
38. Mortada I, Raymond F, Nabha S, David MO, Fares Y, Almawi WY, Sadier NS. Immunotherapies for neurodegenerative diseases. *Front Neurol.* 2021;12:654739. https://doi.org/10.3389/fneur.2021.654739.
39. Shin J, Kim H-J, Jeon B. Immunotherapy targeting neurodegenerative proteinopathies: α-synucleinopathies and tauopathies. *J Mov Disord.* 2020;13:11. https://doi.org/10.14802/jmd.19057.
40. Kwon S, Iba M, Kim C, Masliah E. Immunotherapies for aging-related neurodegenerative diseases – emerging perspectives and new targets. *Neurotherapeutics.* 2020;17:935–54. https://doi.org/10.1007/s13311-020-00853-2.
41. Wang S, Colonna M. Microglia in Alzheimer's disease: A target for immunotherapy. *J Leukoc Biol.* 2019;106:219–27. https://doi.org/10.1002/JLB.MR0818-319R.
42. Manjula R, Anuja K, Alcain FJ. SIRT1 and SIRT2 activity control in neurodegenerative diseases. *Front Pharmacol.* 2021;11:585821. https://doi.org/10.3389/fphar.2020.585821.
43. Gorabi AM, Kiaie N, Barreto GE, Read MI, Tafti HA, Sahebkar A. The therapeutic potential of mesenchymal stem cell-derived exosomes in treatment of neurodegenerative diseases. *Mol Neurobiol.* 2019;56:8157–67. https://doi.org/10.1007/s12035-019-01663-0.
44. Krauss R, Bosanac T, Devraj R, Engber T, Hughes RO. Axons matter: The promise of treating neurodegenerative disorders by targeting SARM1-mediated axonal degeneration. *Trends Pharmacol Sci.* 2020;41:281–93. https://doi.org/10.1016/j.tips.2020.01.006.
45. Vijiaratnam N, Foltynie T. Therapeutic strategies to treat or prevent off episodes in adults with Parkinson's disease. *Drugs.* 2020;80:775–96. https://doi.org/10.1007/s40265-020-01310-2.
46. Mullane K, Williams M. Alzheimer's disease beyond amyloid: Can the repetitive failures of amyloid-targeted therapeutics inform future approaches to dementia drug discovery? *Biochem Pharmacol.* 2020;177:113945. https://doi.org/10.1016/j.bcp.2020.113945.
47. Liu D, Zhang Q, Luo P, Gu L, Shen S, Tang H. Neuroprotective effects of celastrol in neurodegenerative diseases-unscramble its major mechanisms of action and targets. *Aging Dis.* 2022;13.
48. Krzysztoforska K, Mirowska-Guzel D, Widy-Tyszkiewicz E. Pharmacological effects of protocatechuic acid and its therapeutic potential in neurodegenerative diseases: Review on the basis of in vitro and in vivo studies in rodents and humans. *Nutr Neurosci.* 2019;22:72–82. https://doi.org/10.1080/1028415X.2017.1354543.
49. Abdel-Salam OM. Drugs used to treat Parkinson's disease, present status and future directions. *CNS Neurol Disord Drug Targets.* 2008;7:321–42. https://doi.org/10.2174/187152708786441867.
50. Scatena R, Martorana GE, Bottoni P, Botta G, Pastore P, Giardina B. An update on pharmacological approaches to neurodegenerative diseases. *Expert Opin Investig Drugs.* 2007;16:59–72. https://doi.org/10.1517/13543784.16.1.59.

51. Marucci G, Buccioni M, Dal Ben D, Lambertucci C, Volpini R, Amenta F. Efficacy of acetylcholinesterase inhibitors in Alzheimer's disease. *Neuropharmacology*. 2021;190:108352. https://doi.org/10.1016/j.neuropharm.2020.108352.
52. Shaikh NK, Bhangale JO, Agrawal M. Book chapter 10- Novel Formulation Approaches for Targeting Alzheimer's diseases. In book- Neurodegenerative Disorders: From Pathogenesis to Clinical Approaches edited by Agrawal M, Singhal M, Nayak AK, Bhatt S & Shah K, published by NOVA Science Publishers, USA. pp. 215–240; https://doi.org/10.52305/GQSI1551; https://novapublishers.com/shop/neurodegenerative-disorders-from-pathogenesis-to-clinical-approaches/.
53. Verma A, Kumar Waiker D, Bhardwaj B, Saraf P, Shrivastava SK. The molecular mechanism, targets, and novel molecules in the treatment of Alzheimer's disease. *Bioorg Chem*. 2022 Feb;119:105562. doi: 10.1016/j.bioorg.2021.105562.
54. Cummings J. New approaches to symptomatic treatments for Alzheimer's disease. *Mol Neurodegener*. 2021;16:1–13. https://doi.org/10.1186/s13024-021-00424-9.
55. Zec RF, Burkett NR. Non-pharmacological and pharmacological treatment of the cognitive and behavioral symptoms of Alzheimer disease. *NeuroRehabilitation*. 2008;23:425–38.
56. Moreira NCDS, Lima JEBF, Marchiori MF, Carvalho I, Sakamoto-Hojo ET. Neuroprotective effects of cholinesterase inhibitors: Current scenario in therapies for Alzheimer's disease and future perspectives. *J Alzheimer's Dis Rep*. 2022 Apr 18;6(1):177–193. https://doi.org/10.3233/ADR-210061.
57. Tanna T, Sachan V. Mesenchymal stem cells: Potential in treatment of neurodegenerative diseases. *Curr Stem Cell Res Ther*. 2014;9:513–21. https://doi.org/10.2174/1574888x09666140923101110.
58. Yao P, Zhou L, Zhu L, Zhou B, Yu Q. Mesenchymal stem cells: A potential therapeutic strategy for neurodegenerative diseases. *Eur Neurol*. 2020;83:235–41. https://doi.org/10.1159/000509268.
59. Trippier PC, Jansen Labby K, Hawker DD, Mataka JJ, Silverman RB. Target- and mechanism-based therapeutics for neurodegenerative diseases: Strength in numbers. *J Med Chem*. 2013 Apr 25;56(8):3121–47. https://doi.org/10.1021/jm3015926
60. Geldenhuys WJ, Van der Schyf CJ. Designing drugs with multi-target activity: The next step in the treatment of neurodegenerative disorders. *Expert Opin Drug Discov*. 2013;8:115–129. https://doi.org/10.1517/17460441.2013.744746.
61. Rizzo S, Tarozzi A, Bartolini M, Da Costa G, Bisi A, Gobbi S, Belluti F, Ligresti A, Allara M, Monti JP, Andrisano V, Di Marzo V, Hrelia P, Rampa A. 2-Arylbenzofuran-based molecules as multipotent Alzheimer's disease modifying agents. *Eur J Med Chem*. 2012;58:519–32. https://doi.org/10.1016/j.ejmech.2012.10.045.
62. Sorrenti V, Salerno L, Di Giacomo C, Acquaviva R, Siracusa MA, Vanella A. Imidazole derivatives as antioxidants and selective inhibitors of nNOS. *Nitric Oxide*. 2006;14:45–50. https://doi.org/10.1016/j.niox.2005.09.005.
63. Galdeano C, Viayna E, Sola I, Formosa X, Camps P, Badia A, Clos MV, Relat J, Ratia M, Bartolini M, Mancini F, Andrisano V, Salmona M, Minguillon C, Gonzalez-Munoz GC, Rodriguez-Franco MI, Bidon-Chanal A, Luque FJ, Munoz-Torrero D. Huprine-tacrine heterodimers as anti-amyloidogenic compounds of potential interest against Alzheimer's and prion diseases. *J Med Chem*. 2012;55:661–9. https://doi.org/10.1021/jm200840c.
64. Shi X, Zhong X, Deng L, Wu X, Zhang P, Zhang X, Wang G. Mesenchymal Stem Cell-derived Extracellular Vesicle-enclosed microRNA-93 Prevents Hypoxic-ischemic Brain Damage in Rats. *Neuroscience*. 2022;500:12–25. https://doi.org/10.1016/j.neuroscience.2022.06.037.
65. Wang S-S, Jia J, Wang Z. Mesenchymal stem cell-derived extracellular vesicles suppresses iNOS expression and ameliorates neural impairment in Alzheimer's disease mice. *J Alzheimer's Dis*. 2018;61:1005–13. https://doi.org/10.3233/JAD-170848.
66. Yuan Y, Sun J, You T, Shen W, Xu W, Dong Q, Cui M. Extracellular Vesicle-Based Therapeutics in Neurological Disorders. *Pharmaceutics*. 2022; 14(12): 2652. https://doi.org/10.3390/pharmaceutics14122652.
67. Sharma P, Mesci P, Carromeu C, McClatchy DR, Schiapparelli L, Yates JR, Muotri AR, Cline HT. Exosomes regulate neurogenesis and circuit assembly. *Proc Natl Acad Sci*. 2019; 116(32): 16086–16094. https://doi.org/10.1073/pnas.1902513116.
68. El Andaloussi S, Mäger I, Breakefield XO, Wood MJ. Extracellular vesicles: Biology and emerging therapeutic opportunities. *Nat Rev Drug Discov*. 2013;12:347–57. https://doi.org/10.1038/nrd3978.
69. Mendt M, Rezvani K, Shpall E. Mesenchymal stem cell-derived exosomes for clinical use. *Bone Marrow Transplant*. 2019;54:789–92. https://doi.org/10.1038/s41409-019-0616-z.
70. Palanisamy CP, Pei J, Alugoju P, Anthikapalli NVA, Jayaraman S, Veeraraghavan VP, Gopathy S, Roy JR, Janaki CS, Thalamati D, Mironescu M, Luo Q, Miao Y, Chai Y, Long Q. New strategies of

neurodegenerative disease treatment with extracellular vesicles (EVs) derived from mesenchymal stem cells (MSCs). *Theragnostic*. 2023 Jul 16;13(12):4138–65. https://doi.org/10.7150/thno.83066.

71. Zielinska A, Costa B, Ferreira MV, Migueis D, Louros JMS, Durazzo A, Lucarini M, Eder P, Chaud MV, Morsink M, Willemen N, Severino P, Santini A, Souto EB. Nanotoxicology and Nanosafety: Safety-By-Design and Testing at a Glance. *Int J Environ Res Public Health*. 2020 Jun 28;17(13):4657. https://doi.org/10.3390/ijerph17134657.
72. Lamptey RNL, Chaulagain B, Trivedi R, Gothwal A, Layek B, Singh J. A Review of the Common Neurodegenerative Disorders: Current Therapeutic Approaches and the Potential Role of Nanotherapeutics. *International Journal of Molecular Sciences*. 2022; 23(3):1851. https://doi.org/10.3390/ijms23031851.
73. Suk JS, Xu Q, Kim N, Hanes J, Ensign LM. PEGylation as a strategy for improving nanoparticle-based drug and gene delivery. *Adv Drug Deliv Rev*. 2016;99:28–51.
74. Neves A, van der Putten L, Queiroz J, Pinheiro M, Reis S. Transferrin-functionalized lipid nanoparticles for curcumin brain delivery. *J Biotechnol*. 2021;331:108–17.
75. Ji X, Lu W, Wu K, Cho WC. Influencing factors of the pharmacokinetic characters on Nanopharmaceutics. *Pharm Nanotechnol*. 2017;5:24–31.
76. Zhu F-D, Hu Y-J, Yu L, Zhou X-G, Wu J-M, Tang Y, Qin D-L, Fan Q-Z, Wu A-G. Nanoparticles: A hope for the treatment of inflammation in CNS. *Front Pharmacol*. 2021;12:683935.
77. Patyar S, Prakash A, Medhi B. Dual inhibition: A novel promising pharmacological approach for different disease conditions. *J Pharm Pharmacol*. 2011;63:459–71. https://doi.org/10.1111/j.2042-7158.2010.01236.x

13 Electron Spinning and Electron Spraying Techniques and Their Correlation with Neurodegenerative Disorders

Subhasri Bogadi, Gowthamarajan Kuppusamy, Karri V. V. S. Narayana Reddy, and Swati Swagatika Swain
Department of Pharmaceutics, JSS College of Pharmacy, JSS Academy of Higher Education & Research, Ooty, Tamil Nadu, India

13.1 INTRODUCTION

Electron spinning and electron spraying techniques are at the forefront of cutting-edge technologies, transforming the field of microscopy and analytical chemistry (1). Electron spinning, also known as electron spin resonance (ESR) or electron paramagnetic resonance (EPR), is a spectroscopy technique that studies the interactions of unpaired electrons with an external magnetic field. This non-destructive method provides essential insights into the electrical structure of molecules and materials, revealing features at the atomic level with the use of microwave radiation (2). Electron spraying, often known as ESI, is a flexible ionization process used extensively in mass spectrometry. It generates highly charged aerosol droplets from liquid samples using a strong electric field, creating gas-phase ions that can be thoroughly analyzed. Electron spraying is particularly useful in biomolecular analysis since it allows for the exact characterization of proteins and nucleic acids. Despite their different applications, electron spinning and spraying procedures serve the same core purpose: to provide comprehensive structural and molecular information (3). This introduction lays the framework for thoroughly examining these strategies in the context of neurodegenerative diseases. As we progress through the following sections, the unique strengths and capabilities of electron spinning and spraying will become apparent, providing a more nuanced knowledge of their contributions to unraveling the complexity of neurodegenerative processes (4).

Understanding neurodegenerative illnesses is critical in both the medical and scientific fields because of their devastating impact on people, families, and society as a whole. Neurodegenerative illnesses, which include conditions such as Alzheimer's disease, Parkinson's disease, and amyotrophic lateral sclerosis (ALS), are defined by the progressive degradation of neurons, which leads to cognitive decline, motor dysfunction, and, ultimately, lower quality of life (5). Clinical research into neurodegenerative illnesses is critical for creating practical diagnostic tools, treatment techniques, and possible therapies. As the world's population ages, the prevalence of these ailments rises, offering a massive burden to healthcare systems globally. Unraveling the underlying mechanisms and molecular complexities of neurodegenerative illnesses is critical for developing biomarkers that can aid in early detection, allowing for timely therapies to reduce or prevent disease development. Neurodegenerative illnesses provide a unique insight into the complexity of the human brain. Investigating these disorders reveals the distinct pathology of each ailment and common mechanisms that may underpin numerous neurodegenerative processes (6). This

DOI: 10.1201/9781003478652-15

understanding is essential for the development of focused therapy approaches capable of treating multiple illnesses at the same time. Furthermore, investigating neurodegenerative illnesses has far-reaching consequences for understanding fundamental neurobiology. The complex interplay of genetic, environmental, and behavioral factors in various diseases improves understanding of basic brain processes. Insights acquired from studying neurodegeneration can lead to breakthroughs in neuroscience, potentially affecting a broad spectrum of neurological illnesses other than typical neurodegenerative diseases (7).

The importance of investigating neurodegenerative illnesses goes far beyond their direct clinical implications. It addresses a serious public health issue and contributes to a broader scientific understanding of the nervous system's complexities, providing hope for novel interventions and improved quality of life for those suffering from these debilitating disorders (8).

The goals of this chapter are twofold: first, to present a complete overview of the novel electron spinning and electron spraying procedures, and second, to investigate their relationship to neurodegenerative illnesses. The chapter is intended to serve as an essential resource for academics, clinicians, and hobbyists interested in converging modern analytical approaches with the complex realm of neurodegeneration (9). The chapter thoroughly reviews the principles, applications, and comparative strengths of electron spinning and spraying. It aims to illustrate the fundamental mechanisms that underpin these techniques, highlighting their historical evolution and the various scientific disciplines they have influenced. By developing a solid understanding of these approaches, the chapter prepares the groundwork for a focused examination of their applicability in neurodegenerative research (10). In parallel, the chapter discusses the importance of investigating neurodegenerative illnesses. It emphasizes the clinical significance, emphasizing the urgent need for advances in diagnostics, therapies, and preventive efforts.

Furthermore, the scope includes broader implications for our understanding of basic neurobiology and the potential cross-disciplinary influence on various neurological diseases (11, 12). This chapter aims to add to the joint knowledge base by clarifying the objectives and scope, encouraging a better understanding of the delicate synergy between cutting-edge technologies and the intricacies of neurodegenerative illnesses. It aims to pique interest, promote multidisciplinary collaboration, and lay the path for future advances in both analytical approaches and neurological research.

13.2 FUNDAMENTALS OF ELECTRON SPINNING AND ELECTRON SPRAYING

Electron spinning, also known as ESR or EPR, is a complex spectroscopy technique used to investigate the behaviors of unpaired electrons in a variety of materials (13). This approach is beneficial for gaining insights into the electronic structure of molecules, materials, and biological systems. In electron spinning, a sample is exposed to a strong external magnetic field while microwaving. This allows electrons with unpaired spins to absorb energy and move to higher energy states, providing important information about the electrons' local surroundings. By analyzing the resulting spectrum, researchers can better understand the nature of paramagnetic species, revealing atomic and molecular motions within the sample (14).

Electron spraying, often known as ESI, is a flexible ionization process most commonly used in mass spectrometry. This approach allows for examining various substances, particularly biomolecules, such as proteins, peptides, and nucleic acids. Electron spraying creates highly charged aerosol droplets from a liquid sample by applying a strong electric field. These droplets evaporate, forming ions that are then analyzed by mass spectrometers. Electrospray ionization is well-known for producing intact, highly charged ions that accurately represent the molecular composition of complicated biological materials. Its sensitivity and versatility have made it a standard technique in analytical chemistry, considerably advancing biomolecular research and drug discovery (2). Both electron spinning and electron spraying techniques demonstrate their abilities to uncover the secrets of molecular and atomic structures, with each playing a distinct role in furthering scientific understanding across multiple disciplines. While electron spinning provides insight into the electronic complexities of materials, electron spraying excels at detailed analysis of biomolecules, enriching our understanding of the microscopic world (15). ESR, or EPR, is based on the fundamentals of magnetic resonance spectroscopy. At its core, the approach takes advantage of the magnetic properties of electrons with unpaired spins, which are common in paramagnetic materials. When a sample is exposed to an external magnetic field, the unpaired electrons absorb energy through microwave radiation, causing them to move to higher energy states (16). Researchers can generate an ESR spectrum by measuring the energy absorbed, which provides valuable information on the surrounding environment, concentration, and mobility of unpaired electrons. This spectroscopic method helps research a wide range of materials, from inorganic compounds to biological systems, as it provides a potent tool for characterizing electron behavior at the atomic and molecular levels (17). ESI follows electrohydrodynamic and mass spectrometric principles. The procedure begins with generating charged aerosol droplets from a liquid sample using a high electric field. As these droplets evaporate, ions form, carrying a net charge. The generated ions are then sent into a mass spectrometer for examination. The capacity of electrospray ionization to produce intact and highly charged ions is the fundamental breakthrough, making it ideal for analyzing significant macromolecules such as proteins and nucleic acids. The technique's versatility extends to its compatibility with various solvents and sample types, establishing it as a cornerstone of modern analytical chemistry. Overall, electron spraying offers a methodical technique for converting complex liquid samples into ions suited for precision mass spectrometric analysis, allowing for revolutionary insights into the composition and structure of a wide range of biological components. Understanding these fundamental principles and mechanisms is critical for realizing the full potential of electron spinning and electron spraying techniques in unraveling the mysteries of molecular and atomic structures, contributing to advances in scientific study across different fields (18).

ESR has various features that have established it as a potent analytical tool. One significant advantage is its capacity to provide extensive information on the electronic structure of materials with unpaired electrons. This non-destructive approach is beneficial for researching paramagnetic species, providing information about their concentration, movement, and local surroundings. Electron spinning is highly diverse, with applications in various scientific fields such as chemistry, physics, and biology. Furthermore, its capacity to function under ambient circumstances makes it suitable for multiple samples. As a result, researchers may use electron spinning to elucidate the electrical complexities of materials, considerably improving our understanding of atomic and molecular dynamics. While electron spinning is a valuable technology, it has some limits. One significant limitation is its sensitivity, as the approach may struggle to detect paramagnetic species at low concentrations.

Furthermore, the strategy may encounter difficulties in adequately estimating the quantity of unpaired electrons in complicated systems. Sample size can also be a limiting factor, as there must be enough material to give a quantifiable signal. Despite these limits, ongoing advances in instrumentation and methods continue to address and alleviate these issues, increasing electron spinning's overall value and applicability in a wide range of scientific research. ESI is distinguished by its adaptability and wide range of applications, particularly in mass spectrometry. One of its key advantages is its capacity to efficiently ionize large, complex molecules, including proteins, peptides, and nucleic acids. The approach produces intact and highly charged ions, enabling precise mass analysis. Additionally, electron spraying is compatible with several solvents, allowing for examining a wide range of sample types. Its sensitivity and capacity to offer comprehensive structural information make it a vital tool in current analytical chemistry, with applications ranging from proteomics to metabolomics and drug development. Despite its numerous benefits, electron spraying has several drawbacks (1). One such limitation is its ability to generate various charging states of ions, resulting in complicated mass spectra. Furthermore, the approach may be sensitive to experimental conditions, with differences in parameters such as solvent composition and flow rate affecting ionization efficiency. Electrospray ionization may also have difficulty ionizing some substances, and ion suppression difficulties might impair the precision of quantitative tests. Continuous developments and modifications in technology and methodology attempt to solve these constraints, increasing the robustness and dependability of electron spraying in analytical applications.

13.3 APPLICATION OF ELECTRON SPINNING IN NEURODEGENERATIVE RESEARCH

A thorough review of studies using electron ESR in neurodegenerative illness research reveals a diverse range of investigations aiming at unraveling the molecular complexities behind these devastating conditions. This precise spectroscopic approach has helped provide novel insights into the cellular and molecular changes linked with dementia (19). Several studies employed ESR to investigate the role of oxidative stress in neurodegenerative diseases. Oxidative stress, an imbalance between reactive oxygen species (ROS) production and the cellular antioxidant defense system, is a characteristic of neurodegenerative disease. Electron spinning has been used to identify and quantify paramagnetic species that indicate the presence of ROS in neural tissues. These studies have improved our understanding of oxidative damage to cellular structures, emphasizing its role in the etiology of neurodegenerative illnesses such as Alzheimer's and Parkinson's seen in Figure 13.1.

Furthermore, electron spinning has helped us understand the role of free radicals and metal ions in neurodegenerative processes. Spin labeling techniques have been used to investigate the interactions of paramagnetic species within neuronal membranes. This technique has shed light on changes in membrane fluidity, lipid peroxidation, and protein oxidation, providing helpful knowledge regarding how these changes affect neuronal function and survival. Electron spinning has expanded beyond the study of oxidative stress to encompass research into the involvement of transition metal ions such as iron and copper in neurodegenerative illnesses. ESR studies have

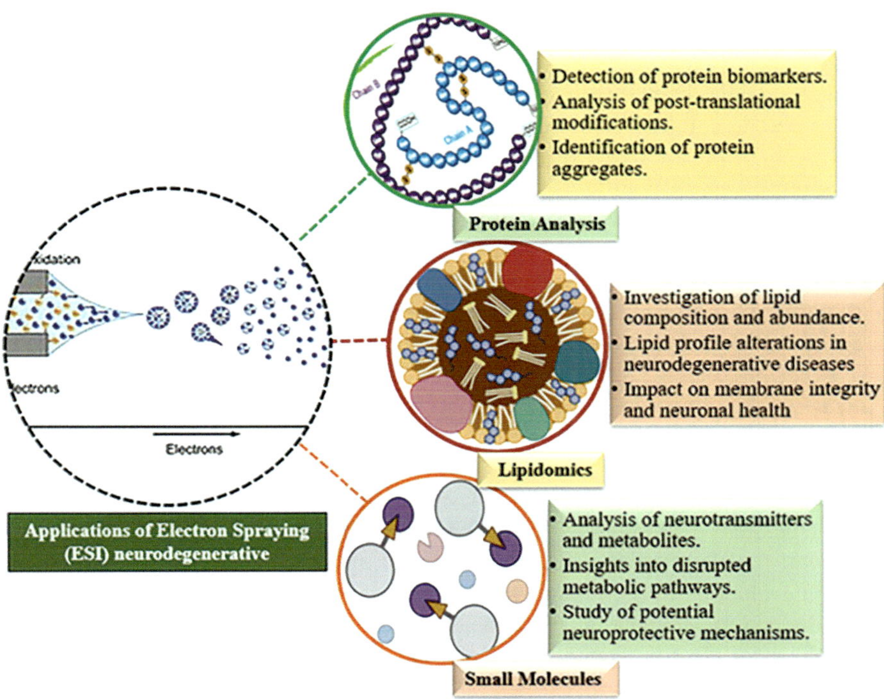

FIGURE 13.1 Application of electron spinning in neurodegenerative research.

investigated the distribution and coordination of these metal ions in the brains of people with neurodegenerative diseases. The discoveries have improved our understanding of metal ion dysregulation and its potential role in illnesses such as Huntington's disease and ALS. The evaluation of works using electron spinning in neurodegenerative illness research emphasizes its critical significance in determining the molecular underpinnings of these complicated diseases. Electron spinning has dramatically increased our understanding of the cellular mechanisms that cause neurodegeneration by offering a window into oxidative stress, free radical dynamics, and metal ion interactions inside neural tissues. These findings show the potential for developing targeted therapeutic techniques and interventions to reduce the impact of neurodegenerative illnesses on individuals and society (20). Electron spinning, commonly known as ESR, has emerged as a valuable tool for studying the complicated cellular structures involved in neurodegeneration. This advanced spectroscopic technique allows for investigating electron dynamics inside biological systems, providing novel insights that contribute to our understanding of the molecular foundations of neurodegenerative illnesses. The study of cell membranes, particularly lipid bilayers, is an essential area of research because they play a critical role in the integrity and functionality of neurons. Electron spinning enables researchers to investigate the organization and fluidity of these membranes by focusing on paramagnetic species. Alterations in membrane structure, such as increased stiffness or disordered lipid packing, are signs of pathogenic alterations linked with neurodegeneration. Using spin labels or probes, researchers can gain precise information about the cellular membrane's sensitivity to oxidative stress, a frequent feature of many neurodegenerative disorders.

Furthermore, electron spinning has helped researchers understand the impact of protein misfolding and aggregation in neural structures. ESR-based studies have explored structural alterations in proteins related to neurodegenerative disorders, such as tau and alpha-synuclein in Alzheimer's and Parkinson's diseases, respectively (19). This method identifies paramagnetic centers inside these proteins, which provides valuable information on their conformational changes and aggregation states. Understanding the structural dynamics of these proteins is crucial for determining the mechanisms that cause neurodegeneration and devising targeted treatment approaches. Furthermore, electron spinning has been used to investigate the involvement of metal ions, specifically iron and copper, in neurodegenerative processes. These ions are necessary for many physiological activities, but their dysregulation has been linked to oxidative stress and neuronal damage. ESR investigations have allowed researchers to look into the distribution and coordination of metal ions inside neural tissues, shedding light on their role in neurodegenerative disorders. In conclusion, investigating how electron spinning provides insights into cellular structures associated with neurodegeneration indicates a significant impact on understanding the molecular complexities of these conditions. By providing a microscopic picture of membrane dynamics, protein structures, and metal ion interactions, electron spinning improves our understanding of the cellular changes that underpin neurodegenerative diseases. This understanding is critical for developing therapeutic options that preserve cellular integrity and slow the course of neurodegenerative illnesses.

13.4 APPLICATION OF ELECTRON SPRAYING IN NEURODEGENERATIVE RESEARCH

Studies using electron spraying, notably ESI, have enriched our understanding of neurodegenerative illnesses. This analytical approach, primarily connected with mass spectrometry, has given researchers a vital tool for studying the molecular and metabolic elements of neurodegeneration (21). A thorough evaluation of these studies indicates a diverse approach elucidating fundamental aspects of cellular and molecular changes associated with distinct neurodegenerative diseases. One significant emphasis of electron spraying research has been the identification of protein biomarkers linked with neurodegenerative diseases. Using ESI with mass spectrometry, researchers could identify and quantify particular proteins implicated in the etiology of diseases such as Alzheimer's and Parkinson's. This technique detects changes in protein expression, post-translational modifications, and protein aggregation, providing crucial insights into the molecular fingerprints associated with neurodegenerative diseases. Identifying such biomarkers is critical for early detection and may open the way for creating tailored therapeutic therapies. Furthermore, electron spraying has helped analyze lipid profile changes linked with dementia. Lipids are essential for the formation and function of cell membranes, and disruption has been linked to various neurodegenerative illnesses. ESI enables a thorough investigation of lipid species, including composition, abundance, and alterations. This knowledge is critical for understanding the impact of lipid metabolism on neuronal health, and it may provide new targets for therapeutic therapies (22).

In addition to proteins and lipids, investigations on electron spraying have examined tiny molecules such as neurotransmitters and metabolites in the context of neurodegenerative illnesses. ESI enables precise measurement of these compounds, providing helpful information about changes in neurotransmitter levels, metabolic pathways, and putative neuroprotective mechanisms. This extensive research improves our understanding of the metabolic changes that occur in the neurodegenerative environment, providing a more complete picture of the disease process. The overview of works using electron spraying in neurodegenerative illness research emphasizes its adaptability and importance in understanding the molecular complexity of these conditions. From protein biomarkers to lipid profiles and small compounds, ESI has been shown to be a game changer, providing vital data that increases our understanding of underlying mechanisms and holds promise for developing focused therapeutic solutions (23).

Electron spraying, specifically ESI, has substantially aided the investigation of the molecular and metabolic aspects underlying neurodegenerative diseases. This new technology, frequently used with mass spectrometry, provides a robust platform for in-depth study, allowing researchers to uncover intricate details that impact our understanding of neurodegeneration at the molecular level. Electron spraying contributes to understanding the protein landscape associated with neurodegenerative diseases. Researchers can use ESI with mass spectrometry to identify and quantify particular proteins involved in disorders such as Alzheimer's, Parkinson's, and Huntington's. This method detects changed protein expression and identifies post-translational changes and the presence of protein aggregates. ESI's thorough protein profiling helps to elucidate the molecular hallmarks of neurodegenerative processes, potentially leading to the development of diagnostic biomarkers and therapeutic targets.

Furthermore, electron spraying has been critical in identifying changes in lipid profiles linked with neurodegeneration (24). Lipids are essential components of neuronal membranes, and their disruption has been linked to various neurodegenerative diseases. ESI allows for thorough investigation of lipid species, providing information about their composition, abundance, and changes. Lipidomics research using electron spraying offers researchers a more detailed understanding of the impact of lipid metabolism on brain health, perhaps uncovering crucial parameters influencing membrane integrity and cellular function.

Furthermore, electron spraying has proven to be an effective method for studying tiny molecules, such as neurotransmitters and metabolites, in the context of neurodegenerative illnesses. ESI allows for exact measurement of these molecules, providing insights into neurotransmitter level changes, metabolic pathway disruptions, and potential compensatory mechanisms. This comprehensive research leads to a more holistic understanding of the biochemical alterations in the neurodegenerative environment, providing insight into the complicated interplay of molecular components. The investigation into how electron spraying adds to our understanding of neurodegeneration's molecular and metabolic elements demonstrates its revolutionary importance in improving research in this area. From unraveling protein complexities to decoding lipid profiles and dissecting small molecule dynamics, electron spraying offers researchers a versatile and comprehensive toolkit, providing valuable insights that deepen our understanding of the molecular complexities underlying neurodegenerative disorders (25) (see Figure 13.2).

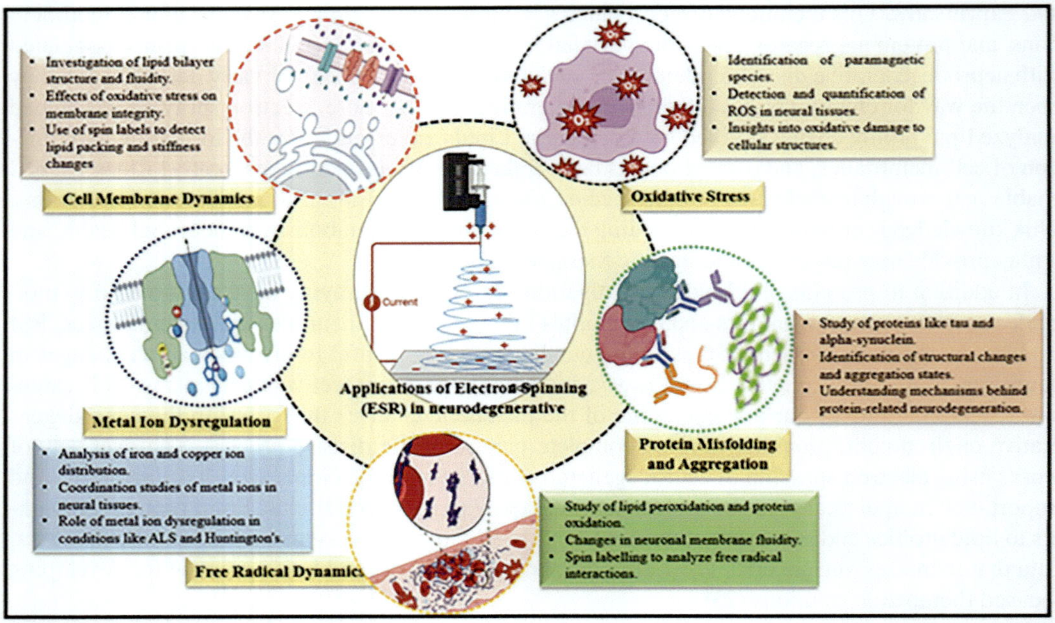

FIGURE 13.2 Application of electron spraying in neurodegenerative research.

13.5 CORRELATION STUDIES: BRIDGING TECHNIQUES AND NEURODEGENERATIVE MARKERS

A detailed evaluation of studies on the relationships between electron spinning and electron spraying data and neurodegenerative markers provides a varied and multifaceted approach to comprehending the complicated complexity of these conditions (26). The combined use of these modern approaches provides a complete picture, greatly aiding in detecting and interpreting genetic signatures associated with neurodegenerative diseases. Several research studies have investigated the relationship between electron spinning, precisely ESR data, and oxidative stress markers in neurodegenerative illnesses. Researchers have used electron spraying techniques to link the presence of certain paramagnetic species observed through ESR with the identification of biomolecules in mass spectrometric investigations. This integrated method has shown links between oxidative stress markers and specific cellular structures, thereby improving our understanding of the role of ROS in neurodegeneration. The correlation investigations have expanded to include the evaluation of protein aggregation, characteristic of many neurodegenerative disorders. Researchers used spinning and spraying techniques to link data on protein structure and aggregation states. Electron spinning (ESR) provides insights into the paramagnetic centers within proteins, while electron spraying and mass spectrometry aid in identifying and quantifying individual proteins implicated in aggregation formation. These correlations help to provide a more complete understanding of the proteinaceous components linked to neurodegenerative indicators.

Furthermore, researchers have examined the relationship between metal ion dysregulation and neurodegenerative processes. Studies have used electron spinning and spraying techniques to match ESR-detected paramagnetic signals, indicating metal ions with mass spectrometric data identifying metal-binding proteins. This dual approach has shed light on the complex link between metal ion homeostasis and proteins involved in neurodegenerative pathways, providing essential insights into possible treatment targets. The integration of spinning and spraying data has not only improved our understanding of specific neurodegenerative indicators. Still, it has also enabled a more comprehensive examination of the molecular landscape associated with these disorders (27). This collaborative strategy can potentially find new biomarkers, decode complex signaling networks, and ultimately improve our ability to diagnose and intervene in neurodegenerative disorders. As technology advances, the combination of spinning and spraying techniques is poised to play a critical role in unraveling the complexities of neurodegenerative illnesses, bringing hope for improved diagnostics and focused therapy solutions. Researchers investigated comprehensively to identify patterns and trends in electron spinning and electron spraying techniques used in neurodegenerative illness research (28). This sophisticated procedure entails combining data from several research studies and identifying important correlations that provide more significant insights into the molecular foundations of these complex illnesses. One notable emerging trend is the link between oxidative stress markers and neurodegenerative processes. Studies using both electron spinning and spraying techniques have consistently shown connections between paramagnetic signals indicating oxidative stress detected via ESR and particular chemical markers discovered via electron spraying mass spectrometry. This repeating relationship emphasizes the importance of oxidative stress in neurodegenerative diseases and its potential as a prominent subject in the search for diagnostic biomarkers and therapeutic strategies. Furthermore, pattern recognition allows for the investigation of protein aggregation processes. Researchers systematically matched electron spinning data, revealing changes in protein structure and paramagnetic centers, with mass spectrometric data, identifying specific proteins implicated in aggregate formation (29). This convergence of information has revealed consistent patterns in the molecular signatures of protein aggregation across many neurodegenerative illnesses, laying the groundwork for identifying common causes and potential intervention locations.

Furthermore, the study of metal ion dysregulation has revealed identifiable patterns through electron spinning and spraying approaches. Correlations between paramagnetic signals indicating metal ions and mass spectrometric data identifying metal-binding proteins have shown persistent tendencies in the role of certain metals in neurodegenerative diseases. This pattern recognition helps to detect the intricacies of metal ion contributions to pathology, opening up possibilities for tailored treatment approaches.

Discovering patterns and trends in these associated datasets is critical for furthering our understanding of neurodegenerative illnesses. These findings help refine existing ideas and act as a stimulus for new research questions and experimental paths. As researchers continue to enhance and expand their analytical approaches, the discovery of patterns and trends through the combined use of spinning and spraying techniques has the potential to reveal unique molecular signatures and the complexities of neurodegenerative processes (30). The use of electron spinning and electron spraying techniques in studying neurodegenerative processes has significant ramifications, providing a comprehensive and precise insight beyond individual molecular components. These consequences echo throughout numerous aspects of neurodegenerative research, altering our understanding of disease causes, diagnostic techniques, and prospective treatments.

The combination of spinning and spraying techniques allows for unique molecular insights into neurodegenerative processes. The spinning data, which highlights paramagnetic signals indicating oxidative stress and metal ion dysregulation, and spraying data, which identifies individual proteins and metabolites, allow for a thorough mapping of molecular events (31). This comprehensive perspective improves our understanding of the complex interplay of different molecular components in neurodegenerative illnesses, offering information on disease initiation, development, and potential intervention locations. The integrated strategy makes identifying reliable biomarkers related to neurodegenerative diseases easier. Researchers can locate chemical markers that consistently characterize various illnesses by identifying trends and correlations between spinning and spraying data (32). These biomarkers show great promise for early detection, illness monitoring, and targeted therapeutics development. The precise and diverse information obtained from these techniques aids in developing a molecular fingerprint that goes beyond typical diagnostic indicators. Spinning and spraying procedures provide nuanced molecular information, allowing for the distinction of various neurodegenerative illnesses (33). Subtypes within more significant illness categories can be identified using unique molecular profiles, allowing for more exact diagnostic categorization. This has implications for adapting treatment techniques to distinct subtypes, recognizing the variability of neurodegenerative illnesses.

Understanding the molecular complexities of spinning and spraying data correlations opens possibilities for discovering medicinal targets. The exact proteins, pathways, and cellular processes involved in neurodegenerative illnesses become apparent, providing a road map for creating tailored therapies (34). This knowledge is critical for developing new medications or repurposing current ones to modify disease-associated cellular pathways. The integrated approach enables dynamic disease progression tracking and therapy efficacy assessments. Changes in molecular markers detected by spinning and spraying procedures can be sensitive indicators of illness progression or responsiveness to therapeutic interventions (35). This helps establish personalized treatment programs and allows for timely therapy adjustment.

To summarize, incorporating spinning and spraying techniques into neurodegenerative research has far-reaching ramifications, substantially altering how we approach the study and treatment of these complicated conditions (36–38). By giving a multidimensional molecular perspective, these approaches pave the way for more detailed knowledge of neurodegenerative processes and actionable insights that can revolutionize future diagnostics and therapeutics.

13.6 CHALLENGES IN APPLICATION

While electron spinning, namely ESR or EPR, is an effective spectroscopic technique, it is not without technical and methodological hurdles (1). Addressing and overcoming these issues is critical for generating accurate and consistent results in various scientific applications. Some major problems include such as achieving adequate sensitivity in ESR can be difficult, especially when working with samples containing low amounts of paramagnetic molecules. The signal-to-noise ratio may become a limiting issue, making it difficult to detect weak signals or minor differences in the sample. Experimental parameters such as modulation amplitude and microwave power must be carefully optimized to improve signal sensitivity while reducing background noise.

To get representative ESR spectra, the sample must be homogeneously distributed. Inhomogeneity, differences in sample composition, or the presence of contaminants can create artifacts and make data interpretation difficult (39). Furthermore, sample size might affect signal strength; larger samples may have broader lines, making it difficult to resolve small spectral details. Special attention must be taken to guarantee sample homogeneity and reduce potential artifacts. And proper calibration is required for quantifying the concentration of paramagnetic species in the sample. Creating a valid calibration curve is difficult due to differences in sample composition, experimental circumstances, and equipment factors. Accurate quantification requires rigorous calibration techniques as well as the use of standard reference materials. High-quality ESR spectra require a uniform magnetic field across the sample volume. Inhomogeneities can expand spectral lines and distort signals (40). Sophisticated magnet designs and meticulous shimming techniques are used to reduce field inhomogeneity. However, these methods may not eradicate the problem. ESR experiments frequently include temperature-dependent observations. Maintaining accurate and steady temperatures throughout tests is difficult yet necessary for producing repeatable results. Temperature fluctuations can alter the linewidth and structure of ESR signals; hence, it is critical to design effective temperature control systems.

Magnetic contaminants in the sample or experimental apparatus can introduce undesired signals and disrupt the analysis of the intended paramagnetic species. Stringent purification processes and rigorous material characterization are required to reduce the impact of contaminants on ESR measurements. When dealing with biological samples, the presence of multiple paramagnetic species, such as metal ions or free radicals, adds complexity to the analysis. Distinguishing and characterizing individual components in complex biological systems pose additional challenges and may require advanced signal processing and modeling techniques. Addressing these technical and methodological hurdles in electron spinning requires experimental optimization, rigorous sample preparation, and apparatus advances. Continuous attempts to improve procedures and develop novel methodologies are needed to push the boundaries of electron spinning applications and broaden their utility in various scientific domains.

Electron spraying, particularly via ESI, has proven indispensable in mass spectrometry, especially for investigating biological systems relevant to neurodegenerative diseases. However, it comes with limitations that researchers must consider, as detailed in Table 13.1 (41). For instance, in complex biological samples, electron spraying might encounter ion suppression issues, leading to decreased sensitivity and potential oversight of crucial compounds. Additionally, it may struggle to ionize hydrophobic molecules like lipids and large macromolecules, limiting its effectiveness in comprehensive analysis. Furthermore, achieving precise quantitative measurements can be challenging due to variations in ionization efficiency and reproducibility across experiments (42).

In response to these challenges, researchers are exploring various strategies. They focus on meticulous optimization of experimental parameters and the exploration of hybrid techniques combining electron spraying with other ionization methods. High-resolution mass spectrometers and isotopic labeling techniques are being employed to enhance precision and accuracy in measurements (43). Moreover, advancements in sample preparation techniques and chromatographic separation methods are helping to mitigate matrix effects and improve sensitivity. Collaboration among multidisciplinary teams and staying abreast of technological advancements are crucial for advancing electron spraying applications in neurodegenerative research. Through these efforts, researchers aim to gain deeper insights into the complexities of neurodegenerative diseases and advance the field of mass spectrometry analysis (44).

13.7 FUTURE PERSPECTIVES AND TECHNOLOGICAL ADVANCEMENTS

Recent advances in electron spinning technologies are transforming materials science and biological studies. One significant trend is the advancement of instrumentation and approaches for studying nano-scale structures (65). Modern ESR spectrometers incorporate more significant magnetic fields, enhanced resonator designs, and advanced pulse sequences, resulting in unparalleled sensitivity and resolution. Furthermore, advances in spin labeling techniques, which use innovative paramagnetic probes, enable more detailed examinations of local surroundings within biological systems. Combining electron spinning with other imaging modalities, such as atomic force and super-resolution microscopy, is also gaining popularity, providing a synergistic approach to unraveling complicated nano-scale structures. These advances make electron spinning a versatile tool for researching varied materials, from nano-particles to biological macromolecules, enabling new paths for scientific exploration (66).

Electron spraying methods, particularly ESI, constantly evolve in response to the desire for increased sensitivity, selectivity, and versatility in mass spectrometry. A notable trend is the development of ambient ionization techniques, which allow for the direct analysis of samples in their natural condition without considerable preparation. Ambient ionization technologies like desorption electrospray ionization (DESI) and laser ablation electrospray ionization (LAESI) allow for real-time surface and tissue investigation. Another rising trend is using machine learning (ML) and artificial intelligence (AI) in data processing, resulting in more efficient and precise identification of complicated chemical patterns (67). Furthermore, advances in ESI source design, such as structures for increased ion transmission and innovative emitter materials, help to improve ionization efficiency and resilience. These trends highlight the ongoing efforts to push the boundaries of electron spraying technologies, making them more potent tools in domains ranging from proteomics and metabolomics to drug discovery.

A significant new trend is the confluence of electron spinning and spraying technologies, which combine their complementary qualities to enable thorough molecular research. Integrating ESR

TABLE 13.1
Challenges and Strategies in Electron Spinning and Spraying Techniques

S. No	Challenges	Electron Spinning	Overcoming Strategies for Electron Spinning	Electron Spraying	Overcoming Strategies for Electron Spraying	References
1	Sensitivity and signal-to-noise ratio	High	Advanced pulse sequences, signal averaging	Moderate to high	Improved source design, enhanced ion transmission	(45, 46)
2	Sample homogeneity and size	Critical	Homogeneous sample preparation, advanced resonator designs	Moderate	Optimization of sample preparation techniques	(47–49)
3	Quantification and calibration	Complex	Standardized calibration protocols, internal standards	Challenging	Use of stable isotope labeling, external calibration	(50)
4	Magnetic field in homogeneity	Significant	Advanced field homogenization techniques	Minimal	Not applicable	(51, 52)
5	Temperature control	Crucial	Precise temperature regulations, cryogenic systems	Important	Maintain stable operating temperatures	(53, 54)
6	Interference from paramagnetic impurities	Prone	Purification techniques, paramagnetic-free solvents	Moderate	Implementation of ion purification steps	(55, 56)
7	Complexities in biological samples	High	Enhanced sample cleanup, advanced extraction methods	Moderate	Improved sample cleanup, limited chromatographic separations	(50, 57)
8	Ion suppression and matrix effects	Not applicable	Not applicable	Significant	Advanced ionization source designs, chromatographic separations	(58, 59)
9	Limited coverage of hydrophobic molecules	Not applicable	Not applicable	High	Optimization of solvent systems for improved coverage	(60, 61)
10	Difficulty in ionizing high-molecular-weight species	Not applicable	Not applicable	Moderate to high	Advanced ionization technique, extensive fragmentation	(47, 62)
11	Limited detection sensitivity for low-abundance species	Not applicable	Not applicable	High	Improved sensitivity through advanced source designs	(50, 63)
12	Challenges in quantitative analysis	Not applicable	Not applicable	Moderate to high	Utilization of Is	(47, 64)

Note: This table provides an overview of challenges associated with electron spinning and spraying techniques and suggests potential strategies to overcome these challenges for each technique and ESI in a hybrid method allows researchers to link paramagnetic information received through spinning with mass spectrometric data obtained via spraying. This collaborative method provides a more comprehensive understanding of molecular structures, dynamics, and interactions, especially in complex biological systems (68). Combining spin labeling techniques and high-resolution mass spectrometry improves biomolecule identification and quantification, providing a more comprehensive understanding of molecular processes. Electron spinning and spraying can open up new dimensions in studying materials and biological systems as these technologies evolve, fostering novel solutions and discoveries across scientific disciplines. The merging of modern electron spinning and spraying technologies can potentially progress neurodegenerative research dramatically. These novel approaches provide a multifaceted approach to examining the molecular complexities involved with neurodegenerative illnesses, revealing findings that potentially redefine our understanding of these complicated conditions. The combination of electron spinning and spraying technologies enables unprecedented precision in molecular characterization of neurodegenerative processes. Electron spinning, which can examine the spin dynamics of paramagnetic species, gives precise information about the structural and environmental properties of molecules involved in neurodegenerative pathways (18). Concurrently, electron spraying in mass spectrometry allows for identifying and quantifying a diverse range of biomolecules, including proteins, lipids, and metabolites. The combined use of these approaches results in a complete and detailed molecular profile, offering information on the complex interplay of many components involved in neurodegenerative illnesses. The combined strategy of electron spinning and spraying aids in the identification of new biomarkers associated with neurodegenerative diseases. By integrating spin resonance data insights with mass spectrometry's extensive molecular information, researchers can identify unique molecular signatures that indicate disease development or specific clinical occurrences. These revealed biomarkers improve diagnostic capacities and serve as prospective targets for creating precise and targeted treatment interventions, representing a substantial advancement in personalized medicine methods for neurodegenerative illnesses. Neurodegenerative illnesses frequently entail complex protein interactions and aggregations. Electron spinning's capacity to clarify paramagnetic centers inside proteins and electron spraying's mass spectrometry capabilities provide a potent tool for unraveling the complexity of protein dynamics in neurodegenerative diseases. Understanding conformational changes, post-translational modifications, and interactions within protein networks offers crucial insights into the mechanisms that drive illness start and progression. This understanding paves the way for therapies targeting specific protein pathways, potentially slowing or stopping neurodegenerative processes. Lipid dysregulation is a hallmark of many neurodegenerative illnesses, and the combined use of electron spinning and spraying technologies allows for a thorough investigation of lipid changes. Electron spinning offers vital information regarding the fluidity and organization of lipid membranes, whereas electron spraying allows for lipidomics analysis, which reveals changes in lipid composition and distribution. This integrated approach advances our understanding of how lipid metabolism affects brain health and may identify new lipid-based treatment targets for neurodegenerative disorders.

To summarize, the potential impact of sophisticated electron spinning and spraying technologies on neurodegenerative research is transformational. These methodologies provide a solid foundation for expanding our understanding of neurodegenerative illnesses through precision, biomarker development, protein interaction unraveling, and extensive lipid research. As these technologies advance, their integrative application has the potential to speed discoveries, paving the path for more targeted and successful therapy options in the complex field of neurodegenerative research.

The future of neurodegenerative research depends on the seamless integration of modern electron spinning and spraying technologies with in vivo imaging and spatialomics methods. Researchers can visualize and map molecular alterations in specific brain regions by combining molecular insights from spinning and streaming techniques with spatial context from imaging methodologies such as positron emission tomography (PET) or magnetic resonance imaging (MRI). This comprehensive method will improve our understanding of the spatial distribution of biomolecules, cellular connections, and pathological alterations, allowing for a more thorough investigation of neurodegenerative illnesses in their natural settings. The advancement of electron spinning and spraying technologies is anticipated to usher in a new age of single-cell analysis in neurodegenerative research. By improving these approaches' spatial resolution and sensitivity, researchers can get insight into the variety of cellular populations in the brain. Single-cell analysis will reveal the molecular markers of individual neurons, glial cells, and other cell types, providing new insight into the cellular diversity that underpins neurodegenerative diseases. This level of granularity is critical for understanding cell-specific responses, detecting early pathogenic alterations, and customizing therapies for specific cell populations. As the volume and complexity of data created by electron spinning and spraying technologies grow, quantum computing will improve data processing. Quantum computing's ability to manage large amounts of information and conduct complex calculations at unprecedented speeds is ideal for decoding delicate chemical interactions. This development will allow researchers to unravel intricate networks, find subtle patterns, and extract significant correlations from the vast information generated by these advanced analytical approaches, boosting the rate of discovery in neurodegenerative research. Integrating AI and ML algorithms will be critical in predictive modeling for neurodegenerative illnesses. AI-driven analysis of electron spinning and spraying data will allow for identifying predictive biomarkers, constructing prognostic models, and investigating new treatment targets. These prediction models will aid in early diagnosis, prognosis assessment, and the development of personalized treatment methods tailored to patients' unique genetic profiles, ushering in a shift toward precision medicine in neurodegenerative research. Future directions in neurodegenerative research will include a deliberate effort to foster global collaborations, share data, and standardize procedures. Establishing consistent data collection, processing, and reporting standards will make it easier to compare and integrate findings from different research projects. This collaborative strategy will use researchers' global experience, promoting a more cohesive and synergistic effort to address the numerous difficulties of neurodegenerative illnesses. Shared datasets and standardized techniques will spark breakthroughs, foster innovation, and speed up the translation of research findings into practical therapeutic applications seen in Table 13.2.

When considering these future directions, integrating cutting-edge technologies, collaborative initiatives, and computational advancements promises to transform our understanding of neurodegenerative disorders, paving the way for more effective diagnostic tools and tailored therapeutic interventions.

13.8 CONCLUSION

Investigating electron spinning and spraying technologies in neurodegenerative research has provided significant findings that promise to transform our understanding of these complex conditions. Electron spinning, particularly ESR, has emerged as an effective method for studying the nanoscale structural and dynamic properties of biomolecules involved in neurodegenerative diseases. This method provides significant insights into paramagnetic centers, providing a unique perspective on the molecular complexities that underpin disorders like Alzheimer's and Parkinson's disease. In addition to electron spinning, ESI has played an important role in mass spectrometry-based studies, allowing the identification and quantification of a wide range of biomolecules. Combining these new approaches has enabled a more holistic approach to neurodegenerative research, offering a whole molecular profile that includes proteins, lipids, and metabolites. This multimodal approach is critical in understanding the complex interplay of molecular components involved in the genesis and progression of neurodegenerative diseases.

The combination of electron spinning and spraying technologies has signaled a new age of accuracy and depth in neurodegenerative research. The ability to connect paramagnetic information received by turning with mass spectrometric data obtained through spraying has

TABLE 13.2
Future Aspects and Trends in Electron Spinning and Spraying Techniques

S. No	Future Aspects and Trends	Electron Spinning	Electron Spraying	References
1	Integration with in vivo imaging	Enhancing the spatial context	Real-time imaging and molecular mapping	(66, 69)
2	Advancements in signal cell analysis	Single-cell structure dynamics	Single-cell proteomics and metabolomics	(67, 70)
3	Quantum computing for data analysis	Handling large datasets	Advanced data analysis and pattern recognition	(71–73)
4	Application of artificial intelligence	Predictive modeling for biomarkers	Machine learning in mass spectrometry	(74, 75)
5	Cross-platform integration	Synergy with neuroimaging and genomics	Multimodal approaches for comprehensive insights	(18, 75)
6	Global collaborations and standardizations	Shared protocols and collaborative research initiatives	Common standards for mass spectrometry	(53)
7	User-friendly platforms	Simplifying data acquisition and analysis	Accessibility and intuitive interfaces	(18, 65)
8	Ethical considerations and patient engagement	Ensuring responsible and transparent practices	Informed consent and ethical data use	(76, 77)
9	Longitudinal studies for disease progression	Unraveling dynamic molecular changes	Tracking biomarker evolution over time	(78)
10	Development of targeted therapies	Identifying the novel therapeutic targets	Tailoring treatments to molecular profiles	(65, 79)
11	Applications of clinical trails	Enhanced evaluation of treatment efficacy	Enriching trails with molecular data	(66, 80)
12	Advancements in biomarkers research	Uncovering prognostic indicators	Exploring novel biomarkers for diagnosis	(67, 80)
13	Integration with wearable sensors	Monitoring molecular changes in real-time	Continues biomarkers tracking in patients	(81, 82)
14	Educational initiatives for research training	Training the next generation of researchers	Building a skilled workforce in mass spectrometry	(65)
15	Exploration of neuroinflammatory processes	Unveiling the molecular dynamics of inflammation	Linking inflammation to molecular signatures	(80)
16	Data-driven insights and computational approaches	Leveraging quantum computing for complex analysis	Implanting AI for enhanced data interpretation	(65, 74)
17	Inclusion of diverse patient populations	Addressing heterogeneity in molecular signatures	Extending applications to underrepresented demographics	(18, 50)

Note: This table provides an overview of future aspects and trends in applying electron spinning and electron spraying techniques, including their integration with emerging technologies and considerations for advancing research and clinical applications.

allowed researchers to unravel complicated protein relationships, uncover novel biomarkers, and investigate lipid modifications with unprecedented precision. This integrative method has the potential to significantly improve diagnostic skills, allowing for targeted therapeutic approaches based on individual patients' biological characteristics – further developments in electron spinning and spraying technologies point to increased sophistication and application in neurodegenerative research. The seamless integration of in vivo imaging, advances in single-cell analysis, quantum computing for data processing, and AI in predictive modeling are all poised to transform the area. Furthermore, worldwide collaborations and data-sharing programs demonstrate a shared commitment to standardization and knowledge development, setting the framework for revolutionary breakthroughs and translating research findings into concrete clinical advances.

The journey into the molecular landscape of neurodegenerative disorders via electron spinning and spraying technologies has not only provided profound insights into the fundamental mechanisms at work but has also laid the groundwork for a future marked by precision, collaboration, and innovation in the pursuit of effective diagnostic and therapeutic strategies for these challenging conditions.

The merging of modern electron spinning and spraying technologies has essential implications for neurodegenerative research, representing a paradigm shift in our approach to understanding and treating these complicated conditions. One of the most significant ramifications is the potential for precise diagnosis. The accurate molecular profiling provided by electron spinning and spraying technologies enables the detection of specific biomarkers linked with neurodegenerative diseases. This precision in diagnostics allows for early diagnosis and intervention, allowing healthcare providers to begin therapeutic efforts at the earliest phases of illness progression when therapies are most effective. These new technologies provide molecular insights that open the door for personalized therapy approaches in neurodegenerative illnesses. Clinicians can adjust therapy techniques to patients' neurodegenerative conditions by identifying their specific molecular markers. This shift to customized medicine marks a substantial change from traditional one-size-fits-all approaches, promising more effective and bearable therapies.

An in-depth investigation of protein interactions, lipid changes, and other molecular characteristics associated with neurodegenerative illnesses opens the door to identifying new therapeutic targets. Electron spinning and spraying technologies enable a thorough grasp of the complex molecular pathways involved in disease progression. This knowledge is crucial for creating targeted therapies that modulate specific molecular components, opening up new drug discovery and development opportunities. Combining electron spinning and spraying technologies considerably increases biomarker studies in neurodegenerative diseases. Researchers may now delve into the complexities of molecular changes, uncovering diagnostic biomarkers and prognostic signs that provide insight into disease development. The revised biomarker landscape helps to construct reliable and effective tools for tracking disease trajectories and assessing therapy efficacy.

The interdisciplinary character of these modern technologies encourages cross-disciplinary partnerships in neurodegenerative research. Researchers from several fields, including chemistry, biology, and medicine, are brought together to capitalize on the synergies between spinning and spraying techniques. This collaborative mindset accelerates discovery by fostering the interchange of ideas, approaches, and insights, contributing to better knowledge of neurodegenerative illnesses.

The massive datasets produced by electron spinning and spraying technologies need a combination of data-driven insights and computational methods. AI and ML algorithms play critical roles in interpreting complicated patterns in large datasets, providing prediction models, and revealing subtle correlations that traditional analytical methods may miss. This computational revolution improves our ability to extract useful information from abundant molecular data, advancing our understanding of neurodegenerative processes. The ramifications of electron spinning and spraying technologies in neurodegenerative research go beyond the lab and influence the future landscape of diagnostics and therapies. These technologies usher in a new era of precision medicine, in which individualized therapies, novel therapeutic targets, and collaborative research projects combine to meet the complex problems of neurodegenerative illnesses.

The transformational potential of electron spinning and spraying technologies in neurodegenerative research necessitates an urgent call to action for additional investigation and discovery. As we stand on the verge of a new era in comprehending these complex illnesses, three critical routes beckon the scientific community. Launch large-scale validation experiments to carefully test the dependability and reproducibility of results achieved using electron spinning and spraying technologies. Standardized protocols and coordinated efforts can assist in establishing the reliability of these techniques in various study contexts and populations. Conduct longitudinal research to understand the dynamic changes in molecular profiles during neurodegenerative diseases. Tracking the evolution of biomarkers and molecular changes over time can provide significant insights into disease trajectories, allowing for the identification of early diagnostic signals and critical windows for intervention.

Integrate electron spinning and spraying technology into clinical trial designs for neurodegenerative treatments. Incorporating these new approaches into clinical trials can provide molecular data, allowing for a more detailed evaluation of therapy efficacy and an understanding of the molecular mechanisms driving therapeutic responses or failures. Invest in creating user-friendly platforms and analytical tools to make electron spinning and spraying technologies more accessible to a broader range of scientists and physicians. Streamlining data collecting and analysis processes will make these techniques more accessible, encouraging widespread adoption and collaboration. Investigate cross-platform integration with other cutting-edge technologies, such as neuroimaging, genomics, and wearable sensors. A comprehensive strategy that integrates data from numerous sources can provide a more complete picture of neurodegenerative illnesses, allowing for the generation of multidimensional profiles for better diagnosis and therapy monitoring. Direct your attention to the involvement of neuroinflammatory processes in neurodegenerative diseases using electron spinning and spraying technologies. Unraveling the molecular dynamics of neuroinflammation may reveal new therapeutic targets and ways for modifying the immune response in neurodegenerative diseases. Increase the use of these technologies in diverse patient populations, especially those with underrepresented demographics and unusual neurodegenerative illnesses. Investigating molecular fingerprints in different populations can provide a more comprehensive knowledge of the variety in these complicated settings. Launch educational programs to train the next generation of researchers, doctors, and analytical scientists in electron spinning and spraying technology. Building a professional workforce capable of realizing the full potential of these tools is critical for furthering neurodegenerative research and putting scientific discoveries into therapeutic practice.

Address the ethical concerns surrounding using sophisticated molecular technology in neurodegenerative research. Encourage patient involvement programs on ethical procedures, informed consent, and open communication. Ensuring these technologies are used ethically and responsibly is critical for establishing confidence with patients and the wider community.

Collaboration and communal efforts emerge as the key to success in this call to action. By moving forward on these fronts, the scientific community will realize the full promise of electron spinning and spraying technologies, ushering in a new era of understanding, diagnosing, and targeting intervention for neurodegenerative illness. The time has come for a collaborative effort to address the difficulties and exploit the opportunities afforded by these ground-breaking technologies.

ABBREVIATIONS

AI	artificial intelligence
ALS	Amyotrophic lateral sclerosis
APCI	Air pressure chemical ionization
DESI	Desorption electrospray ionization
EPR	Electron paramagnetic resonance
ESI	Electrospray ionization
ESI	Electrospray ionization
ESR	Electron spin resonance
GC	gas chromatography
iTRAQ	isobaric tags for relative and absolute quantification
LAESI	laser ablation electrospray ionization
LC	liquid chromatography
MALDI	Matrix-assisted laser desorption/ionization
ML	Machine learning
MRI	Magnetic resonance imaging
MS	Mass spectrometry
NMR	Nuclear magnetic resonance
PET	Positron emission tomography
ROS	Reactive oxygen species
SILAC	Isotope labeling by amino acids in cell culture
TOF	time-of-flight

REFERENCES

1. Panigrahi A, Patro SK. Review on neuro degenerative disorders–with reference to ethinomedicinal and novel strategies. *Research Journal of Pharmacy and Life Sciences: Volume*. 2023 Jan;4(1):34–57.
2. Dos Santos DM, Correa DS, Medeiros ES, Oliveira JE, Mattoso LH. Advances in functional polymer nanofibers: From spinning fabrication techniques to recent biomedical applications. *ACS Applied Materials & Interfaces*. 2020 Sep 16;12(41):45673–701.
3. Tang L, Wang C, Tian S, Zhang Z, Yu Y, Song D, Zhang Z. Label-free and ultrasensitive detection of butyrylcholinesterase and organophosphorus pesticides by Mn (II)-based ESRspectroscopy with a zero background signal. *Analytical Chemistry*. 2022 Nov 4;94(46):16189–95.
4. Badarinath AV, Rao KM, Chetty CM, Ramkanth ST, Rajan TV, Gnanaprakash K. A review on in-vitro antioxidant methods: Comparisions, correlations and considerations. *International Journal of PharmTech Research*. 2010 Jan 1;2(2):1276–85.
5. Banik D, Bhattacharjee RR. Nanotechnology-based point-of-care diagnostics and therapeutics for neurological disorders. In *Functionalized Smart Nanomaterials for Point-of-Care Testing* 2023 Oct 24 (pp. 211–233). Singapore: Springer Nature Singapore.
6. Pihlstrøm L, Wiethoff S, Houlden H. Genetics of neurodegenerative diseases: An overview. *Handbook of Clinical Neurology*. 2018 Jan 1;145:309–23.
7. Musgrove RE, Jewell SA, Di Monte DA. Overview of neurodegenerative disorders and susceptibility factors in neurodegenerative processes. In *Environmental Factors in Neurodevelopmental and Neurodegenerative Disorders* 2015 Jan 1 (pp. 197–210). Academic Press.
8. Hofmann-Apitius M, Ball G, Gebel S, Bagewadi S, De Bono B, Schneider R, Page M, Kodamullil AT, Younesi E, Ebeling C, Tegner J. Bioinformatics mining and modelling methods for the identification of disease mechanisms in neurodegenerative disorders. *International Journal of Molecular Sciences*. 2015 Dec 7;16(12):29179–206.
9. Kirkitadze MD, Bitan G, Teplow DB. Paradigm shifts in Alzheimer's disease and other neurodegenerative disorders: The emerging role of oligomeric assemblies. *Journal of Neuroscience Research*. 2002 Sep 1;69(5):567–77.
10. Hachinski V, Iadecola C, Petersen RC, Breteler MM, Nyenhuis DL, Black SE, Powers WJ, DeCarli C, Merino JG, Kalaria RN, Vinters HV. National Institute of Neurological Disorders and

Stroke–Canadian stroke network vascular cognitive impairment harmonization standards. *Stroke*. 2006 Sep 1;37(9):2220–41.
11. Pratt J, De Vito G, Segurado R, Pessanha L, Dolan J, Narici M, Boreham C. Plasma neurofilament light levels associate with muscle mass and strength in middle-aged and older adults: Findings from GenoFit. *Journal of Cachexia, Sarcopenia and Muscle*. 2022 Jun;13(3):1811–20.
12. Zlokovic BV, Gottesman RF, Bernstein KE, Seshadri S, McKee A, Snyder H, Greenberg SM, Yaffe K, Schaffer CB, Yuan C, Hughes TM. Vascular contributions to cognitive impairment and dementia (VCID): A report from the 2018 National Heart, Lung, and Blood Institute and National Institute of Neurological Disorders and Stroke Workshop. *Alzheimer's & Dementia*. 2020 Dec;16(12):1714–33.
13. Heimann RB. *Plasma-Spray Coating: Principles and Applications*. John Wiley & Sons; 2008 Jul 11.
14. Lukáš D, Sarkar A, Martinova L, Vodsed'álková K, Lubasová D, Chaloupek J, Pokorný P, Mikeš P, Chvojka J, Komárek M. Physical principles of electrospinning (electrospinning as a nano-scale technology of the twenty-first century). *Textile Progress*. 2009 May 29;41(2):59–140.
15. Ratheesh G, Tian L, Venugopal JR, Ezhilarasu H, Sadiq A, Fan TP, Ramakrishna S. Role of medicinal plants in neurodegenerative diseases. *Biomanufacturing Reviews*. 2017 Dec;2:1–6.
16. Braghirolli DI, Steffens D, Pranke P. Electrospinning for regenerative medicine: A review of the main topics. *Drug Discovery Today*. 2014 Jun 1;19(6):743–53.
17. Liu H, Zhou Y, Chen S, Bu M, Xin J, Li S. Current sustained delivery strategies for the design of local neurotrophic factors in treatment of neurological disorders. *Asian Journal of Pharmaceutical Sciences*. 2013 Oct 1;8(5):269–77.
18. Yan B, Zhang Y, Li Z, Zhou P, Mao Y. Electrospun nanofibrous membrane for biomedical application. *SN Applied Sciences*. 2022 Jun;4(6):172.
19. Floyd RA, Neto HC, Zimmerman GA, Hensley K, Towner RA. Nitrone-based therapeutics for neurodegenerative diseases: Their use alone or in combination with lanthionines. *Free Radical Biology and Medicine*. 2013 Sep 1;62:145–56.
20. Zheng H, Gal S, Weiner LM, Bar-Am O, Warshawsky A, Fridkin M, Youdim MB. Novel multifunctional neuroprotective iron chelator-monoamine oxidase inhibitor drugs for neurodegenerative diseases: In vitro studies on antioxidant activity, prevention of lipid peroxide formation and monoamine oxidase inhibition. *Journal of Neurochemistry*. 2005 Oct;95(1):68–78.
21. Franklin R, Hare M, Beckman JS. Determining copper and zinc content in superoxide dismutase using electron capture dissociation under native spray conditions. *Methods in Molecular Biology*. 2022 Jan 1;2500:201–10.
22. Strnad Š, Pražienková V, Sýkora D, Cvačka J, Maletínská L, Popelová A, Vrkoslav V. The use of 1, 5-diaminonaphthalene for matrix-assisted laser desorption/ionization mass spectrometry imaging of brain in neurodegenerative disorders. *Talanta*. 2019 Aug 15;201:364–72.
23. Li B, Zhang G, Tahirbegi IB, Morten MJ, Tan H. Monitoring amyloid-β 42 conformational change using a spray-printed graphene electrode. *Electrochemistry Communications*. 2021 Feb 1;123:106927.
24. Nidheesh T, Salim C, Rajini PS, Suresh PV. Antioxidant and neuroprotective potential of chitooligomers in Caenorhabditis elegans exposed to monocrotophos. *Carbohydrate Polymers*. 2016 Jan 1;135:138–44.
25. Lakkappa N, Krishnamurthy PT, Yamjala K, Hwang SH, Hammock BD, Babu B. Evaluation of antiparkinson activity of PTUPB by measuring dopamine and its metabolites in *Drosophila melanogaster*: LC–MS/MS method development. *Journal of Pharmaceutical and Biomedical Analysis*. 2018 Feb 5;149:457–64.
26. Ravid R. Biobanks for biomarkers in neurological disorders: The Da Vinci bridge for optimal clinicopathological connection. *Journal of the Neurological Sciences*. 2009 Aug 15;283(1–2):119–26.
27. Bjorkli C, Sandvig A, Sandvig I. Bridging the gap between fluid biomarkers for Alzheimer's disease, model systems, and patients. *Frontiers in Aging Neuroscience*. 2020 Sep 2;12:272.
28. Pajer K, Andrus BM, Gardner W, Lourie A, Strange B, Campo J, Bridge J, Blizinsky K, Dennis K, Vedell P, Churchill GA. Discovery of blood transcriptomic markers for depression in animal models and pilot validation in subjects with early-onset major depression. *Translational Psychiatry*. 2012 Apr;2(4):e101.
29. Koníčková D, Menšíková K, Tučková L, Hényková E, Strnad M, Friedecký D, Stejskal D, Matěj R, Kaňovský P. Biomarkers of neurodegenerative diseases: Biology, taxonomy, clinical relevance, and current research status. *Biomedicines*. 2022 Jul 21;10(7):1760.
30. Nilsson J, Cousins KA, Gobom J, Portelius E, Chen-Plotkin A, Shaw LM, Grossman M, Irwin DJ, Trojanowski JQ, Zetterberg H, Blennow K. Cerebrospinal fluid biomarker panel of synaptic dysfunction in Alzheimer's disease and other neurodegenerative disorders. *Alzheimer's & Dementia*. 2023 May;19(5):1775–84.

31. Battaglia RA, Beltran AS, Delic S, Dumitru R, Robinson JA, Kabiraj P, Herring LE, Madden VJ, Ravinder N, Willems E, Newman RA. Site-specific phosphorylation and caspase cleavage of GFAP are new markers of Alexander disease severity. *Elife*. 2019 Nov 4;8:e47789.
32. Graham SF, Chevallier OP, Roberts D, Holscher C, Elliott CT, Green BD. Investigation of the human brain metabolome to identify potential markers for early diagnosis and therapeutic targets of Alzheimer's disease. *Analytical Chemistry*. 2013 Feb 5;85(3):1803–11.
33. Mitra S, Gera R, Siddiqui WA, Khandelwal S. Tributyltin induces oxidative damage, inflammation and apoptosis via disturbance in blood–brain barrier and metal homeostasis in cerebral cortex of rat brain: An in vivo and in vitro study. *Toxicology*. 2013 Aug 9;310:39–52.
34. Spickett CM, Wiswedel I, Siems W, Zarkovic K, Zarkovic N. Advances in methods for the determination of biologically relevant lipid peroxidation products. *Free Radical Research*. 2010 Oct 1;44(10):1172–202.
35. Marmor-Kollet H, Siany A, Kedersha N, Knafo N, Rivkin N, Danino YM, Moens TG, Olender T, Sheban D, Cohen N, Dadosh T. Spatiotemporal proteomic analysis of stress granule disassembly using APEX reveals regulation by SUMOylation and links to ALS pathogenesis. *Molecular Cell*. 2020 Dec 3;80(5):876–91.
36. Ahmed RM, Devenney EM, Irish M, Ittner A, Naismith S, Ittner LM, Rohrer JD, Halliday GM, Eisen A, Hodges JR, Kiernan MC. Neuronal network disintegration: Common pathways linking neurodegenerative diseases. *Journal of Neurology, Neurosurgery & Psychiatry*. 2016 May 11; 87(11):1234–41.
37. Eidelberg D. Metabolic brain networks in neurodegenerative disorders: A functional imaging approach. *Trends in Neurosciences*. 2009 Oct 1;32(10):548–57.
38. Seeley WW, Crawford RK, Zhou J, Miller BL, Greicius MD. Neurodegenerative diseases target large-scale human brain networks. *Neuron*. 2009 Apr 16;62(1):42–52.
39. Ayala-Fuentes JC, Chavez-Santoscoy RA. Nanotechnology as a key to enhance the benefits and improve the bioavailability of flavonoids in the food industry. *Foods*. 2021 Nov 5;10(11):2701.
40. Ghorbani M, Aleman RS. Electro-spinning and electro-spraying as innovative approaches in developing of a suitable food vehicle for polyphenols-based functional ingredients. In *Bioactive Compounds-Biosynthesis, Characterization and Applications* 2021 Jul 6. IntechOpen.
41. Tong X, Ga L, Bi LG, Ai J. Wearable electrochemical sensors based on nanomaterials for healthcare applications. *Electroanalysis*. 2023 Feb;35(2):e202200228.
42. Gunnarsson LG, Bodin L. Occupational exposures and neurodegenerative diseases – a systematic literature review and meta-analyses. *International Journal of Environmental Research and Public Health*. 2019 Feb;16(3):337.
43. Sun J, Ludvigsson JF, Ingre C, Piehl F, Wirdefeldt K, Zagai U, Ye W, Fang F. Hospital-treated infections in early-and mid-life and risk of Alzheimer's disease, Parkinson's disease, and amyotrophic lateral sclerosis: A nationwide nested case-control study in Sweden. *PLoS Medicine*. 2022 Sep 15;19(9):e1004092.
44. Santibanez M, Bolumar F, Garcia AM. Occupational risk factors in Alzheimers disease: A review assessing quality of published epidemiological studies. *Occupational and Environmental Medicine*. 2007 May 24;64(11):723–32.
45. Hang Y, Boryczka J, Wu N. Visible-light and near-infrared fluorescence and surface-enhanced Raman scattering point-of-care sensing and bio-imaging: A review. *Chemical Society Reviews*. 2022;51(1):329–75.
46. Luo H, Jie T, Zheng L, Huang C, Chen G, Cui W. Electrospun nanofibers for cancer therapy. *Bio-Nanomedicine for Cancer Therapy*. 2021:163–90.
47. Esfanjani AF, Jafari SM. Biopolymer nano-particles and natural nano-carriers for nano-encapsulation of phenolic compounds. *Colloids and Surfaces B: Biointerfaces*. 2016 Oct 1;146:532–43.
48. Miller A, Chia S, Toprakcioglu Z, Hakala T, Schmid R, Feng Y, Kartanas T, Kamada A, Vendruscolo M, Ruggeri FS, Knowles TP. Enhanced surface nanoanalytics of transient biomolecular processes. *Science Advances*. 2023 Jan 13;9(2):eabq3151.
49. Rostamabadi H, Falsafi SR, Rostamabadi MM, Assadpour E, Jafari SM. Electrospraying as a novel process for the synthesis of particles/nano-particles loaded with poorly water-soluble bioactive molecules. *Advances in Colloid and Interface Science*. 2021 Apr 1;290:102384.
50. Tripathi KM, Kim T, Losic D, Tung TT. Recent advances in engineered graphene and composites for detection of volatile organic compounds (VOCs) and non-invasive diseases diagnosis. *Carbon*. 2016 Dec 1;110:97–129.
51. Thakkar S, Misra M. Electrospun polymeric nanofibers: New horizons in drug delivery. *European Journal of Pharmaceutical Sciences*. 2017 Sep 30;107:148–67.
52. Xu Q, Gao X, Zhao S, Liu YN, Zhang D, Zhou K, Khanbareh H, Chen W, Zhang Y, Bowen C. Construction of bio-piezoelectric platforms: From structures and synthesis to applications. *Advanced Materials*. 2021 Jul;33(27):2008452.

53. Barman J, Tirkey A, Batra S, Paul AA, Panda K, Deka R, Babu PJ. The role of nanotechnology based wearable electronic textiles in biomedical and healthcare applications. *Materials Today Communications*. 2022 Jul 16:104055.
54. Kundu A, Shetti NP, Basu S, Mondal K, Sharma A, Aminabhavi TM. Versatile carbon nanofiber-based sensors. *ACS Applied Bio Materials*. 2022 Aug 30;5(9):4086–102.
55. Yeo LY, Chang HC, Chan PP, Friend JR. Microfluidic devices for bioapplications. *Small*. 2011 Jan 3;7(1):12–48.
56. Garkal A, Bangar P, Rajput A, Pingale P, Dhas N, Sami A, Mathur K, Joshi S, Dhuri S, Parikh D, Mutalik S. Long-acting formulation strategies for protein and peptide delivery in the treatment of PSED. *Journal of Controlled Release*. 2022 Oct 1;350:538–68.
57. Kim SJ, Kim MG, Kim J, Jeon JS, Park J, Yi HG. Bioprinting methods for fabricating in vitro tubular blood vessel models. *Cyborg and Bionic Systems*. 2023 Aug 1;4:0043.
58. McLaughlin N, Bielinski TM, Tressler CM, Barton E, Glunde K, Stumpo KA. Pneumatically sprayed gold nano-particles for mass spectrometry imaging of neurotransmitters. *Journal of the American Society for Mass Spectrometry*. 2020 Aug 25;31(12):2452–61.
59. Simeoni MC, Pellegrini M, Sergi M, Pittia P, Ricci A, Compagnone D. Analysis of polyphenols in the Lamiaceae family by matrix solid-phase dispersion extraction followed by ultra-high-performance liquid chromatography–tandem mass spectrometry determination. *ACS Omega*. 2018 Dec 18;3(12):17610–6.
60. Zare M, Ramakrishna S. Current progress of electrospun nanocarriers for drug delivery applications. In *Proceedings* 2020 Dec 1 (Vol. 4, p. 8790).
61. Mathew AM, Bhuvanendran S, Nair RS, Radhakrishnan AK. Exploring the anti-inflammatory activities, mechanism of action and prospective drug delivery systems of tocotrienol to target neurodegenerative diseases. *F1000Research*. 2023 Mar 27;12(338):338.
62. Cook KD. To the Graduate Council: I am submitting herewith a dissertation written by Maolian Chen entitled "Characterization of Aβ Aggregates Using Hydrogen/Deuterium Exchange Mass Spectrometry." I have examined the final electronic copy of this dissertation for form and content and recommended that it be accepted in partial fulfillment for the degree of.
63. Wei X, Herbst A, Ma D, Aiken J, Li L. A quantitative proteomic approach to prion disease biomarker research: Delving into the glycoproteome. *Journal of Proteome Research*. 2011 Jun 3;10(6):2687–702.
64. Natarajan D, Ye Z, Wang L, Ge L, Pathak JL. Rare earth smart nanomaterials for bone tissue engineering and implantology: Advances, challenges, and prospects. *Bioengineering & Translational Medicine*. 2022 Jan;7(1):e10262.
65. Muthuraman A, Kaur J. Antimicrobial nanostructures for neurodegenerative infections: Present and future perspectives. In *Nanostructures for Antimicrobial Therapy* 2017 Jan 1 (pp. 139–167). Elsevier.
66. Binda A, Murano C, Rivolta I. Innovative therapies and nanomedicine applications for the treatment of Alzheimer's disease: A state-of-the-art (2017–2020). *International Journal of Nanomedicine*. 2020 Aug 14:6113–35.
67. Hussain Z, Thu HE, Amjad MW, Hussain F, Ahmed TA, Khan S. Exploring recent developments to improve antioxidant, anti-inflammatory and antimicrobial efficacy of curcumin: A review of new trends and future perspectives. *Materials Science and Engineering: C*. 2017 Aug 1;77:1316–26.
68. Shangguan W, Li S, Cao L, Wei M, Wang Z, Xu H. Electrospinning and nanofibers: Building drug delivery systems and potential in pesticide delivery. *Materials Today Communications*. 2022 Sep 8:104399.
69. Maiolo L, Polese D, Convertino A. The rise of flexible electronics in neuroscience, from materials selection to in vitro and in vivo applications. *Advances in Physics: X*. 2019 Jan 1;4(1):1664319.
70. Kumar R, Rauti R, Scaini D, Antman-Passig M, Meshulam O, Naveh D, Ballerini L, Shefi O. Graphene-based nanomaterials for neuroengineering: Recent advances and future prospective. *Advanced Functional Materials*. 2021 Nov;31(46):2104887.
71. Adam T, Gopinath SC. Nanosensors: Recent perspectives on attainments and future promise of downstream applications. *Process Biochemistry*. 2022 Jun 1;117:153–73.
72. Chugh V, Basu A, Kaushik NK, Kaushik A, Mishra YK, Basu AK. Smart nanomaterials to support quantum-sensing electronics. *Materials Today Electronics*. 2023 Dec 1;6:100067.
73. Sarma ID, Bhowmick D, Bhilkar P, Sharma R, Chaudhary RG, Sarma ID, Chaudhary RG. Nanotechnology: Fundamental aspects and biomedical and technological applications. *Nanobiomaterials: Perspectives for Medical Applications in the Diagnosis and Treatment of Diseases*. 2023 Jun 5;145:1–8.
74. Sinha R, Shukla P. Protein engineering for improved health: Technological perspectives. *Current Protein and Peptide Science*. 2019 Sep 1;20(9):856–60.
75. Moulahoum H, Ghorbanizamani F, Beduk T, Beduk D, Ozufuklar O, Celik EG, Timur S. Emerging trends in nanomaterial design for the development of point-of-care platforms and practical applications. *Journal of Pharmaceutical and Biomedical Analysis*. 2023 Aug 1:115623.

76. Kiparissides C, Kammona O. Nanotechnology advances in diagnostics, drug delivery, and regenerative medicine. In *The Nano-Micro Interface: Bridging the Micro and Nano Worlds* 2015 Jan 21 (pp. 311–40). Weinheim: Wiley-VCH.
77. Keservani RK, Kesharwani RK, Sharma AK, editors. *Advances in Novel Formulations for Drug Delivery* 2023 Mar 14. John Wiley & Sons.
78. Öhman F, Hassenstab J, Berron D, Schöll M, Papp KV. Current advances in digital cognitive assessment for preclinical Alzheimer's disease. *Alzheimer's & Dementia: Diagnosis, Assessment & Disease Monitoring*. 2021;13(1):e12217.
79. Jampílek J, Kráľová K, Novák P, Novák M. Nanobiotechnology in neurodegenerative diseases. *Nanobiotechnology in Neurodegenerative Diseases*. 2019:65–138.
80. Silva Adaya D, Aguirre-Cruz L, Guevara J, Ortiz-Islas E. Nanobiomaterials' applications in neurodegenerative diseases. *Journal of Biomaterials Applications*. 2017 Feb;31(7):953–84.
81. Shajari S, Kuruvinashetti K, Komeili A, Sundararaj U. The emergence of AI-based wearable sensors for digital health technology: A review. *Sensors*. 2023 Nov 29;23(23):9498.
82. Nag A, Alahi ME, Mukhopadhyay SC. Recent progress in the fabrication of graphene fibers and their composites for applications of monitoring human activities. *Applied Materials Today.* 2021 Mar 1; 22:100953.

14 Neurodegenerative Disease and Its Therapeutic Approaches

Formulation Approach for Overcoming the Barriers during the Treatment of Neurodegenerative Disorder

Koushik Yetukuri,[1] *M.S. Umashankar,*[1]
Sandeep Kanna,[2] *and Rama Rao Nadendla*[2]
[1]Department of Pharmaceutics, SRM Institute of Science and Technology, Kattankulathur, Chennai, Tamil Nadu, India
[2]Department of Pharmaceutics, Chalapathi Institute of Pharmaceutical Sciences, Lam, Guntur, Andhra Pradesh, India

14.1 INTRODUCTION

The central nervous system (CNS) is protected by specific barriers that control molecular and cellular interaction between the brain and systemic circulation. The blood–brain barrier (BBB) is the most restrictive, largely consisting of brain microvascular endothelial cells (BMECs) linked by tight junctions (TJs) and adherens junctions (AJs). This barrier is necessary for preserving CNS homeostasis by restricting the ingress of xenobiotics, poisons, and immune cells, while enabling the transfer of vital nutrients [1].

In pathological circumstances, including neurodegenerative illnesses (e.g., Alzheimer's disease (AD), Parkinson's disease (PD)), stroke, and CNS infections, the integrity of the BBB is impaired, leading to heightened permeability and dysregulated transport pathways. This disturbance enables the entry of neurotoxic chemicals, immune cells, and pro-inflammatory mediators, ultimately worsening neuronal injury and disease development [2].

The growing global incidence of neurodegenerative illnesses, attributed to prolonged life expectancy, presents considerable socio-economic and healthcare challenges. Current pharmaceutical therapies generally provide symptomatic alleviation without significantly altering disease progression. A primary challenge in the development of effective neurotherapeutics is the transport of drugs across the BBB, as most therapeutic agents do not attain sufficient brain penetration [3].

Nanoparticles (NPs) have arisen as a multifaceted and promising platform for drug delivery to address these limitations. NP-based formulations can improve drug stability, extend circulation duration, and enable targeted distribution to specific brain areas. Preclinical investigations have shown that intravenously delivered NPs, especially those functionalized with surfactants or targeting ligands, exhibited improved BBB penetration and increased therapeutic efficacy. A comprehensive

DOI: 10.1201/9781003478652-16

understanding of BBB modifications in pathological conditions is crucial for enhancing NP-based approaches to ensure accurate, effective, and prolonged medication delivery to neurodegenerative disorder (ND) locations [4].

14.2 CHALLENGES IN TREATING NDS

Despite improvements in our knowledge of NDs, there are still a number of obstacles to overcome in order to provide efficient treatments:

BBB Restriction

- Most medications' entrance is restricted by the BBB, which lowers the effectiveness of treatment.
- The BBB restricts the entry of most drugs, hence diminishing therapy efficacy.

Lack of Disease-Modifying Therapies

- Present therapies (e.g., cholinesterase inhibitors, dopamine agonists) merely offer symptomatic alleviation.
- Many potential neuroprotective drugs do not penetrate the BBB at therapeutic quantities.
- Therapies effectively impede or reverse illness advancement.

Complex and Multifactorial Pathophysiology

- Single-target drugs may not address the various underlying causes of neurodegeneration.
- Requirement for multi-targeted and combinatorial drugs.

Late Diagnosis and Irreversible Neuronal Damage

- Symptoms often arise after substantial neuronal damage.
- Preliminary biomarkers for diagnosis remain at the developmental stage.

Challenges in Clinical Trials

- Elevated failure rates in clinical trials (~99% failure rate in the development of AD drugs).
- Challenges in patient recruitment for longitudinal research.

Personalized Medicine and Heterogeneity

- Genetic and environmental variables influence illness variability.
- Requirement for precision drug strategies customized for individual patients.

NDs provide a considerable worldwide health challenge, with rising frequency attributed to aging populations. The intricacy of their pathogenesis, characterized by protein misfolding, oxidative stress, and neuroinflammation, complicates treatment. Contemporary treatments emphasize symptomatic alleviation rather than disease change [5]. Addressing treatment obstacles, specifically the BBB and constraints in neural regeneration, need innovative formulation strategies and sophisticated drug delivery devices.

14.3 BBB: GENERAL CONCEPT AND MECHANISMS OF PASSAGE

The CNS is safeguarded by a variety of physiological barriers that govern molecular and cellular interaction between the brain and systemic circulation. These encompass the BBB, the blood–cerebrospinal fluid barrier, the blood–retinal barrier, and the blood–spinal cord barrier. The BBB is the most comprehensive and stringent, acting as a highly selective barrier that preserves CNS homeostasis and obstructs the ingress of neurotoxic agents, infections, and circulating immune cells [6].

14.3.1 Structural and Functional Composition of the BBB

The BBB consists mostly of brain BMECs, which are linked by TJs and AJs, creating a continuous, impermeable monolayer (Figure 14.1). The non-fenestrated endothelium is augmented by pericytes, astrocytes, and the basement membrane, which collectively enhance barrier integrity, signal transmission, and neurovascular coupling [7].

- **Endothelial Cells**: These specialized cells constitute the principal structural element of the BBB, demonstrating minimal pinocytotic activity and an absence of fenestrations to limit passive diffusion. The expression of TJs and AJs, consisting of proteins including claudins, occludin, junctional adhesion molecules, and zonula occludens proteins, enhances barrier function and restricts paracellular permeability.
- **Pericytes**: Pericytes, located in the vascular basement membrane, regulate capillary blood flow, endothelial proliferation, and vascular remodeling. These contractile cells regulate vascular tone via actin-myosin interactions and play a role in maintaining the BBB.
- **Astrocytes**: Astrocytic end-feet encase almost the entire capillary surface, improving BBB functionality by regulating TJ integrity, neuronal signaling, and blood flow. Astrocytes are essential for synapse maintenance, neuroprotection, and metabolic support.
- **Basement Membrane**: This extracellular matrix structure offers mechanical support and acts as a framework for cellular interactions that affect endothelium stability, pericyte function, and immune cell signaling.

14.3.2 Mechanisms of Molecular Transport across the BBB

Notwithstanding its restrictive characteristics, the BBB allows the selective transit of vital nutrients, ions, and signaling molecules via a combination of paracellular and transcellular transport systems (Figure 14.2). The equilibrium between these channels dictates BBB permeability in both healthy and pathological states [8].

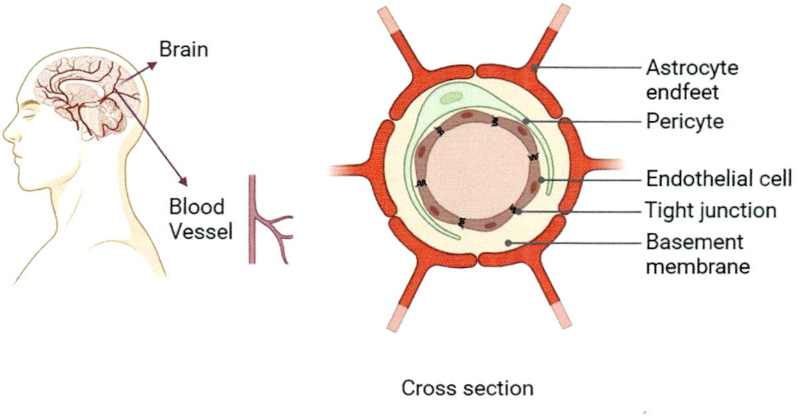

FIGURE 14.1 Blood–brain barrier cross-section.

Paracellular Transport:

- Transport of tiny hydrophilic molecules and ions via interendothelial gaps, governed by the integrity of TJs and electrochemical gradients.
- Generally constrained by the elevated expression of TJ proteins, which impede solute movement.

Transcellular Transport:

- **Passive Diffusion**: Lipophilic substances, like as oxygen and carbon dioxide, diffuse across the endothelium membrane passively, without the necessity of energy expenditure.
- **Carrier-Mediated Transport**: Essential nutrients, like glucose and amino acids, necessitate specialized transporters, such as glucose transporter-1 (GLUT-1) for glucose absorption [9].
- **Receptor-Mediated Transport**: Substantial biomolecules (e.g., insulin, transferrin) employ distinct endothelium receptors to enable transcytosis into the brain parenchyma.
- **Adsorptive-Mediated Endocytosis**: Ionized molecules engage with endothelial membrane constituents to enable transport.
- **Caveolae-Mediated Transcytosis**: Formation of diminutive vesicles that enclose cargo molecules, facilitating their transit across the endothelium.

Efflux Transport Systems:

- As active efflux pumps, ATP-binding cassette (ABC) transporters, such P-glycoprotein (P-gp) and breast cancer resistance protein, return potentially hazardous materials to the bloodstream, including neurotoxic compounds and some drugs [10].
- This method poses a significant difficulty for CNS drug delivery because many therapeutic compounds are efficiently removed before reaching their targeted sites. [11].

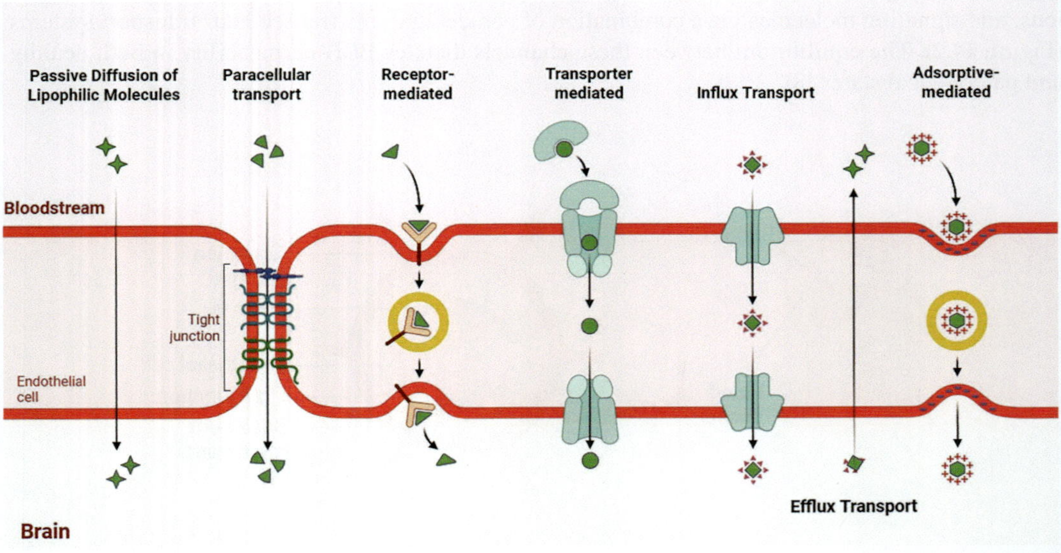

FIGURE 14.2 Molecular transport across the BBB.

14.3.3 Implications for Drug Delivery and Therapeutic Strategies

The BBB's highly selective nature complicates neurotherapeutic administration, especially for hydrophilic drugs and big macromolecules [12]. Several techniques are being investigated to increase BBB permeability and improve CNS drug targeting (Figure 14.3), including:

- **NP-Based Drug Delivery**: Engineered liposomes, polymeric NPs, and exosomes can encapsulate therapeutic substances, facilitating regulated release and improved brain penetration.
- **Receptor-Targeted Strategies**: Utilizing receptor-mediated transcytosis (e.g., targeting transferrin or insulin receptors) to enhance drug delivery.
- **Modulation of Efflux Transporters**: Inhibiting ABC transporters to diminish drug clearance and augment CNS drug retention.
- **BBB Disruption Methods**: Temporarily augmenting BBB permeability using targeted ultrasound, osmotic agents, or pharmacological modulators.

A comprehensive understanding of BBB physiology and transport mechanisms is crucial for devising novel techniques to deliver neuroprotective and regenerative chemicals that would otherwise be unable to traverse this vital barrier [13].

14.4 NANOTECHNOLOGY-BASED FORMULATION STRATEGIES FOR BBB DRUG DELIVERY

Formulation strategies based on nanotechnology are essential for improving drug delivery across the BBB. These methodologies enhance medication stability, bioavailability, and targeted delivery, mitigating the problems associated with conventional therapy [14]. Utilizing nanoscale carriers enhances drug penetration into the brain, facilitates regulated release, and diminishes systemic side effects, providing new approaches for the treatment of neurological illnesses [15]. Several nanocarrier systems have been explored for BBB drug delivery (Figure 14.4), including the following:

- Lipid-based nanoparticles (LNPs)
- Polymer-based nanoparticles (PNPs)
- Nanogels
- Dendrimers

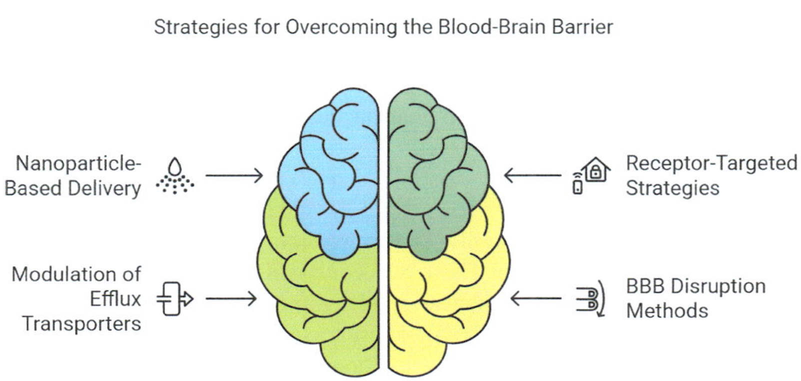

FIGURE 14.3 Strategies for overcoming blood–brain barrier.

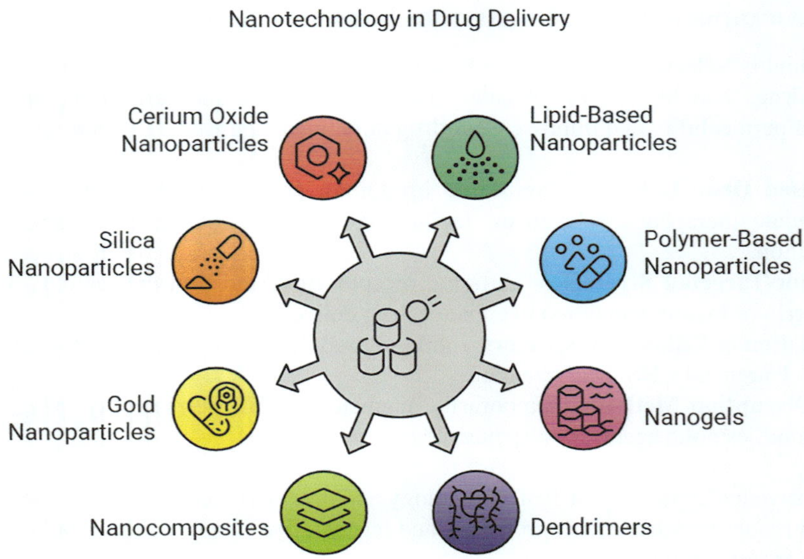

FIGURE 14.4 Nanotechnology-based formulation techniques for BBB.

- Nanocomposites
- Gold nanoparticles (AuNPs)
- Silica nanoparticles (SiNPs)
- Cerium oxide nanoparticles

14.4.1 Lipid-Based NPs

LNPs have garnered considerable interest as a sophisticated drug delivery mechanism, especially for brain targeting. Their biocompatibility, biodegradability, and safety provide them with appropriate carriers for medicinal medicines [16]. Their primary advantage is in their structural resemblance to the BBB, which facilitates their capacity to traverse this protective barrier and efficiently administer medications to the CNS. LNPs consist mostly of phospholipids, fatty acids, monoglycerides, triglycerides, fatty alcohols, and waxes, endowing them with distinctive physicochemical characteristics [17].

14.5 TYPES OF LNPS

There are three major classes of LNPs used in pharmaceutical applications:

- **Liposomes:** Vesicular formations consisting of one or many lipid bilayers encasing an aqueous core. They are extensively researched for medication delivery because of their capacity to encapsulate both hydrophilic and hydrophobic compounds.
- **Solid Lipid Nanoparticles (SLNs):** These NPs consist of solid lipids at ambient and physiological temperatures, providing enhanced stability and regulated drug release.
- **Nanostructured Lipid Carriers (NLCs):** These are altered solid lipid NPs with a largely liquid lipid matrix, which improves drug loading capacity and diminishes crystallinity, therefore inhibiting drug ejection.

These LNPs have been thoroughly investigated for CNS medication delivery in the treatment of neurodegenerative illnesses, including AD, PD, multiple sclerosis (MS), anxiety, and depression [18]. Their adaptability facilitates the effective encapsulation of both hydrophilic and hydrophobic pharmaceuticals, enhancing pharmacokinetics and biodistribution [19].

14.6 CHALLENGES AND LIMITATIONS

Notwithstanding their potential, certain problems impede the extensive manufacture and clinical application of LNPs:

- **Low Stability**: LNPs are susceptible to aggregation, resulting in diminished shelf life.
- **Complex Manufacturing Processes**: Extensive production needs advanced processes like high-pressure homogenization or ultrasonication [20].
- **Polymorphic Transitions**: The lipids within the NPs may experience unforeseen structural alterations, impacting medication release characteristics.
- **Drug Leakage During Storage**: Encapsulated pharmaceuticals may gradually disperse, diminishing their efficacy [21].

14.7 FUTURE INSIGHTS

Notwithstanding these constraints, LNPs – especially SLNs and NLCs – persist in exhibiting potential as diagnostic and therapeutic modalities for NDs. Their capacity to improve drug bioavailability, enable regulated drug release, and assist in BBB penetration renders them an invaluable asset in contemporary CNS drug delivery research [22]. Continued progress in surface modification, ligand conjugation, and hybrid NP formulations may augment their effectiveness, facilitating wider utilization in clinical environments.

14.7.1 Polymer-Based NPs

PNPs have emerged as highly effective drug delivery systems owing to their prolonged shelf life, improved biocompatibility, and biodegradability, facilitating regulated, targeted, and sustained drug release. These characteristics are particularly advantageous in the treatment of NDs, where precise drug delivery to specific brain regions is crucial for therapeutic efficacy. PNPs can encapsulate a diverse array of therapeutic agents, including both hydrophobic and hydrophilic drugs, providing substantial benefits over traditional drug delivery systems, especially in surmounting biological barriers such as the BBB [23].

PNPs can be produced from both synthetic and natural polymers, each providing unique benefits for stability, drug encapsulation efficiency, and biodegradability. Synthetic polymers, including polylactic acid, polyglycolic acid, and polyacrylates, are frequently utilized for their meticulous regulation of particle size and drug release kinetics. These polymers are frequently selected for their biocompatibility, long-term stability, and the capacity to customize their physicochemical features, including hydrophilicity/hydrophobicity and breakdown rate. Conversely, natural polymers like chitosan, alginate, and albumin are preferred due to their intrinsic biodegradability, non-toxicity, and bioavailability, rendering them appropriate for physiological applications, including brain-targeted distribution [24].

Poly(lactic-co-glycolic acid) (PLGA) is one of the most thoroughly researched and clinically promising polymeric carriers for NDs. PLGA is a copolymer of lactic acid and glycolic acid, and its distinctive characteristics – namely, biodegradability, biocompatibility, and regulated degradation into non-toxic byproducts (lactic acid and glycolic acid) – render it an optimal choice for drug delivery, especially to the CNS. PLGA NPs demonstrate considerable permeability through the BBB, presenting a substantial challenge in the treatment of NDs such as AD. Owing to their nanoscale dimensions, PLGA NPs can be designed to traverse the BBB, facilitating more effective medication administration directly to the brain. This attribute is essential for the neuroprotective efficacy of PLGA NPs in AD treatment, since these particles may administer both hydrophilic and lipophilic medications with enhanced accuracy and prolonged release profiles, minimizing systemic toxicity and improving therapeutic results in the brain [25].

The architecture of PLGA NPs facilitates the integration of diverse therapeutic agents, including neuroprotective chemicals, anti-amyloid drugs, and anti-inflammatory medications, aimed at addressing the fundamental pathophysiological mechanisms of neurodegenerative illnesses. Moreover, PLGA NPs can be modified with targeting ligands, like peptides, antibodies, or aptamers, which augment their capacity to traverse the BBB and selectively target certain cells or regions in the brain, therefore enhancing medication efficacy and reducing off-target effects. Moreover, PLGA NPs can be customized to regulate the release rate of encapsulated pharmaceuticals, so providing a prolonged therapeutic impact, which is especially advantageous in the management of chronic ailments such as AD [26].

PLGA-based NPs constitute a promising approach to address the challenges of drug delivery to the brain, presenting substantial potential for treating NDs by enhancing drug bioavailability, improving targeted delivery, and ensuring sustained release of therapeutic agents. These NPs are not only suitable for delivering small molecules but also capable of encapsulating larger biomolecules such as genes, proteins, and RNA, thereby broadening their applicability in sophisticated treatments for conditions such as AD, PD, and amyotrophic lateral sclerosis [27].

14.8 CHALLENGES AND LIMITATIONS

PNPs provide significant therapeutic delivery potential but encounter many challenges:

- **BBB Penetration**
  Notwithstanding their diminutive dimensions, PNPs have difficulties in traversing the BBB effectively, particularly with hydrophilic or bulky compounds. Enhancing PNPs with specific targeting ligands can optimize delivery; nonetheless, challenges persist in chronic conditions such as AD.
- **Stability and Drug Release Control**
  PNPs exhibit sensitivity to storage conditions, potentially resulting in aggregation and premature drug release. Ensuring uniform drug release kinetics over prolonged durations is crucial for the therapy of chronic diseases.
- **Toxicity and Biocompatibility**
  Although mostly biocompatible, the long-term safety of PNPs in the brain remains ambiguous. The buildup of polymeric substances or detrimental breakdown byproducts may induce toxicity or inflammatory reactions.
- **Manufacturing Complexity and Scalability**
  The synthesis of PNPs is intricate and expensive, posing difficulties for large-scale manufacture. Variability in particle size and drug encapsulation between batches can influence consistency and therapeutic results.
- **Particle Size and Polydispersity**
  Consistent particle size is essential for efficient BBB penetration and precise distribution. Polydispersity may result in fluctuations in pharmacokinetics, complicating the attainment of uniform therapeutic outcomes.
- **Immunogenicity and Clearance**
  PNPs may elicit immunological responses or be rapidly eliminated by the immune system, diminishing their bioavailability and constraining their therapeutic efficacy.

- **Regulatory and Clinical Approval**
  PNPs encounter regulatory obstacles due to the comprehensive safety and toxicity assessments necessary for licensing. Translating preclinical discoveries to human applications and adhering to clinical trial criteria pose considerable obstacles.

14.9 FUTURE DIRECTIONS

The prospects for PNPs in the treatment of neurodegenerative illnesses are optimistic, with anticipated breakthroughs in multiple domains. This encompasses enhanced functionalization for BBB penetration and tailored distribution, alongside the creation of intelligent drug delivery systems that release pharmaceuticals in reaction to certain stimuli [28]. Efforts will additionally concentrate on augmenting PNP stability and guaranteeing regulated medication release over time. Enhancements in biocompatibility, scalable manufacturing methods, and the integration of PNPs with other therapies, such as gene therapy, present significant promise for individualized treatment. Continuous research and clinical trials are essential for converting these advances into effective therapies for disorders such as AD and PD.

14.9.1 Nanogels in Drug Delivery for Neurodegenerative Diseases

Nanogels have emerged as a highly promising drug delivery system owing to their distinctive physicochemical features, which render them particularly effective for targeted therapy in NDs. These nanoscale carriers demonstrate improved drug loading capacity, significant stability, core-shell structures, and reactivity to environmental stimuli, all of which boost their effectiveness in drug delivery [29]. Their capacity to traverse biological barriers, particularly the BBB, distinguishes them from traditional drug delivery systems, providing a non-invasive method for addressing CNS problems.

14.9.1.1 Unique Structural and Functional Properties of Nanogels

Nanogels generally consist of hydrophilic polymer network particles capable of encapsulating and delivering diverse therapeutic agents, including antisense oligonucleotides and small-molecule drugs. The particles are produced by crosslinking polyethylene glycol with polyethyleneimine, forming a three-dimensional colloidal structure. Owing to their diminutive size (often <200 nm), nanogels can efficiently penetrate physiological barriers, such as the BBB, enabling the precise delivery of pharmaceuticals to the brain [30]. The architecture of nanogels is exceptionally adaptable, permitting various configurations including hollow, hairy, multilayered, and core-shell structures, each offering unique benefits based on the medication and target location.

14.9.1.2 Nanogel Modification for Targeted Drug Delivery

One notable feature of nanogels is their capacity to be designed to react to particular environmental stimuli, including pH, temperature, and ionic strength. This responsiveness facilitates the regulated and precise delivery of therapeutic substances to designated locations inside the body [31]. Moreover, surface changes, such as the incorporation of targeted ligands and magnetic structures, enhance the ability of nanogels to traverse the BBB more effectively. These alterations guarantee that the drug-loaded nanogels can attach to specific receptors on target brain cells, hence improving the accuracy and efficacy of drug delivery [32].

14.9.1.3 Applications in Neuroprotective Drug Delivery

Nanogels are being engineered for the administration of several neuroprotective pharmaceuticals, essential for the management of NDs, including AD, PD, and Huntington's disease. This category of medications encompasses nucleic acids (including DNA- and RNA-based therapies), peptides, nerve growth factors, and antioxidants, all of which are essential for safeguarding neurons from degeneration, mitigating inflammation, and enhancing cell survival [33]. The utilization of nanogels for the delivery of therapeutic drugs guarantees enhanced stability, bioavailability, and controlled release, which are crucial for the efficient treatment of neurodegenerative illnesses.

14.10 CHALLENGES AND LIMITATIONS

Although nanogels exhibit considerable potential in drug delivery, numerous hurdles must be resolved prior to their extensive clinical application. The technique is in its nascent phase, and the formulation for CNS drug delivery necessitates comprehensive preclinical and clinical investigations to address challenges, including scalability, long-term stability, and regulatory approval. Comprehending the pharmacokinetics and toxicity profiles of nanogels in vivo is essential for guaranteeing their safety and efficacy in human therapies [34]. Furthermore, optimizing surface changes and investigating innovative drug combinations will be crucial to augment therapeutic outcomes and enhance the efficacy of nanogels.

14.11 LONG-TERM PERSPECTIVES

The prospective application of nanogels in the treatment of neurodegenerative illnesses seems promising, particularly due to innovations in targeted drug delivery and controlled release systems. Enhancing surface changes and engineering responsive nanogels can yield more accurate and effective treatments. Hybrid systems integrating various drugs may improve outcomes further [35]. Nonetheless, issues such as scalability, stability, and toxicity must be resolved through ongoing research and clinical trials prior to the widespread application of nanogels in therapeutic environments.

14.11.1 Dendrimers in Drug Delivery for Neurodegenerative Diseases

Dendrimers are a category of extremely branched, nanoscale polymers characterized by their exact chemical structure. These precisely delineated structures, distinguished by a core-shell configuration, enable the engineering of dendrimers for particular uses, especially in the targeted delivery of pharmaceuticals and genetic material to the brain. Dendrimers have emerged as useful carriers for the treatment of NDs, including AD, PD, and prion disorders, owing to their diminutive size, customizable surface functional groups, and elevated surface area [36].

14.12 UNIQUE PROPERTIES OF DENDRIMERS

The architecture of dendrimers provides certain benefits for pharmacological administration in NDs:

- **Size and Surface Functionalization**: Dendrimers often measure under 10 nm in diameter, facilitating effective penetration into the brain. Their surface can be readily altered with diverse functional groups, rendering them appropriate for transporting various therapeutic agents, including pharmaceuticals, genes, or proteins [37]. This adaptability improves their capacity to traverse the BBB and localize specific areas within the brain.
- **Core-Shell Architecture**: The core-shell architecture of the dendrimer can encapsulate therapeutic chemicals, safeguarding them from degradation and facilitating controlled release. This configuration facilitates the attachment of several Drugs or targeting ligands to the surface, enhancing the accuracy of drug delivery [38].

14.13 LIMITATIONS AND CHALLENGES

Notwithstanding their promise, dendrimers encounter numerous obstacles in clinical translation:

- **High Production Costs**: The intricate and costly manufacturing of dendrimers impedes their extensive utilization in clinical environments. The production costs must be diminished for dendrimers to become feasible for extensive therapeutic applications.
- **Long-Term Exposure and Toxicity**: The long-term safety of dendrimers in humans is inadequately established. Dendrimers are typically regarded as harmless; nonetheless, their potential toxicity with prolonged usage, particularly in the CNS, need additional research. The accumulation of dendrimers in tissues and their interactions with biological systems require investigation to ascertain their safety during prolonged usage [37].

14.13.1 Dendritic Polyglycerol Derivatives: A Safer Alternative

Dendritic polyglycerol (dPG) derivatives have emerged as a viable alternative to mitigate some limitations inherent in standard dendrimers. dPG dendrimers are recognized for their biocompatibility and reduced toxicity relative to other dendrimer varieties. dPG microspheres have been effectively loaded with medicinal compounds, including dimethylfumarate and curcumin, exhibiting regulated release and minimal in vitro toxicity. These dendrimers exhibit promise in the treatment of MS, providing a safer and more cost-effective alternative for medication delivery [39].

14.14 STRATEGIC DIRECTIONS

Dendrimers exhibit considerable promise as nanocarriers for ND treatment; yet, their clinical application remains nascent. Current research aims to enhance production techniques, save expenses, and guarantee safety through extensive longitudinal investigations. Future innovations in dendrimer technology, particularly the creation of dPG-based dendrimers, may address existing obstacles, providing a safer and more economical approach for the treatment of neurodegenerative illnesses [40]. Nonetheless, extensive biocompatibility assessments and clinical trials are necessary to fully harness the therapeutic potential of dendrimers in clinical applications.

14.14.1 Nanocomposites in Drug Delivery for Neurodegenerative Diseases

Nanocomposites are materials synthesized by amalgamating various components to enhance their qualities. These materials typically consist of a matrix and a reinforcing phase, which may take numerous forms, including tubes, threads, layers, or spheres. Nanocomposites are classified according to their structure: particulate composites (containing particles inside the matrix), laminated composites (comprising thin bonded layers), and fibrous composites (including fibers within a matrix). When polymers serve as the matrix, the composite is referred to as a polymer-based composite [41].

14.14.1.1 Polymer-Based Composites for Biomedical Applications

Polymer-based composites have gained significance in the biological sectors owing to their capacity for customization for particular purposes. These materials are frequently biodegradable and bioresorbable, rendering them appropriate for application in medical devices, tissue engineering, and drug delivery systems. These composites have the advantage of amalgamating the superior qualities of distinct materials, yielding a product that frequently exhibits enhanced attributes relative to the separate components [42]. The interaction between the polymer matrix and reinforcing elements, such as fibers or NPs, is crucial in defining the composite's performance.

14.14.1.2 Use of Nanocomposites for Neurodegenerative Diseases

In NDs such as AD and PD, several therapeutic approaches have been suggested to mitigate cognitive impairments and inhibit the aggregation of detrimental protein deposits. One technique entails utilizing neuroprotective proteins, like as Hsp70 and Hsp27, to degrade these detrimental aggregates. Nonetheless, transporting these proteins to the brain poses a hurdle due to the BBB, which limits their translocation [43]. Systemic protein injections frequently necessitate elevated dosages or several administrations, heightening the likelihood of adverse effects and diminishing patient adherence.

Nanocomposite systems, especially those integrating inorganic clay, have been investigated to enhance the delivery of neuroprotective proteins. An inorganic clay-based nanocomposite has been proposed to augment the brain penetration and efficacy of donepezil, a medication for AD, by facilitating its traversal over the BBB and enhancing its pharmacokinetics within the brain [42].

14.15 CHALLENGES AND LIMITATIONS OF NANOCOMPOSITES IN DRUG DELIVERY

1. **BBB Penetration:** A significant hurdle in utilizing nanocomposites for neurodegenerative illnesses is traversing the BBB. The barrier substantially restricts the quantity of therapeutic drugs that can access the brain, necessitating creative approaches for efficient administration.
2. **Scalability and Production Costs:** The production of nanocomposites, particularly polymer-based and inorganic clay composites, can be expensive and intricate. Increasing output while ensuring consistency and quality poses a substantial challenge for clinical translation.
3. **Long-Term Safety and Toxicity:** Although nanocomposites present intriguing therapeutic potential, their long-term safety, encompassing potential toxicity and immunological responses, requires further exploration prior to extensive clinical application [44].
4. **Regulatory Approval:** The innovative characteristics of nanocomposite-based Drugs may result in regulatory obstacles that hinder their clinical approval. Comprehensive preclinical and clinical investigations are necessary to satisfy safety and efficacy criteria.
5. **Patient Compliance:** The requirement for elevated dosages or many treatments to attain therapeutic levels may diminish patient adherence. Furthermore, the intricacy of nanocomposite systems may result in difficulties in management and oversight.

14.16 PROSPECTIVE DEVELOPMENTS

Nanocomposites possess significant potential for enhancing drug delivery to the brain in the management of NDs. Ongoing investigation into polymer-based and inorganic nanocomposite systems is crucial for optimizing drug administration, improving therapeutic results, and addressing the challenges presented by the BBB. Innovations in nanocomposite design may offer novel pathways for enhanced treatments, minimizing the necessity for elevated dosages and increasing patient outcomes in the management of AD, PD, and other NDs [45].

14.16.1 AuNPs in Drug Delivery for Neurodegenerative Diseases

AuNPs have garnered considerable interest in biomedical research owing to their distinctive physicochemical characteristics, including biocompatibility, facile surface modification, and adjustable size and morphology. These characteristics render AuNPs appropriate for many applications, including drug administration, diagnostic imaging, and therapeutic interventions, especially in the treatment of NDs like AD and PD [46].

14.16.1.1 Properties of AuNPs

AuNPs exhibit distinct features that enhance their efficacy in biomedical applications:

- **Size and Surface Area**: Their diminutive dimensions (often ranging from 1 to 100 nm) and extensive surface area facilitate interactions with diverse biomolecules, rendering AuNPs optimal for drug encapsulation and targeted administration.
- **Biocompatibility**: Gold is a biologically inert substance, minimizing the likelihood of negative immunological reactions in medical applications.
- **Surface Modifiability**: AuNPs can be readily functionalized with diverse biomolecules (e.g., peptides, antibodies, ligands) to selectively target certain cells or tissues, thereby augmenting their therapeutic potential, particularly for targeted drug administration [47].
- **Optical Properties**: Gold NPs display distinctive optical characteristics, including surface plasmon resonance, which can be utilized in diagnostic imaging and photothermal therapy.

14.16.1.2 AuNPs for Drug Delivery across the BBB

A major problem in the treatment of neurodegenerative illnesses is the efficient transport of therapeutic drugs across the BBB. AuNPs have emerged as a viable solution to this problem, as their diminutive size and surface changes enable them to penetrate the BBB and administer medications directly to the brain [48].

14.16.1.3 Surface Functionalization for BBB Penetration

Surface functionalization is essential for improving the capacity of AuNPs to traverse the BBB. Researchers have altered the surface of AuNPs with ligands, peptides, and antibodies that preferentially bind to receptors present on the endothelial cells of the BBB. For example:

- **Transferrin Receptor Targeting**: Transferrin, a protein that enables iron transport across the BBB, has been utilized to functionalize gold NPs, facilitating their passage over the barrier via receptor-mediated transcytosis [49].
- **Cell-Penetrating Peptides (CPPs)**: CPPs can be coupled to AuNPs to enhance direct penetration of the BBB through endocytosis, hence enhancing drug delivery efficiency.

14.16.1.4 Applications in Neurodegenerative Diseases

AuNPs have been examined for their potential in the delivery of therapeutic drugs for the treatment of neurodegenerative illnesses [50]. Some of their applications include:

- **Targeted Drug Delivery**: AuNPs can facilitate the targeted delivery of several therapeutic agents, such as small molecules, peptides, and nucleic acids, directly to the brain. This method is particularly advantageous in conditions such as AD and PD, where traditional drug delivery systems encounter difficulties in traversing the BBB.
- **Gene Therapy**: AuNPs can also be utilized for gene delivery in NDs. Their capacity to encapsulate and safeguard genetic material from degradation facilitates effective transport to brain cells, rendering them intriguing candidates for gene therapy approaches.
- **Thermal and Radiation Therapy**: Gold NPs demonstrate significant photothermal characteristics. Upon exposure to near-infrared light, AuNPs can produce heat, which may be utilized for the thermal ablation of amyloid plaques in AD or the mitigation of neuroinflammation in PD.

14.17 CHALLENGES AND LIMITATIONS OF AUNPS IN NEURODEGENERATIVE DISEASES

Although AuNPs exhibit significant potential, some hurdles persist regarding their application in drug delivery for NDs:

Toxicity and Biocompatibility: Notwithstanding their biocompatibility, prolonged exposure to AuNPs may result in a buildup in multiple organs and possible toxicity. Additional research on their biodegradability and long-term safety is essential.

1. **Scalability**: The mass manufacture of AuNPs with uniform quality and specified attributes continues to be a problem. Furthermore, the expense associated with the production of high-quality AuNPs can be excessively substantial, constraining their clinical utilization.
2. **BBB Penetration Efficiency**: Despite advancements in surface functionalization techniques enhancing the capacity of AuNPs to traverse the BBB, the attainment of efficient, targeted, and regulated drug delivery to precise regions within the brain continues to pose a problem.
3. **Regulatory Approval**: The incorporation of NPs in drug delivery systems necessitates thorough preclinical and clinical assessments to ascertain their safety, effectiveness, and regulatory compliance, potentially postponing their clinical implementation [51].

14.18 FUTURE IMPLICATIONS

AuNPs exhibit significant promise in the treatment of neurodegenerative illnesses, since improvements in surface modification techniques facilitate BBB penetration and specific targeting. To guarantee clinical success, enhanced emphasis on biocompatibility, safety, and production efficiency is required. Furthermore, the integration of AuNPs with alternative treatment modalities, such as gene therapy or photothermal therapy, may augment their efficacy [52]. As research advances, AuNPs may provide novel therapies for AD, PD, and other NDs.

14.18.1 SiNPs in Drug Delivery for Neurodegenerative Diseases

SiNPs have emerged as a very adaptable and promising medium for drug delivery, especially in the treatment of NDs. Their distinctive physicochemical characteristics, such as elevated surface area, biocompatibility, and adjustable surface chemistry, render SiNPs optimal candidates for targeted drug delivery systems capable of addressing significant challenges in the treatment of NDs like AD and PD [53].

14.18.1.1 Structure and Functionalization of SiNPs

SiNPs generally comprise an amorphous silica core that offers a stable and porous framework, adept at encapsulating diverse medicinal ingredients. The core can be additionally functionalized with several chemical groups, including amino, carboxyl, or hydroxyl groups, to augment stability, boost drug loading, and facilitate particular interactions with biological components [54]. This customization enables the engineering of SiNPs to fulfill the precise requirements of a certain treatment, encompassing controlled release and targeted delivery to designated tissues or cells.

14.18.1.2 Advantages in Drug Delivery

A key advantage of SiNPs in drug administration is their elevated surface-to-volume ratio, facilitating substantial drug loading capacity. SiNPs are optimal for administering medications that necessitate delayed or sustained release. The adjustable surface characteristics allow for the modification of SiNPs to enhance solubility, augment stability in biological fluids, and diminish aggregation or premature release of encapsulated pharmaceuticals [55].

14.18.1.3 Overcoming the BBB

The BBB is a significant impediment in the treatment of neurodegenerative illnesses, as it limits the entry of therapeutic drugs into the brain. SiNPs demonstrate significant potential in surmounting this obstacle owing to their diminutive size, biocompatibility, and capacity for surface modification. Researchers have devised multiple strategies to augment the capacity of SiNPs to traverse the BBB, such as coating the particles with targeting ligands (e.g., peptides, antibodies) that promote receptor-mediated endocytosis, or altering the surface with hydrophobic groups to enhance BBB penetration [56].

14.19 THERAPEUTIC APPLICATIONS

SiNPs have been extensively investigated for the administration of many therapeutic drugs employed in NDs.

These include:

- **Antioxidants and Neuroprotective Agents**: SiNPs can transport substances like curcumin, resveratrol, and other antioxidants, recognized for their capacity to mitigate oxidative stress, a significant factor in dementia.
- **Gene Therapy**: SiNPs can be modified to transport nucleic acids, such as siRNA or genes that encode neuroprotective proteins, providing prospective gene therapy approaches for conditions including AD and PD.
- **Amyloid-β and Tau Targeting**: In AD, SiNPs have been utilized to provide Drugs aimed at amyloid plaques or tau tangles, which are associated with disease pathology.
- **Dopamine Delivery**: In PD, silicon nanoparticles (SiNPs) have been investigated for the administration of dopamine or its precursors to reestablish the neurochemical equilibrium in the brain.

14.19.1 Safety and Biocompatibility

SiNPs are typically regarded as biocompatible and non-toxic, rendering them a secure choice for drug administration. Nonetheless, their long-term safety and possible toxicity require additional investigation. The biodegradability of SiNPs is a crucial consideration, as their accumulation in the body over time may result in detrimental effects [57]. Current research examines the in vivo behavior of SiNPs, encompassing their distribution, clearance, and potential immunogenic or cytotoxic reactions.

14.20 CHALLENGES AND LIMITATIONS

Notwithstanding their potential, SiNPs have numerous obstacles and constraints in drug delivery for NDs.

1. **Scalability and Manufacturing**: The mass manufacture of SiNPs with uniform quality and attributes can be difficult. Advanced manufacturing procedures must be adjusted to guarantee scalability while preserving reproducibility.
2. **BBB Penetration Efficiency**: Although SiNPs demonstrate potential in traversing the BBB, their efficacy and capacity to target specific regions of the brain require further enhancement.

3. **Long-Term Safety**: The long-term biocompatibility and biodegradability of SiNPs require comprehensive evaluation, particularly in light of the possibility of chronic exposure during treatment.
4. **Targeting Specificity**: Despite the potential for functionalizing SiNPs with targeting moieties, attaining high specificity for particular brain areas or specific cell types inside the brain is a significant hurdle.

14.21 ANTICIPATED DEVELOPMENTS

SiNPs exhibit potential in drug delivery for neurological disorders. Current research seeks to augment their BBB permeability, refine targeting specificity, and optimize surface changes for improved drug release characteristics. Progress in scalable production methods and comprehensive safety assessments is essential for clinical translation [58]. Ongoing advancements in nanotechnology position SiNPs to transform the treatment of neurodegenerative illnesses through targeted, sustained, and more efficient drug delivery methods, thereby enhancing patient outcomes.

14.21.1 Cerium Oxide Nanoparticles in Drug Delivery for Neurodegenerative Diseases

Cerium oxide nanoparticles (CeONPs), or nanoceria, have garnered considerable interest owing to their distinctive antioxidant characteristics and capacity to neutralize free radicals. The features are mainly due to the reversible oxidation states of cerium (Ce^{3+} and Ce^{4+}), allowing CeONPs to efficiently neutralize reactive oxygen species and reactive nitrogen species (RNS). The radical-scavenging capability of CeONPs renders them interesting candidates for the treatment of several neurological illnesses characterized by oxidative stress as a significant contributing factor [58].

14.22 PROPERTIES AND MECHANISM OF ACTION

CeONPs demonstrate a remarkable ability to alleviate oxidative damage by diminishing free radicals, which are significantly associated with NDs such as AD, PD, and MS. Their diminutive size (often under 5 nm) facilitates their traversal of the BBB, a crucial attribute for administering therapeutic drugs to the brain, a significant obstacle in the treatment of CNS illnesses. Upon entering the brain, CeONPs engage with neural cells, where they efficiently sequester RNS and mitigate their detrimental effects [59].

The capacity of NPs to regulate the oxidative milieu in the brain renders them significant in the treatment of NDs, wherein inflammation and oxidative stress contribute to neuronal injury. Furthermore, CeONPs can regenerate between Ce^{3+} and Ce^{4+} oxidation states, hence maintaining a prolonged antioxidant activity.

14.23 BIOMEDICAL APPLICATIONS

Besides their antioxidant properties, CeONPs have been investigated for drug delivery applications, especially in NDs. The particles can be modified with diverse targeting ligands to enhance their selectivity for neurons, facilitating more accurate drug delivery. They have demonstrated the ability to augment the stability of bioactive molecules, including anti-inflammatory medications and neuroprotective drugs, therefore enhancing their therapeutic efficacy [60].

14.24 CHALLENGES AND LIMITATIONS

CeONPs exhibit considerable potential for the treatment of neurodegenerative illnesses; however, numerous hurdles and restrictions must be resolved prior to their widespread clinical application [61].

1. **Stability Issues:** Ensuring the long-term stability of CeONPs presents a significant problem. NPs may agglomerate or disintegrate under specific conditions, resulting in diminished efficacy and unpredictability in their activity inside biological systems.
2. **Biocompatibility Concerns:** The long-term interactions between CeONPs and biological tissues remain inadequately comprehended. Although they demonstrate antioxidant effects, considerable research is necessary to assess their biocompatibility, especially with prolonged use in the human body. Prolonged exposure may result in inadvertent immune system reactions or cellular harm.
3. **Toxicity Potential:** Despite the recognized radical-scavenging capabilities of CeONPs, their in vivo toxicity continues to be a significant concern. The diminutive size and extensive surface area of NPs may result in their accumulation in organs, potentially inducing toxicity in healthy cells and tissues. Additional toxicity studies are required to ascertain safe dosages and long-term consequences [62].
4. **Size and Surface Modification:** The dimensions and surface charge of CeONPs must be meticulously regulated to guarantee their successful traversal of the BBB and attainment of the target site in the brain without inducing adverse effects. Inadequate surface functionalization may result in non-specific binding or diminished therapeutic efficacy.
5. **Clinical Translation and Safety:** Notwithstanding the encouraging preclinical findings, the conversion of CeONPs into human therapies poses difficulties, particularly the necessity for stringent clinical trials to evaluate their long-term safety and effectiveness. Their regulatory approval process is intricate and necessitates much proof to validate their safety profile.
6. **Production Costs:** The synthesis of CeONPs necessitates meticulous regulation of particle dimensions, morphology, and surface properties, potentially elevating production expenses. To ensure the economic viability of these NPs, it is essential to establish scalable and cost-efficient manufacturing procedures [63].

14.25 FUTURE APPROACHES

The prospects for CeONPs in the treatment of neurodegenerative illnesses are favorable, with potential to enhance therapeutic approaches. The current study will likely concentrate on enhancing their dimensions, surface modification, and stability to augment their biocompatibility and therapeutic effectiveness. Advancements in targeted delivery systems, including surface changes to improve BBB penetration, may boost the efficacy of CeONPs in accessing the brain [64]. Furthermore, additional clinical investigations are required to ascertain their long-term safety and efficacy in human trials. Ongoing improvements suggest that CeONPs may be crucial in mitigating oxidative stress and providing innovative therapeutic alternatives for neurological illnesses.

14.26 CONCLUSION

Recent advancements in drug delivery systems for NDs highlight the promising potential of nanotechnology-based strategies, including LNPs, polymer-based NPs, nanogels, nanocomposites, dendrimers, and metal-based NPs (e.g., cerium oxide and gold). These novel strategies seek to surmount the BBB and enhance therapeutic effectiveness. These techniques possess the potential for more successful therapies by improving drug absorption, accurately targeting damaged brain regions, and minimizing side effects. Nonetheless, additional research, encompassing preclinical and clinical trials, is essential to refine these systems for safer, long-term therapeutic interventions in neurodegenerative illnesses.

REFERENCES

1. Cipolla MJ. *The Cerebral Circulation*. San Rafael (CA): Morgan & Claypool Life Sciences; 2009.
2. Choonara YE, Pillay V, Du Toit LC, Modi G, Naidoo D, Ndesendo VMK, Sibambo SR. Trends in the molecular pathogenesis and clinical therapeutics of common neurodegenerative disorders. *Int J Mol Sci*. 2009;10(6):2510–2557.
3. Goel P, Chakrabarti S, Goel K, Bhutani K, Chopra T, Bali S. Neuronal cell death mechanisms in Alzheimer's disease: An insight. *Front Mol Neurosci*. 2022;15:937133.
4. Taliyan R, et al. Nanocarrier mediated drug delivery as an impeccable therapeutic approach against Alzheimer's disease. *J Controlled Release*. 2022;343:528–550.
5. Obermeier B, Daneman R, Ransohoff RM. Development, maintenance and disruption of the blood–brain barrier. *Nat Med*. 2013;19:1584–1596.
6. Abbott NJ, Patabendige AAK, Dolman DEM, Yusof SR, Begley DJ. Structure and function of the blood–brain barrier. *Neurobiol Dis*. 2010;37:13–25.
7. Boyer-Di Ponio J, El-Ayoubi F, Glacial F, Ganeshamoorthy K, Driancourt C, Godet M, et al. Instruction of circulating endothelial progenitors in vitro towards specialized blood–brain barrier and arterial phenotypes. *PLoS One*. 2014;9.
8. Huber JD, Witt KA, Hom S, Egleton RD, Mark KS, Davis TP. Inflammatory pain alters blood–brain barrier permeability and tight junctional protein expression. *Am J Physiol Heart Circ Physiol*. 2001;280:H1241–H1248.
9. Wolburg H, Lippoldt A. Tight junctions of the blood–brain barrier: Development, composition and regulation. *Vasc Pharmacol*. 2002;38:323–337.
10. Persidsky Y, Ramirez SH, Haorah J, Kanmogne GD. Blood–brain barrier: Structur al components and function under physiologic and pathologic conditions. *J NeuroImmune Pharmacol*. 2006;1:223–236.
11. Choi YK, Kim K-W. Blood–neural barrier: Its diversity and coordinated cell-to cell communication. *BMB Rep*. 2008;41:345–352.
12. Re F, Gregori M, Masserini M. Nanotechnology for neurodegenerative disorders. *Maturitas*. 2012;73(1):45–51.
13. Choi H, et al. Strategies for targeted delivery of exosomes to the brain: Advantages and challenges. *Pharmaceutics*. 2022;14(3):672.
14. Wei X, et al. Brain tumor-targeted drug delivery strategies. *Acta Pharm Sinica B*. 2014;4(3):193–201.
15. Lu W. Adsorptive-mediated brain delivery systems. *Curr Pharm Biotechnol*. 2012;13(12):2340–2348.
16. Gordillo-Galeano A, Mora-Huertas CE. Solid lipid nanoparticles and nanostructured lipid carriers: A review emphasizing on particle structure and drug release. *Eur J Pharm Biopharm*. 2018;133:285–308.
17. ud Din F, Zeb A, Shah KU. Development, in-vitro and in-vivo evaluation of ezetimibe-loaded solid lipid nanoparticles and their comparison with marketed product. *J Drug Deliv Sci Technol*. 2019;51: 583–590.
18. Maqsood S, et al. Levosulpiride-loaded nanostructured lipid carriers for brain delivery with antipsychotic and antidepressant effects. *Life Sci*. 2022;311:121198.
19. Gul M, et al. Formulation optimization, in vitro and in vivo evaluation of agomelatine-loaded nanostructured lipid carriers for augmented antidepressant effects. *Colloids Surf B*. 2022;216:112537.
20. Bibi M, et al. Cilostazol-loaded solid lipid nanoparticles: Bioavailability and safety evaluation in an animal model. *J Drug Deliv Sci Technol*. 2022;74:103581.
21. Puri A, et al. Lipid-based nanoparticles as pharmaceutical drug carriers: From concepts to clinic. *Crit Reviews™ Therapeutic Drug Carrier Syst*. 2009;26(6).
22. Sohail S. Nanotheranostics: The future remedy of neurological disorders. In *Cancer Nanotheranostics: Volume 2*. Springer Nature Switzerland AG; 2021, 117–154.
23. Khan MM, et al. Statistical optimization of co-loaded rifampicin and pentamidine polymeric nanoparticles for the treatment of cutaneous Leishmaniasis. *J Drug Deliv Sci Technol*. 2023;79:104005.
24. Kakkar V, et al. Curcumin and its nanoformulations as therapeutic for Alzheimer's disease. *Nanobiotechnol Neurodegenerative Dis*. 2019:343–67.

25. Aalinkeel R, et al. Neuroprotective effects of a biodegradable poly(lactic-co-glycolic acid)-ginsenoside Rg3 nanoformulation: A potential nanotherapy for Alzheimer's disease? *J Drug Target*. 2018;26(2):182–93.
26. Zhao L, et al. The effect of poly(lactic-co-glycolic acid) (PLGA) coating on the mechanical, biodegradable, bioactive properties and drug release of porous calcium silicate scaffolds. *Biomed Mater Eng*. 2012;22(5):289–300.
27. Mathew A, et al. Curcumin loaded-PLGA nanoparticles conjugated with Tet-1 peptide for potential use in Alzheimer's disease. *PLoS ONE*. 2012;7(3):e32616.
28. Tiwari SK, et al. Curcumin-loaded nanoparticles potently induce adult neurogenesis and reverse cognitive deficits in Alzheimer's disease model via canonical Wnt/β-catenin pathway. *ACS Nano*. 2014;8(1):76–103.
29. ud Din F, et al. Novel dual-reverse thermosensitive solid lipid nanoparticle-loaded hydrogel for rectal administration of flurbiprofen with improved bioavailability and reduced initial burst effect. *Eur J Pharm Biopharm*. 2015;94:64–72.
30. Vigani B, et al. Recent advances in the development of in situ gelling drug delivery systems for non-parenteral administration routes. *Pharmaceutics*. 2020;12(9).
31. Xiao Y, et al. Over-activation of TLR5 signaling by high-dose flagellin induces liver injury in mice. *Cell Mol Immunol*. 2015;12(6):729–742.
32. Xing R, et al. Development, characterization, and evaluation of SLN-loaded thermoresponsive hydrogel system of topotecan as biological macromolecule for colorectal delivery. *BioMed Research International*. 2021;2021:1–14.
33. Sarkar A, et al. Nanoparticles as a carrier system for drug delivery across blood brain barrier. *Curr Drug Metable*. 2017;18(2):129–137.
34. Patel V, Chavda V, Shah J. Nanotherapeutics in Neuropathologies: Obstacles, challenges and recent advancements in CNS targeted drug delivery systems. *Curr Neuropharmacol*. 2021;19(5):693–710.
35. Liu J, et al. A self-assembled α-synuclein nanoscavenger for Parkinson's disease. *ACS Nano*. 2020; 14(2):1533–1549.
36. Singh K. Dendrimer-drug conjugates in drug delivery and targeting. *Pharm Nanatechnol*. 2015; 3(4):239–260.
37. Klajnert B, et al. Influence of dendrimer's structure on its activity against amyloid fibril formation. *Biochem Biophys Res Commun*. 2006;345(1):21–28.
38. Pérez-Carrión MD, Posadas I. Dendrimers in neurodegenerative diseases. *Processes*. 2023;11(2):319.
39. Milowska K, et al. In vitro PAMAM, phosphorus and viologen-phosphorus dendrimers prevent rotenone-induced cell damage. *Int J Pharm*. 2014;474(1–2):42–49.
40. da Silva PV, de Queiroz AAA. Long term multiple sclerosis drug delivery using dendritic polyglycerol flower-like microspheres. *J Biomater Sci Polym Ed*. 2020;31(2):188–206.
41. Yu H, et al. Novel composite double-layered dressing with improved mechanical properties and wound recovery for thermosensitive drug, *Lactobacillus brevis*. *Compos Part B: Eng*. 2021;225:109276.
42. Kabiri K, Zohuriaan-Mehr M. Porous superabsorbent hydrogel composites: Synthesis, morphology and swelling rate. *Macromol Mater Eng*. 2004;289:653–661.
43. Shameem MM, et al. A brief review on polymer nanocomposites and its applications. *Materials Today: Proceedings*. 2021;45:2536–2539.
44. Popovic N, Brundin P. Therapeutic potential of controlled drug delivery systems in neurodegenerative diseases. *Int J Pharm*. 2006;314(2):120–126.
45. Singh AK, et al. Inorganic clay nanocomposite system for improved cholinesterase inhibition and brain pharmacokinetics of donepezil. *Drug Dev Ind Pharm*. 2020;46(1):8–19.
46. Sela H, et al. Spontaneous penetration of gold nanoparticles through the blood brain barrier (BBB). *J Nanobiotechnol*. 2015;13:1–9.
47. Ruff J, et al. The effects of gold nanoparticles functionalized with ß-amyloid specific peptides on an in vitro model of blood-brain barrier. *Nanomedicine*. 2017;13(5):1645–1652.
48. Betzer O, et al. The effect of nanoparticle size on the ability to cross the blood-brain barrier: An in vivo study. *Nanomed (Lond)*. 2017;12(13):1533–1546.
49. Hou K, et al. Chiral gold nanoparticles enantioselectively rescue memory deficits in a mouse model of Alzheimer's disease. *Nat Commun*. 2020;11(1):4790.
50. Zhou Y, et al. Simple colorimetric detection of amyloid β-peptide (1–40) based on aggregation of gold nanoparticles in the presence of copper ions. *Small*. 2015;11(18):2144–2149.
51. Nayab DE, Din Fu, Ali H, et al. Nano biomaterials based strategies for enhanced brain targeting in the treatment of neurodegenerative diseases: An up-to-date perspective. *J Nanobiotechnol* 2023;21:477.

52. Saraiva C, Praçab C, Ferreira R, et al. Nanoparticle-mediated brain drug delivery: Overcoming blood–brain barrier to treat neurodegenerative diseases. *Journal of Controlled Release*. 2016;235:34–47.
53. Zhong C, et al. The application, neurotoxicity, and related mechanism of silica nanoparticles. *Neurotox Nanomaterials Nanomed*. 2017:227–257.
54. Selvarajan V, Obuobi S, Ee PLR. Silica nanoparticles - a versatile tool for the treatment of bacterial infections. *Front Chem*. 2020;8:602.
55. Nday CM, et al. Quercetin encapsulation in modified silica nanoparticles: Potential use against Cu(II)-induced oxidative stress in neurodegeneration. *J Inorg Biochem*. 2015;145:51–64.
56. Karimzadeh M, Rashidi L, Ganji F. Mesoporous silica nanoparticles for efficient rivastigmine hydrogen tartrate delivery into SY5Y cells. *Drug Dev Ind Pharm*. 2017;43(4):628–636.
57. Singh AK, et al. Lipid-coated MCM-41 mesoporous silica nanoparticles loaded with berberine improved inhibition of acetylcholine esterase and amyloid formation. *ACS Biomaterials Science & Engineering*. 2021;7(8):3737–3753.
58. Dowding JM, et al. Cerium oxide nanoparticles protect against Aβ-induced mitochondrial fragmentation and neuronal cell death. *Cell Death Differ*. 2014;21(10):1622–1632.
59. Cimini A, et al. Antibody-conjugated PEGylated cerium oxide nanoparticles for specific targeting of Aβ aggregates modulate neuronal survival pathways. *Acta Biomater*. 2012;8(6):2056–2067.
60. Kuhn HG, Dickinson-Anson H, Gage FH. Neurogenesis in the dentate gyrus of the adult rat: Age-related decrease of neuronal progenitor proliferation. *J Neurosci*. 1996;16(6):2027–2033.
61. Palmer TD, Takahashi J, Gage FH. The adult rat hippocampus contains primordial neural stem cells. *Mol Cell Neurosci*. 1997;8(6):389–404.
62. Murrell W, et al. Expansion of multipotent stem cells from the adult human brain. *PLoS ONE*. 2013;8(8):e71334.
63. Zhang G-L, Zhu Z-H, Wang Y-Z. Neural stem cell transplantation therapy for brain ischemic Stroke: Review and perspectives. *World J Stem Cells*. 2019;11(10):817.
64. Santos T, et al. Nanomedicine approaches to modulate neural stem cells in brain repair. *Trends Biotechnol*. 2016;34(6):437–439.

15 Food and Nutrition as a Medicine Class for the Prevention and Treatment of Neurodegenerative Disorder

Vetriselvan Subramaniyan,[1] Muralikrishnan Dhanasekaran,[2] Mukesh Kumar,[3] Mahendran Sekar,[4] Neeraj Kumar Fuloria,[5] Shivkanya Fuloria,[5] Gowthamarajan Kuppusamy,[6] and Karri V. V. S. Narayana Reddy[6]

[1]Sir Jeffery Cheah Sunway Medical School, Faculty of Medical and Life Sciences, Sunway University, Malaysia

[2]Department of Drug Discovery & Development, Harrison College of Pharmacy, Auburn University, Walker building, Auburn, AL, USA

[3]Botany and Microbiology Gurukul Kangri University Haridwar, Uttarakhand, India

[4]School of Pharmacy, Monash University Malaysia, Jalan Lagoon Selatan, Bandar Sunway, Subang Jaya, Selangor

[5]Faculty of Pharmacy, AIMST University, Bukit Air Nasi, Bedong, Kedah, Malaysia

[6]Department of Pharmaceutics, JSS College of Pharmacy, JSS Academy of Higher Education & Research, Ooty, Tamil Nadu, India

15.1 INTRODUCTION

Neurodegenerative disorders include Alzheimer's disease, Parkinson's disease, Huntington's disease, and amyotrophic lateral sclerosis, which have now become one of the increasing public health challenges worldwide. These conditions are characterized by progressive degeneration of the nervous system, leading to severe cognitive, motor, and functional impairments. The increasing prevalence of neurodegenerative diseases, particularly with aging populations, has spurred an urgent need for novel therapeutic strategies. Whereas drugs have been developed with a goal to decelerate or prevent disease advancement and have had minimal success, it is increasingly acknowledged that nutrition has a major role in both prevention and management. Nutrition, over the past couple of decades, has arisen as one of the key modulators of brain health [1]. It is, therefore, with reason that the "food as medicine" concept has started to attract momentum, whereby dietary interventions are now perceived as part of the prevention and therapeutic intervention for neurodegenerative disorders. The use of functional foods, bioactive compounds, and nutritional strategies may represent a complementary means to promote brain health, mitigate neurodegenerative processes, and delay the course of neuropsychiatric disease. This chapter will review new advances in nutrition and food with neuroprotective potential and new therapeutic approaches related to neurodegenerative disease, considering in detail the latest nutritional interventions, the influence of bioactive substances, the axis between gut–brain, and the possible nutrigenetics in disease management [2].

DOI: 10.1201/9781003478652-17

15.1.1 Neuroprotection and Cognitive Health

Role of Functional Foods. Functional foods are defined as foods providing health benefits beyond basic nutrition. These foods contain bioactive compounds, including antioxidants, polyphenols, vitamins, minerals, and fatty acids, which have been shown to exert protective effects on the brain. A growing body of evidence suggests that functional foods may play an important role in promoting cognitive function and protecting the brain from neurodegenerative diseases. Among all other functional foods promoting cognitive health, the most prominent example might be the Mediterranean diet [3]. This Mediterranean diet contains fruits, vegetables, whole grains, legumes, nuts and seeds, olive oil, among other components that together help diminish cognitive decline by offering a potential low risk of age-related cognitive dysfunction and neurodegenerative disease. This contains a highly bioavailable source of antioxidants along with healthy fats, mainly omega-3 fatty acids, proving effective in inflammation and oxidative stress suppression and resulting neuronal damage, one major cause leading toward neurodegenerative phenomena. Other functional foods, such as turmeric containing curcumin and green tea containing catechins, have also been reported to have neuroprotective effects. This is attributed to their anti-inflammatory and antioxidant activities. Of these, curcumin has been undergoing clinical trials, wherein it was shown to reduce amyloid plaque formation, one of the hallmarks of Alzheimer's disease. Omega-3 fatty acids found in fatty fish and flaxseed have also been associated with improved cognitive function, lower risk of Alzheimer's disease, and slowing of dementia in older adults. As shown in Figure 15.1, neuroprotection by functional foods points to the potential of dietary interventions in the prevention and treatment of neurodegenerative diseases. As research continues to uncover the mechanisms by which these foods benefit the brain, there is hope that functional foods could one day complement conventional therapies in managing neurodegenerative disorders [4].

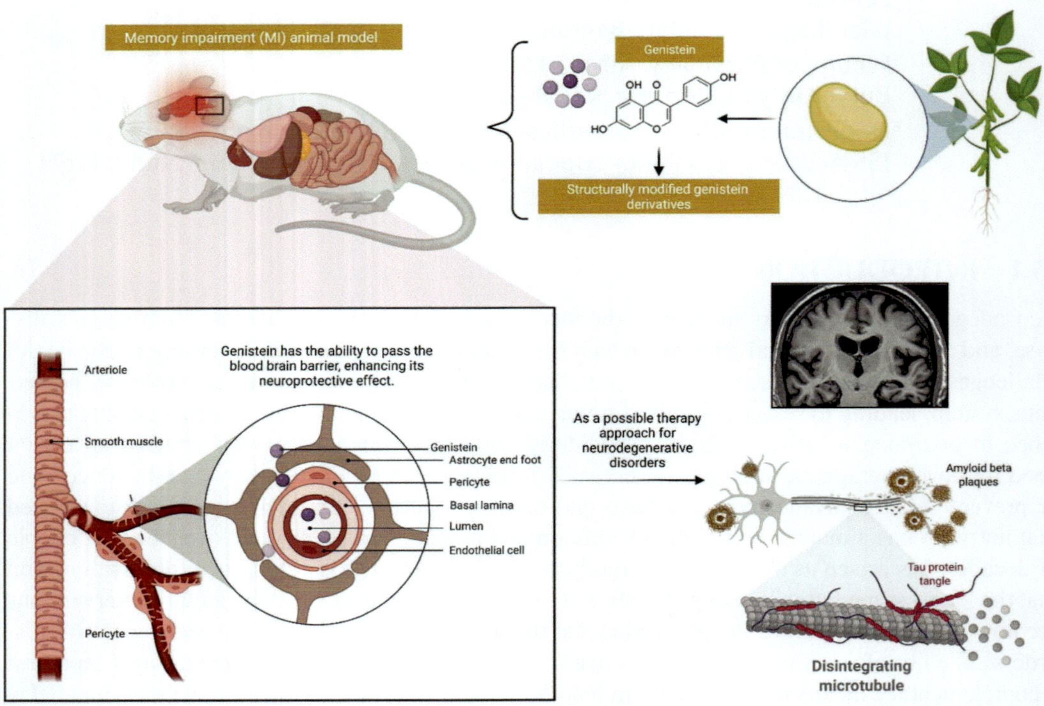

FIGURE 15.1 A potential natural lead molecule for the design and development of novel drugs to treat memory impairment (copyright © 2022; Fuloria and associates). The Creative Commons Attribution License (CC BY) governs the distribution of this open-access article.

15.1.2 Nutritional Approaches to Neurodegenerative Disease Management and Prevention

The goal of nutritional strategies for neurodegenerative disease prevention and treatment is to provide a diet that preserves brain function and lowers disease burden factors. The human brain is extremely sensitive to nutritional stress, and dietary factors have an impact on a variety of physiological processes, including oxidative stress, inflammation, and several signaling pathways that are directly linked to neurodegeneration. Reducing consumption of foods that cause inflammation is a crucial dietary strategy. Neurodegenerative disorders are largely caused by chronic inflammation, and processed diets, trans fats, and refined carbohydrates can exacerbate oxidative stress and inflammation [5]. Conversely, anti-inflammatory diets that include foods high in fiber, antioxidants, and healthy fats can help prevent neurodegeneration in the brain. Because it controls blood sugar, the low-glycemic index diet is crucial for controlling neurodegenerative illnesses. Type 2 diabetes and insulin resistance increase the risk of Alzheimer's disease and other cognitive problems since they are typically associated with elevated blood sugar levels. Diets high in whole grains and legumes, for example, may help reduce these risks by promoting stable blood sugar levels. Optimizing consumption of vitamins and minerals that are critical for brain function, such as magnesium, folic acid, vitamin D, and B vitamins, particularly B6 and B12, is another aspect of nutritional strategy. These nutrients have gained attention due to their important involvement in neuroprotection, neurotransmitter synthesis, and cognitive processes. A diet rich in all of these vital nutrients is crucial because a lack of these vitamins has been linked to a higher risk of cognitive impairment [6].

15.1.3 Bioactive Food Ingredients: How They Affect Brain Health

Flavonoids, carotenoids, and polyphenols are examples of bioactive foods that have recently drawn attention because of their enormous potential in application tactics for maintaining brain function and combating neurodegenerative diseases. Through a variety of mechanisms, including antioxidant activity, inflammatory cytokine pathway modulation, and neuroprotective gene regulation, these phytochemical substances impact the brain. Dark chocolate, citrus fruits, and berries are among the foods that contain them. They also have anti-inflammatory and antioxidant qualities. Numerous studies have shown that flavonoids enhance cognitive performance, lower neuroinflammation, and guard against oxidative stress-induced neuronal damage [7]. For instance, in animal models of Alzheimer's disease, blueberry anthocyanins have been shown to enhance memory and reduce oxidative damage. By shielding neural tissues from oxidative damage, carotenoids – such as lutein and zeaxanthin – found in leafy greens and other vibrant vegetables also promote brain health. According to certain research, these substances enhance cognitive performance and may reduce the risk of cognitive decline and age-related macular degeneration. Numerous polyphenols, such as ellagic acid from pomegranates and resveratrol from grapes and red wine, have shown promise in lowering the production of amyloid plaque and altering brain cell communication pathways in an effort to enhance cognitive performance. The potential for food-derived compounds to function as therapeutic agents in the prevention and treatment of neurodegenerative disorders is highlighted by these bioactive phytochemicals [8].

15.1.4 Dietary Influence on Neurodegeneration and Cognitive Decline: The Gut–Brain Axis

Recent studies on the gut–brain axis – the relationship between the gut and the brain – have revealed that the gut microbiota plays a significant role in regulating brain activity and that disruptions in the microbiome are linked to neurodegenerative diseases like Parkinson's and Alzheimer's. Cognitive decline risk can be decreased, and brain health can be impacted by diets that support a healthy gut microbiota [9]. Prebiotic and probiotic meals, fermented foods, high-fiber fruits and vegetables, and whole grains are all essential components of a healthy gut microbiota diet. These nutrients support the beneficial bacteria in the stomach, which can create compounds that affect brain activity by modulating neurotransmitters, causing neuroprotection, and reducing inflammation. Additionally, diets rich in polyphenols – found in green tea and berries – modify gut microbiota cognition and neuroprotection [10].

15.1.5 The Future of Neurodegenerative Disease Treatment with Personalized Nutrition and Novel Nutritional Therapies

As a new benchmark for nutritional therapies intended to prevent or slow the progression of neurodegenerative diseases, personalized nutrition takes into account individual differences in genotype, metabolome, and health state. The significance of personalized nutrition as a paradigm shift in the treatment of neurodegenerative diseases stems from the possibility that genetic predispositions give people their unique metabolism, interaction, and activity in response to dietary modification or bioactive chemicals [11]. Ketogenic diets, intermittent fasting, and supplements with specific nutrients that promote neuroprotection are examples of emerging dietary interventions that may slow the progression of neurodegenerative illnesses. By improving mitochondrial function and lowering neuroinflammation, the ketogenic diet – a high-fat, low-carb diet – is showing promise in the treatment of Parkinson's and Alzheimer's illnesses. Personalized nutrition may provide even more potent and focused treatments for individuals at risk of neurodegenerative disorders as our knowledge of the intricate relationship between nutrition, genetics, and brain health expands. Promising approaches to the management, prevention, and therapy of neurodegenerative diseases include diet and nutrition. It is possible to preserve the brain, reduce the progression of disease, and enhance cognitive performance by utilizing functional foods, bioactive chemicals, and focused dietary techniques. Food and nutrition may play a significant role in the medical treatment of neurodegenerative illnesses in the future, thanks to the growth of research on the relationship between nutrition and brain function and advancements in tailored nutrition. Indeed, it seems that food can be used as medication and gives patients suffering from these debilitating conditions fresh hope as we continue to understand the mechanisms underlying these nutritional therapies [12].

15.2 THE FUNCTION OF FUNCTIONAL FOODS IN COGNITIVE HEALTH AND NEUROPROTECTION

Today, one of the biggest health burdens in the world is neurodegenerative disorders, including Huntington's, Parkinson's, and Alzheimer's. Progressive neuronal degeneration in these illnesses results in memory loss, motor impairment, poor cognitive function, and ultimately a lower quality of life. These illnesses are becoming more prevalent as the global population ages, which puts an increasing amount of strain on healthcare systems. The majority of pharmacological treatments only slow down or even reverse neurodegeneration, though some are available. All of this has driven researchers to look for other therapeutic techniques, and in this context, functional foods and nutrition have emerged as one of the most popular strategies for preserving brain health and preventing neurodegenerative illnesses [13].

Figure 15.2 illustrates the functional foods are, by definition, foods that have bioactive components that provide health advantages over and beyond their nutritional content. Because of their protective effects on the brain, foods in this category may help prevent and manage neurodegenerative illnesses. They are thought to have beneficial effects on brain health because they include vitamins, minerals, antioxidants, omega-3 fatty acids, and other phytochemicals that are not found in typical meals. Recent research has highlighted the significant potential of neuroprotection with functional meals in enhancing cognitive function and lowering inflammation – neuroprotection against degeneration. Some of the main mechanisms that contribute to neuroprotective effects exerted by functional foods are antioxidant activity, anti-inflammation, regulation of gene expression, and modulation of signaling pathways that control neuronal survival. These are involved in countering the underlying processes that result in neurodegeneration, which include oxidative stress, inflammation, and mitochondrial dysfunction [14].

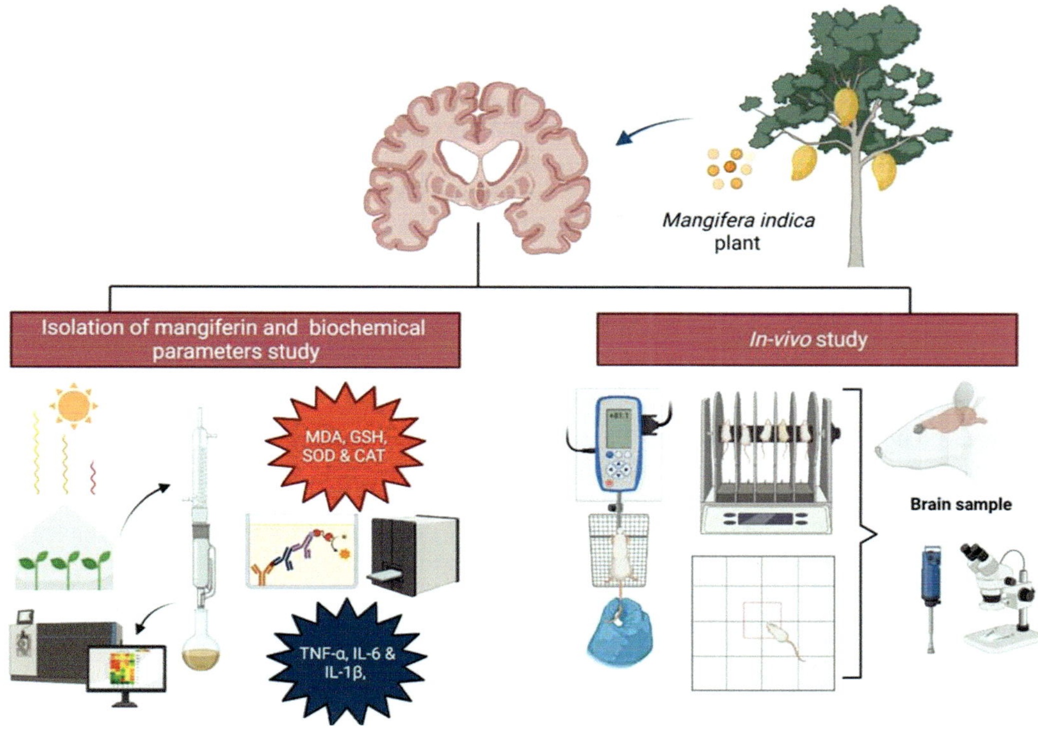

FIGURE 15.2 Mangiferin's neuroprotective effectiveness against 3-nitropropionic acid-induced symptoms resembling Huntington's disease (Copyright © 2023). Lum and associates, The Creative Commons Attribution License (CC BY), governs the distribution of this open-access article.

15.2.1 Activity of Antioxidants

One of the main causes of neurodegeneration is oxidative stress, which arises from an imbalance between the body's ability to neutralize reactive oxygen species (ROS) via antioxidants and the production of ROS. Cell lesions, including lipid, protein, and DNA-triage neuronal cell death, may be caused by the high load of ROS. Foods high in active antioxidants may help counteract the effects of free radicals, resulting in less oxidative damage and a neuroprotective impact against neurodegeneration. The majority of the functional foods on the list, including cruciferous vegetables, dark chocolate, berries, and green tea, are rich in various antioxidants, including flavonoids, polyphenols, and carotenoids. These substances have been linked to reduced oxidative stress in the brain, which protects neurons. Blueberry anthocyanins, for example, have been shown to enhance memory and shield brain cells from oxidative stress. Moreover, green tea polyphenols – most notably, the EGCG – have the potential to support robust antioxidant defenses, maintain cognitive function, and shield neurons from oxidative damage [15].

15.2.2 Effects on Inflammation

Chronic inflammation of the brain, or neuroinflammation, is thought to be a key factor in the etiology of neurodegenerative illnesses. Amyloid plaque buildup in Alzheimer's disease and alpha-synuclein aggregation in Parkinson's disease can both trigger an inflammatory response. Long-term neuroinflammation can lead to cell death, synapse loss, and neuronal dysfunction. Anti-inflammatory functional foods have the potential to prevent neurodegeneration and lessen neuroinflammation.

Omega-3 fatty acids, which may be found in plant sources like flaxseeds and chia seeds and in fatty fish like salmon, are widely known for their anti-inflammatory properties. It has been demonstrated that omega-3 fatty acids, particularly EPA and DHA, improve neuronal health and lessen inflammation in the brain. Increasing omega-3 fatty acid intake has also been linked to a delayed decrease in Alzheimer's disease and other forms of dementia in older adults, according to a number of studies. By blocking pro-inflammatory cytokines, these dietary ingredients – particularly turmeric, which includes the active ingredient curcumin – have also been shown to lessen inflammation. Since curcumin readily penetrates the blood–brain barrier, it has drawn a lot of interest because of its potent anti-inflammatory qualities, which make it an excellent choice for neuroprotective treatment. Green tea and ginger are two more foods that have anti-inflammatory compounds that may help regulate inflammation in the brain and prevent neurodegenerations [16].

15.2.3 Alteration of Signaling Pathways and Gene Expression

By altering gene expression and signals linked to survival and repair, functional meals are thought to enhance brain health. Functional food-derived phytochemicals have the ability to trigger a number of pathways that increase neurotrophic factors that support neuron development, survival, and differentiation. Brain-derived neurotrophic factor (BDNF) is a crucial protein in neuronal growth and synaptic plasticity conditions that are closely linked to cognitive performance. Certain functional foods, such as flavonoids, resveratrol, and curcumin, have been shown to increase BDNF expression in order to carry out their neuroprotective activity. Red wine and red grapes contain resveratrol. It works through the sirtuin pathway, which improves mitochondrial activity and lengthens cell life. Neurodegenerative illnesses whose course is marked by mitochondrial dysfunction benefit from the enhanced mitochondrial function [17].

15.2.4 Alteration in Mitochondrial Activity

The cell's powerhouse, mitochondria, is where energy used by neurons is generated. Mitochondrial malfunction may cause neuronal damage and ultimately cell death in neurodegenerative disorders. Neurodegeneration may be avoided and brain health preserved by functional diets that promote mitochondrial function. The ketogenic diet, for instance, is a high-fat, low-carb diet that has been shown to enhance mitochondrial function and even cognitive performance in patients with Parkinson's and Alzheimer's illnesses. Additional mitochondrial health-promoting functional foods include those high in coenzyme Q10 (CoQ10), such as spinach and broccoli, which enhance mitochondrial activity and ward off neurodegenerative illnesses. Eating meals that improve the brain can lower the risk of neurodegenerative disorders and greatly improve cognitive performance [18]. Omega-3 fatty acids from fatty fish like salmon and mackerel encourage neurogenesis and lower inflammation, while berries, which are high in flavonoids and anthocyanins, fight oxidative stress and aid with memory recall. Essential vitamins, including folate, which lowers homocysteine levels associated with cognitive decline, are found in leafy greens and cruciferous vegetables. Rich in vitamin E and polyphenols, nuts and seeds are essential for brain health since they have neuroprotective properties. Strong neuroprotective, anti-inflammatory, and antioxidant qualities of these functional foods support their function in maintaining brain integrity and possibly averting diseases like Parkinson's and Alzheimer's. Including these nutrient-dense foods in everyday diets could be a key component of integrated prevention and cognitive lifespan as research progresses [19].

15.3 DIETARY TECHNIQUES FOR NEURODEGENERATIVE DISEASE MANAGEMENT AND PREVENTION

Neurodegenerative diseases, such as amyotrophic lateral sclerosis, Parkinson's disease, Alzheimer's disease, and Huntington's disease, are progressive conditions that inevitably cause neuron degeneration. Neurodegenerative disorders can cause movement dysfunction, cognitive decline, and a substantial reduction in quality of life. As a result, these conditions are regarded as one of the biggest public health issues due to the aging of the global population. Despite the availability of certain pharmaceutical treatments, their limited effectiveness so far has sparked interest in nutritional methods that could be considered modifiable risk factors for the prevention and management of neurodegenerative diseases. Certain nutrients and functional meals have been shown to promote brain health and slow the rate of neurodegeneration [20].

15.3.1 Antioxidants and Nutrients That Reduce Inflammation

These diseases have been widely regarded as the primary cause of neurodegenerative processes: oxidative stress, which damages neurons and brain tissues, is caused by an imbalance between the body's antioxidant defenses and ROS, while chronic inflammation speeds up the process of neuronal damage and is frequently observed in the majority of neurodegenerative disorders. Consequently, diets high in antioxidants will provide noticeable neuroprotection (Figure 15.3). Antioxidants like vitamins E and C, flavonoids, polyphenols, and carotenoids are densely packed in fruits, vegetables, nuts, seeds, and plant-based diets. Berries, such as blackberries, blueberries, raspberries, and strawberries, are known to contain anthocyanins, which are potent antioxidants and anti-inflammatory substances. In addition to being rich in antioxidants, green leafy vegetables like spinach, kale, and broccoli contain folate, a nutrient that is critical for cognitive function. Fatty fish like salmon, mackerel, and sardines contain docosahexaenoic acid, an omega-3 fatty acid that is essential for reducing neuroinflammation. These polyunsaturated fats guard against neurodegenerative illnesses like Parkinson's and Alzheimer's and maintain the integrity of the membranes of neurons [21].

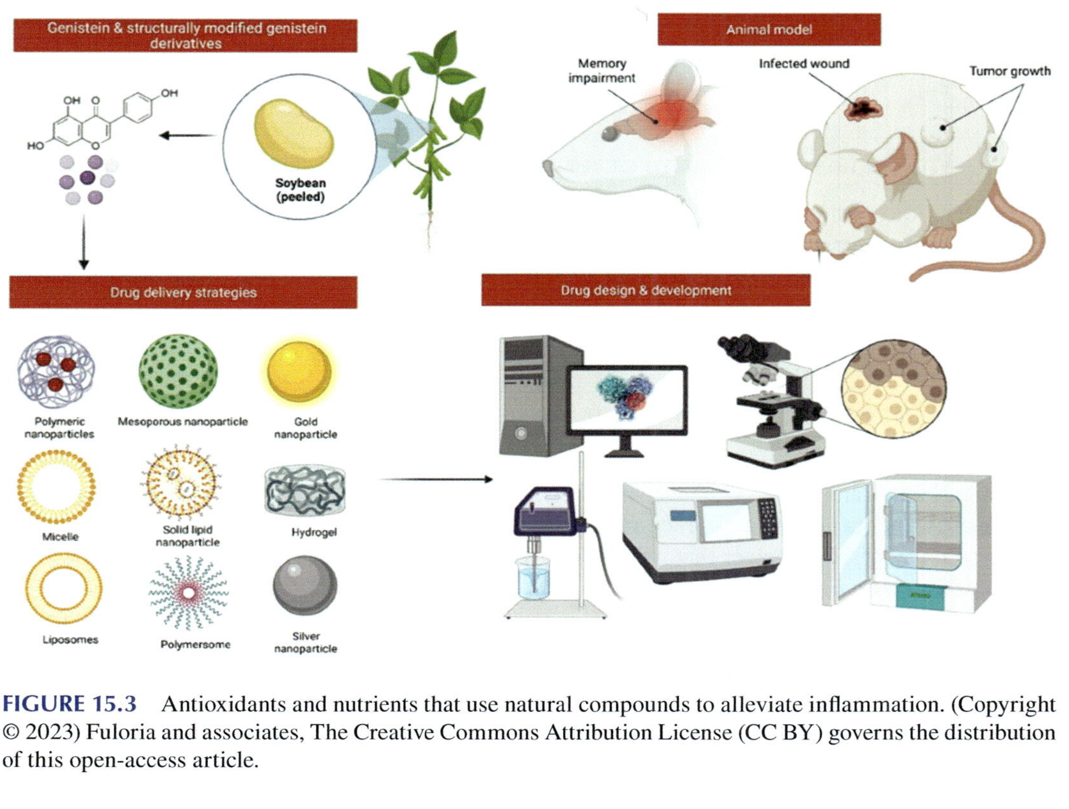

FIGURE 15.3 Antioxidants and nutrients that use natural compounds to alleviate inflammation. (Copyright © 2023) Fuloria and associates, The Creative Commons Attribution License (CC BY) governs the distribution of this open-access article.

15.3.2 Flavonoids and Polyphenols

Plants include chemical molecules called polyphenols, which are crucial for brain health and have antioxidant qualities. Foods that contain green tea, dark chocolate, grapes, apples, and citrus fruits are rich in these chemicals. Apples, onions, and capers contain quercetin, a flavonoid with anti-inflammatory qualities that reduces the amyloid-beta plaques that are common in Alzheimer's disease. Resveratrol, another polyphenol that is included in red wine and berries, has neuroprotective properties by lowering inflammation and oxidative stress while encouraging the synthesis of BDNF, which promotes the formation of new neurons [22].

15.3.3 Dietary Fiber and the Gut–Brain Axis

Recent studies on the gut–brain axis link neurodegenerative illnesses to gut health. Neurodegeneration may result from systemic inflammation brought on by an imbalance in the gut microbiota. The secret to the best possible brain health is dietary practices that positively affect the gut microbiota. Prebiotic foods, like garlic, onions, and bananas, and probiotic meals, like yogurt, kefir, and fermented vegetables, nourish the good bacteria in the gut and lower inflammation. Whole grains, legumes, fruits, and vegetables provide a high-fiber diet that encourages the growth of good bacteria and healthy gut flora, both of which may have a positive impact on brain function [23].

15.3.4 Intermittent Fasting and Calorie Limitation

Intermittent fasting and calorie restriction are two age-related disease-delaying and lifespan-extending practices that have drawn a lot of attention, notably for neurodegenerative illnesses. Calorie restriction enhances mitochondrial function and activates cellular pathways that promote autophagy, the elimination of damaged cells. It has been shown that intermittent fasting, which alternates between eating and fasting times, improves cognitive function and lowers oxidative stress and inflammation. According to recent studies, intermittent fasting may promote brain plasticity and aid in the removal of amyloid-beta plaques, two characteristics frequent in neurodegenerative disorders [24].

15.3.5 Vitamin D and the Health of the Brain

The brain requires vitamin D, and a lack of it is associated with a higher risk of neurodegenerative diseases. To preserve neurons, vitamin D lowers neuroinflammation and regulates immune activity. Sunlight, fortified dairy products, and fatty fish are examples of natural sources. As a result, maintaining cognitive function and preventing neurodegeneration requires adequate vitamin D levels [25].

15.3.6 Tailored Nutritional Strategies

Perhaps the most fascinating of these is individualized nutrition, which is based on a person's lifestyle, microbiome, and genotype. Gene variations, such as the APOE ε4 allele, are known to increase the risk of Alzheimer's disease, which raises the possibility that individualized nutrition, through dietary adjustment, could lower that risk. Additionally, individualized diets that optimize the gut–brain axis can enhance cognitive function and provide a successful preventative and therapeutic approach for neurodegenerative illnesses. As a result, nutritional strategies are highly effective in managing and preventing neurodegenerative illnesses. To lower the risk of neurodegeneration and promote brain function, special attention will be paid to gut health, anti-inflammatory nutrients, antioxidant foods, and customized nutrition. Vitamins, fiber, polyphenols, and omega-3 fatty acids will all play a significant part in lowering inflammation, oxidative stress, and cognitive decline.

According to current studies, nutrition may continue to be the mainstay of care for neurodegenerative diseases and present a chance to preserve and/or improve cognitive function [26].

15.4 BIOACTIVE SUBSTANCES IN FOOD: HOW THEY AFFECT BRAIN HEALTH

Natural, non-nutrient chemicals known as bioactive compounds in food have the potential to improve health and lower the risk of diseases, including neurodegenerative conditions. By modifying several biological processes, including neuronal function, inflammation, oxidative stress, and neuronal cell survival, these substances – which include polyphenols, flavonoids, carotenoids, alkaloids, and peptides – may have strong effects on brain health. To promote neuroprotection, suppress neuroinflammation, and stop oxidative neurodegeneration, the majority of them modulate a variety of cellular signaling pathways. The majority of bioactives fall into the polyphenolic categories, which are widely present in fruits, vegetables, tea, and red wine. As a result, polyphenols are a significant class of compounds that are linked to preserving brain health through their roles as neurogenesis promoters, antioxidants, and anti-inflammatory agents. As scavengers of ROS, which are extremely hazardous chemicals that are continuously created throughout metabolism and whose elevated level is known to induce oxidative stress, quercetin and epicatechin show particular efficacy [27]. In neurodegenerative illnesses like Parkinson's and Alzheimer's, oxidative stress is a significant factor. Additionally, pro-inflammatory mediators, including cytokines and reactive nitrogen species, which are enhanced during chronic neuroinflammation, are suppressed by polyphenols. Polyphenols function neuroprotectively against neuronal injury and enhance general cognitive abilities by modifying the activation of signaling pathways. Additionally, certain polyphenols increase BDNF, a protein that promotes neuronal growth, survival, and differentiation. This suggests improved cognitive function and may reduce the risk of neurodegenerative diseases.

Other important foods that are bioactive include carotenoids like zeaxanthin and lutein. Eggs, corn, and dark leafy greens all contain them. These bioactive substances are strong antioxidants that enhance cognitive function and shield the brain from oxidative damage. It has been demonstrated that lutein builds up in the brain and is linked to enhanced processing speed and memory. By regulating the activation of glial cells involved in the brain's immunological response, carotenoids also lessen neuroinflammation. Carotenoids help maintain the health of neurones and may help stop age-related cognitive decline by reducing oxidative stress and inflammation. By improving synaptic plasticity, regulating neuroinflammatory activity in the central nervous system, and exhibiting metabolic pathway regulation, bioactive peptides – which are generated from dietary proteins – may have neuroprotective benefits as modulators of neurotransmission. Numerous neuronal activities are regulated by those pathways, suggesting that those peptides may actively participate as a therapeutic element in either prevention or treatment. Bioactive food ingredients promote brain health in a variety of ways. These substances can prevent or postpone the beginning of neurodegenerative disorders, improve cognitive function, and shield neurones from harm by functioning as antioxidants, anti-inflammatory agents, and modulators of neurotrophic factors. To ensure lifelong brain health and lower the risk of cognitive decline, a diet rich in bioactive foods – fruits, vegetables, whole grains, and other plant-based foods – may offer the best protection [28].

15.5 THE GUT–BRAIN AXIS: THE IMPACT OF DIET ON COGNITIVE DECLINE AND NEURODEGENERATION

One of the intricate two-way communication channels between the central nervous system and the gastrointestinal tract is the gut–brain axis. Numerous mechanisms, such as the vagus nerve, immune system, neurotransmitters, signaling molecules, and hormones, can mediate this intercommunication. In fact, recent research has linked the gut microbiome – a diverse collection of microorganisms that live in the gut – to brain function, particularly in relation to neurodegenerative illnesses and cognitive decline. In this regard, the gut microbiota plays a key role in a number of regulatory functions, including immune response modulation, the synthesis of neuroactive compounds, and blood–brain barrier integrity maintenance, all of which have a major impact on neurodegenerative events. Up until now, neurological diseases, including multiple sclerosis, Parkinson's disease, and Alzheimer's disease, have been linked to dysbiosis, or an imbalance in the gut microbiota. A "leaky gut" or intestinal permeability is favored by a sick microbiome, which permits toxic compounds from bacteria and inflammatory cytokines to enter the circulation and even reach the brain [9]. Accordingly, these might cause microglia to become activated, which would lead to neuroinflammation, a typical symptom of neurodegenerative diseases. In addition, SCFAs like butyrate, which are produced when gut bacteria ferment dietary fiber, may be impacted by dysbiosis in the gut microbiota. Because of its anti-inflammatory properties, SCFAs help maintain the blood–brain barrier, which protects the brain from harmful substances. SCFAs are therefore widely regarded as one of the crucial elements that support the maintenance of optimal brain health. Dietary practices may have an impact on brain health since they have a significant impact on gut microbiota. The production of SCFAs by gut bacteria is promoted by a diet rich in fiber, which is mostly found in fruits, vegetables, whole grains, and legumes. In addition to aiding in the reduction of gastrointestinal inflammation, these chemicals may potentially have neuroprotective effects via modulating inflammatory responses in the brain. Fermented foods like yogurt, kefir, kimchi, and sauerkraut not only contain fiber but also probiotics since they introduce good bacteria into the digestive system. Probiotics have been demonstrated to reduce neuroinflammation, enhance cognitive performance, and even lessen the symptoms of anxiety and despair that frequently accompany neurodegenerative disorders. They also help to restore microbial equilibrium.

Furthermore, by supporting gut and brain health, antioxidants, polyphenols, and omega-3 fatty acids can affect the gut–brain axis. Foods high in antioxidants, such as almonds, berries, and dark leafy greens, lessen oxidative stress, which can harm neurones and the stomach lining. Walnuts, flaxseeds, and fatty fish are good sources of omega-3 fatty acids, which have anti-inflammatory properties in the brain and intestines. These nutrients help prevent cognitive decline and neurodegeneration by promoting synaptic plasticity, lowering neuroinflammation, and repairing and maintaining neuronal membranes. The gut microbiota plays a key role in regulating brain health, making the gut–brain axis a crucial mechanism in the onset and progression of neurodegenerative disorders. Dietary variables may have a positive or negative impact on neurodegeneration and cognitive functioning by influencing oxidative stress, inflammatory processes, or the synthesis of neuroactive chemicals. The risk of neurodegenerative illnesses may be reduced and cognitive decline significantly lessened if an enabling gut microbiome is achieved through diet, which includes foods high in fiber, fermented foods, antioxidants, and omega-3 fatty acids. The gut–brain axis must be taken into account when developing dietary plans for brain health, according to new research [29].

15.6 NEW DIETARY THERAPIES FOR NEURODEGENERATIVE DISEASES AND CUSTOMIZED NUTRITION

An emerging strategy is personalized nutrition, in which nutritional interventions are customized based on lifestyle, health, and genetic background. It actually entails a customized strategy that takes into account each person's unique metabolic profile, gut microbiota, predispositions, and dietary response. This entails individualized dietary pattern modification in neurodegenerative illnesses to reduce the progression of the disease, prevent its onset, and improve overall brain health. Genetic variations, such as those in the APOE gene, increase an individual's risk of developing Alzheimer's disease and a number of other illnesses. Because those with the APOE ε4 allele are more likely to experience cognitive decline, it makes sense to modify nutrition based on a customized genetic approach, which would be crucial. Dietary modifications that focus on these genetic risk factors may lessen the effects of these differences, enhancing brain function and lowering the chance of neurodegenerative illnesses [30].

The focus of emerging dietary therapies is on adding bioactive substances that show promise for neuroprotection, such as antioxidants, polyphenols, omega-3 fatty acids, and certain vitamins and minerals. To lessen oxidative stress and inflammation in the brain, personalized nutrition may place an emphasis on consuming foods that are anti-inflammatory, such as those that contain polyphenols, like berries, green tea, and dark chocolate. Flaxseeds, walnuts, and fatty fish are good sources of omega-3 fatty acids, especially EPA and DHA, which have been shown to have neuroprotective effects by lowering inflammation and preserving the integrity of neuronal membranes. This comprises a personalized diet that contains the kind and quantity of substances that a given person's biological and metabolic demands require, meeting the brain's requirements and lowering the risk of neurodegeneration. In addition to providing insight into the possibilities for personalized nutrition in optimizing the gut microbiome, the gut–brain axis has come to be recognized as playing a vital role in the development of neurodegenerative diseases. The gut–brain axis may be modulated and neuroinflammation reduced by dietary practices that would preserve a healthy gut microbiota, such as consuming fermented foods, fiber-rich meals, and probiotic and prebiotic supplements. Personalized nutrition may be crucial in improving gut–brain communication pathways and providing novel treatment options for those at risk of neurodegenerative illnesses as microbiome research advances [31].

In conclusion, there is great potential for the prevention and treatment of neurodegenerative illnesses with personalized nutrition and recently developed dietary interventions. Personalized dietary methods might therefore optimize brain health and potentially slow the onset or progression of disorders like Alzheimer's, Parkinson's, and Huntington's by taking into account genetic, metabolic, and microbiome perspectives. With the ongoing advancements in nutritional science, genetics, and microbiome research, personalized nutrition may become increasingly important in integrated care strategies for neurodegenerative disorders, allowing for even more focused treatment and management of these problems.

15.7 CONCLUSION

The expanding discipline of nutritional neuroscience highlights the promise of utilizing food as a therapeutic agent to prevent and control neurodegenerative diseases. Functional foods, bioactive substances, and tailored nutrition techniques exhibit potential in regulating neuroinflammation, oxidative stress, and gut–brain connections, providing supplementary options to traditional therapies. As research advances, incorporating diet into the management of neurodegenerative diseases may become fundamental for maintaining cognitive function and enhancing patient outcomes.

REFERENCES

1. Tsalamandris G, Hadjivassiliou M, Zis P. The role of nutrition in neurological disorders. *Nutrients*. 2023;15(22):4713. https://doi.org/10.3390/nu15224713.
2. van Zonneveld SM, van den Oever EJ, Haarman BCM, Grandjean EL, Nuninga JO, van de Rest O, Sommer IEC. An anti-inflammatory diet and its potential benefit for individuals with mental disorders and neurodegenerative diseases - a narrative review. *Nutrients*. 2024;16(16):2646. https://doi.org/10.3390/nu16162646.
3. Patel V, Akimbekov NS, Grant WB, Dean C, Fang X, Razzaque MS. Neuroprotective effects of magnesium: Implications for neuroinflammation and cognitive decline. *Front Endocrinol (Lausanne)*. 2024;15:1406455. https://doi.org/10.3389/fendo.2024.1406455.
4. Mattson MP, Leak RK. The hormesis principle of neuroplasticity and neuroprotection. *Cell Metable*. 2024;36(2):315–337. https://doi.org/10.1016/j.cmet.2023.12.022.
5. Badaeva AV, Danilov AB, Clayton P, Moskalev AA, Karasev AV, Tarasevich AF, Vorobyeva YD, Novikov VN. Perspectives on neuronutrition in prevention and treatment of neurological disorders. *Nutrients*. 2023;15(11):2505. https://doi.org/10.3390/nu15112505.

6. Barati S, Fabrizio C, Strafella C, Cascella R, Caputo V, Megalizzi D, Peconi C, Mela J, Colantoni L, Caltagirone C, Termine A, Giardina E. Relationship between nutrition, lifestyle, and neurodegenerative disease: Lessons from *ADH1B*, *CYP1A2* and *MTHFR*. *Genes (Basel).* 2022;13(8):1498. https://doi.org/10.3390/genes13081498.
7. Samtiya M, Aluko RE, Dhewa T, Moreno-Rojas JM. Potential health benefits of plant food-derived bioactive components: An overview. *Foods*. 2021;10(4):839. https://doi.org/10.3390/foods10040839.
8. Seo Y, Gang G, Kim HK, Kim Y, Kang S, Kim H, Lee SG, Go GW. Effect of MIND diet on cognitive function in elderly: A narrative review with emphasis on bioactive food ingredients. *Food Sci Biotechnol*. 2023;33(2):297–306. https://doi.org/10.1007/s10068-023-01465-0
9. Park KJ, Gao Y. Gut-brain axis and neurodegeneration: Mechanisms and therapeutic potentials. *Front Neurosci*. 2024;18:1481390. https://doi.org/10.3389/fnins.2024.1481390.
10. Koblinsky ND, Power KA, Middleton L, Ferland G, Anderson ND. The role of the gut microbiome in diet and exercise effects on cognition: A review of the intervention literature. *J Gerontol A Biol Sci Med Sci*. 2023;78(2):195–205. https://doi.org/10.1093/gerona/glac166.
11. Di Renzo L, Gualtieri P, Romano L, Marrone G, Noce A, Pujia A, Perrone MA, Aiello V, Colica C, De Lorenzo A. Role of personalized nutrition in chronic-degenerative diseases. *Nutrients*. 2019;11(8):1707. https://doi.org/10.3390/nu11081707.
12. Palimariciuc M, Balmus IM, Gireadă B, Ciobica A, Chiriță R, Iordache AC, Apostu M, Dobrin RP. The quest for neurodegenerative disease treatment-focusing on Alzheimer's disease personalised diets. *Curr Issues Mol Biol*. 2023;45(2):1519–1535. https://doi.org/10.3390/cimb45020098.
13. Atlante A, Amadoro G, Bobba A, Latina V. Functional foods: An approach to modulate molecular mechanisms of Alzheimer's disease. *Cells*. 2020;9(11):2347. https://doi.org/10.3390/cells9112347.
14. Puri S, Shaheen M, Grover B. Nutrition and cognitive health: A life course approach. *Front Public Health*. 2023;11:1023907. https://doi.org/10.3389/fpubh.2023.1023907.
15. Houldsworth A. Role of oxidative stress in neurodegenerative disorders: A review of reactive oxygen species and prevention by antioxidants. *Brain Commun*. 2024;6(1):fcad356. https://doi.org/10.1093/braincomms/fcad356.
16. Sun Y, Koyama Y, Shimada S. Inflammation from peripheral organs to the brain: How does systemic inflammation cause neuroinflammation? *Front Aging Neurosci*. 2022;14:903455. https://doi.org/10.3389/fnagi.2022.903455.
17. Yip HYK, Papa A. Signaling pathways in cancer: Therapeutic targets, combinatorial treatments, and new developments. *Cells*. 2021;10(3):659. https://doi.org/10.3390/cells10030659.
18. Clemente-Suárez VJ, Redondo-Flórez L, Beltrán-Velasco AI, Ramos-Campo DJ, Belinchón-deMiguel P, Martinez-Guardado I, Dalamitros AA, Yáñez-Sepúlveda R, Martín-Rodríguez A, Tornero-Aguilera JF. Mitochondria and brain disease: A comprehensive review of pathological mechanisms and therapeutic opportunities. *Biomedicines*. 2023;11(9):2488. https://doi.org/10.3390/biomedicines11092488.
19. Lee J, Boo JH, Ryu H. The failure of mitochondria leads to neurodegeneration: Do mitochondria need a jump start? *Adv Drug Deliv Rev*. 2009;61(14):1316–1323. https://doi.org/10.1016/j.addr.2009.07.016.
20. Roy JRS, Janaki CS, Thalamati D, Mironescu M, Luo Q, Miao Y, Chai Y, Long Q. New strategies of neurodegenerative disease treatment with extracellular vesicles (EVs) derived from mesenchymal stem cells (MSCs). *Theranostics*. 2023;13(12):4138–4165. https://doi.org/10.7150/thno.83066.
21. Dehzad MJ, Ghalandari H, Nouri M, Askarpour M. Antioxidant and anti-inflammatory effects of curcumin/turmeric supplementation in adults: A GRADE-assessed systematic review and dose-response meta-analysis of randomized controlled trials. *Cytokine*. 2023;164:156144. https://doi.org/10.1016/j.cyto.2023.156144.
22. Grabska-Kobyłecka I, Szpakowski P, Król A, Książek-Winiarek D, Kobyłecki A, Głąbiński A, Nowak D. Polyphenols and their impact on the prevention of neurodegenerative diseases and development. *Nutrients*. 2023;15(15):3454. https://doi.org/10.3390/nu15153454.
23. Um CY, Peters BA, Choi HS, Oberstein P, Beggs DB, Usyk M, Wu F, Hayes RB, Gapstur SM, McCullough ML, Ahn J. Grain, gluten, and dietary fiber intake influence gut microbial diversity: Data from the food and microbiome longitudinal investigation. *Cancer Res Commun*. 2023;3(1):43–53. https://doi.org/10.1158/2767-9764.CRC-22-0154.
24. Hansen B, Roomp K, Ebid H, Schneider JG. Perspective: The impact of fasting and caloric restriction on neurodegenerative diseases in humans. *Adv Nutr*. 2024;15(4):100197. https://doi.org/10.1016/j.advnut.2024.100197.
25. Navale SS, Mulugeta A, Zhou A, Llewellyn DJ, Hyppönen E. Vitamin D and brain health: An observational and Mendelian randomization study. *Am J Clin Nutr*. 2022;116(2):531–540. https://doi.org/10.1093/ajcn/nqac107.

26. Thomas J, Thomas CJ, Radcliffe J, Itsiopoulos C. Omega-3 fatty acids in early prevention of inflammatory neurodegenerative disease: A focus on Alzheimer's disease. *Biomed Res Int*. 2015;2015:172801. https://doi.org/10.1155/2015/172801.
27. Del Río C, Segura-Carretero A. Neuroprotection with bioactive compounds. *Nutrients*. 2023;15(21):4612. https://doi.org/10.3390/nu15214612.
28. Sorrenti V, Burò I, Consoli V, Vanella L. Recent advances in health benefits of bioactive compounds from food wastes and by-products: Biochemical aspects. *Int J Mol Sci*. 2023 Jan 19;24(3):2019. https://doi.org/10.3390/ijms24032019
29. Fekete M, Lehoczki A, Major D, Fazekas-Pongor V, Csípő T, Tarantini S, Csizmadia Z, Varga JT. Exploring the influence of gut-brain axis modulation on cognitive health: A comprehensive review of prebiotics, probiotics, and symbiotics. *Nutrients*. 2024;16(6):789. https://doi.org/10.3390/nu16060789.
30. Zhang T, Liu Z, Mi Y. Editorial: Nutritional interventions on age-related neurodegenerative diseases. *Front Aging Neurosci*. 2023;15:1215586. https://doi.org/10.3389/fnagi.2023.1215586.
31. Ellouze I, Sheffler J, Nagpal R, Arjmandi B. Dietary patterns and Alzheimer's disease: An updated review linking nutrition to neuroscience. *Nutrients*. 2023;15(14):3204. https://doi.org/10.3390/nu15143204.

16 Flavonoids
Role in Neurodegenerative Studies

Swathi Nalla,[1] Nagaraju Bandaru,[2] Nayudu Teja,[3] and Vadabam Valasa Reddy[4]

[1]Department of Pharmacology, Malla Reddy Pharmacy College, Maisammaguda, Hyderabad

[2]Department of Pharmacology, School of Pharmaceutical Sciences (SOPS), Sandip University, Nasik, Maharashtra

[3]Department of Pharmaceutics, JSS College of Pharmacy, JSS Academy of Higher Education and Research, The Nilgiris, Tamil Nadu, India

[4]Department of Pharmaceutical Chemistry, Adarsa College of Pharmacy, G. Kothapalli, Andhra Pradesh, India.

16.1 NEURODEGENERATIVE DISEASES

Alzheimer's disease (AD), amyotrophic lateral sclerosis (ALS), Parkinson's disease (PD), and Huntington's disease (HD) are all manifestations of NDDs that are defined by the inevitable degradation of particular neuronal clusters. NDDs are a diverse, devastating, and incurable category of degenerative conditions. They are distinguished by a prolonged and cumulative loss of dopaminergic neurons, which in turn results in a steady and growing impairment of particular activities of the central nervous system (CNS). This loss of neuronal cells is the defining feature of these conditions (1). Once neurodegeneration has begun, it can only be slowed down; it cannot be completely stopped once it has started. This class of disorders presents a severe therapeutic challenge because of this.

In most cases, the subcortical areas are the first to experience the beginning stages of neurodegeneration, followed by the cortical areas as the disorder advances. The vital regions of the brain that are affected by neuronal death are a cause of the particular disorder, like striatal regions in PD, hippocampal and cortical regions in AD, striatal and cortical regions in HD, and spinal motor neurons and cortical regions in ALS (2). The rise in life expectancy that has been seen over the course of the past century has resulted in the creation of age-related neurological disorders that provide modern societies in semi and advanced industrialized nations with unique issues and call for active study. The age range of typical patients affected by NDDs is between 50 and 70 years old.

16.1.1 Overview of Neurodegenerative Disorders

Unfortunately, the only treatment option for many NDDs is palliative care, despite the fact that they have catastrophic consequences for the patient. The combination of high costs for formal care, an ever-increasing number of individuals who need it, and a disproportionate dependence on informal care is a significant challenge for society at large. New insights into the role of genetic and epigenetic factors, pharmaceuticals, aging, and infectious agents in the development of neurodegenerative illnesses have emerged as a result of advancements in biomedical research and informatics. The rising expense of healthcare is directly correlated to the rising prevalence of NDDs, which in turn is caused by the rapidly aging population (3).

In Europe, for instance, persons over 65 years of age currently make up around 16% of the population, but that number is predicted to climb to 25% by the year 2030. Dementia is the most

DOI: 10.1201/9781003478652-18

prevalent neurodegenerative disease, affecting over 7 million people throughout the globe. More than 10 million individuals worldwide are living with PD, while the prevalence varies greatly across demographics, including gender, age, and location. In the United States, an estimated USD 14.4 billion is spent annually on PD treatment, and it is predicted that this sum will quadruple by 2040. In Italy, however, the average cost per patient is just € 8,340. The market for PD therapies was worth $3.99 billion in 2016, and it may reach to worth of $5.69 billion by 2022, according to the Global Forecasts to 2022 (4). Because of the exponential rise in the average age of the world's population, AD has become a serious public health concern.

The prevalence of AD is reported to increase every 5 years beyond the age of 65, and it is envisaged that a greater proportion of the population will have AD in the USA, and will rise to 14 million by 2050 from 5.5 million in 2018. By contrast, by 2050, there will be around 135 million individuals with AD, an increase of 100 million from the number of persons with AD now (35 million). If the potential majors are not pursued, it's painful to contemplate the human toll and the rising costs associated with managing the disease's symptoms. Motor neurons in the brainstem, motor cortex, and spinal cord degenerate, causing ALS symptoms that worsen over time and ultimately prove deadly. ALS affects around 1–2 people per 100,000 people per year, and its incidence rate is between 6 and 8 per 100,000 people annually. Ninety-five to a majority (95%) of patients have a random etiology, whereas 5–10% have a hereditary one. Approximately 60% of ALS cases are found to be inherited. Patients with ALS exhibit spasms, twitching, speech difficulties, and weakness as symptoms of the disease (5). On average, two people with ALS live for about 19 months after being diagnosed and 30 months after the disease first shows up.

16.1.1.1 Important Diseases of NDDs

a. AD: The most prevalent NDD is AD, characterized by a slow but steady decline in mental faculties along with other symptoms of neurodegeneration, including altered behavior and psychological states. The hippocampus and cerebral cortex, in particular, suffer extensive neuronal, synaptic, and neurotransmitter loss in AD. AD is characterized by neuronal degeneration, extracellular senile plaques laden with the peptide amyloid (A), and neurofibrillary tangles laden with a hyperphosphorylated variant of the tubulin protein tau (6).
b. PD: The second most frequent kind of NDD is PD, and it is defined by a gradual deterioration of motor and nonmotor functions (bradykinesia, muscular stiffness, resting tremor, and postural instability due to the dearth of catecholaminergic neurons in the substantia nigra (SN). Motor symptoms of PD are connected to the loss of pigmented cells in the SN and the subsequent decrease in dopamine synthesis in the striatum. Lewy bodies and protein aggregates, including ubiquitin and alpha-synuclein, are also present in the cytoplasm of dying nerve cells and serve as a diagnostic marker of this disorder.
c. HD: Chorea is the most prominent motor sign of HD, although the disorder also causes personality shifts, mental disorders, and cognitive decline, such as dementia. Long projection neurons located in both brain and striatum are lost and die off at an abnormally young age, defining this pathology. HD is characterized by the loss of striatal medium spiny GABAergic neurons, an inherited disorder that is inherited in a dominant fashion. The caudate nucleus, globus pallidus, and putamen all gradually atrophy due to this neurodegeneration. The CAG repeats expansion in exon 1 of the gene producing the protein huntingtin is the genetic basis for HD.
d. ALS: As the disease progresses, it weakens and wastes away the skeletal muscles, eventually leading to paralysis and death between 2 and 5 years after diagnosis. Nerve cells in the brainstem, spinal cord's ventral horn, and the motor cortex gradually deteriorate, leading to ALS. Nevertheless, the preponderance of individuals have adequate superoxide dismutase 1 (SOD1) activity despite the fact that about 20% of patients have a mutation in the SOD1 gene (7).

16.1.1.2 Typical Characteristics of NDDs

Each of the NDDs is characterized by its own unique set of signs and symptoms, neurological effects, and biological root causes. The pathophysiology of NDDs is complicated and multifaceted since NDDs represent a diverse group of diseases. It has been speculated that cellular features such as aberrant protein dynamics, breakdown of the neuronal Golgi apparatus, oxidative stress, molecular chaperones, disruption of cellular/axonal transit, choline deficiency, modifications in ion channels, the neuroinflammatory system, deficits of neurotrophins, and neurotoxicity all make a contribution to neuronal loss (Figure 16.1). Multifactorial, with hereditary and environmental components interacting, describes the pathoethiology of NDDs well (8).

The pathophysiological hallmarks of AD in the brain are the extracellular amyloid-beta (A) plaques and NFT development, and the destruction of synaptic connections among neurons crucial in learning and memory. Frontotemporal dementia accounts for 20% of all cases of dementia with an early onset and is characterized by a progressive loss of neurons in the temporal and frontal lobes, resulting in a wide multitude of clinical symptoms that can be broken down into three distinct groups: behavioral dysfunctions variant, progressive nonfluent aphasia and semantic dementia. Forty percent of FTD patients have a favorable family history, indicating a significant genetic component. The second most prominent kind of progressive adult-onset ND, PD, primarily affects the motor system and is characterized by tremors, muscular stiffness, bradykinesia, postural impairment, and speech and cognitive deficits. Misfolded-synuclein proteins accumulate abnormally in neurons as Lewy bodies, leading to degeneration of dopamine-generating neurons of the pars compacta in 4the SN of the middle brain. ALS results from motor neuron degeneration in the brain stem, cerebral cortex, and spinal cord, all of which are involved in the regulation of the patient's voluntary muscles. Neuronal degeneration is associated with a number of different processes, including protein inclusion aggregates, oxidative stress, mitochondrial dysfunction, apoptosis, and glutamate toxicity (9).

16.1.2 Mechanisms of Neurodegeneration

Neurodegeneration refers to the progressive loss of structure or function of neurons, which may eventually lead to their death. This phenomenon underlies a wide range of neurological disorders,

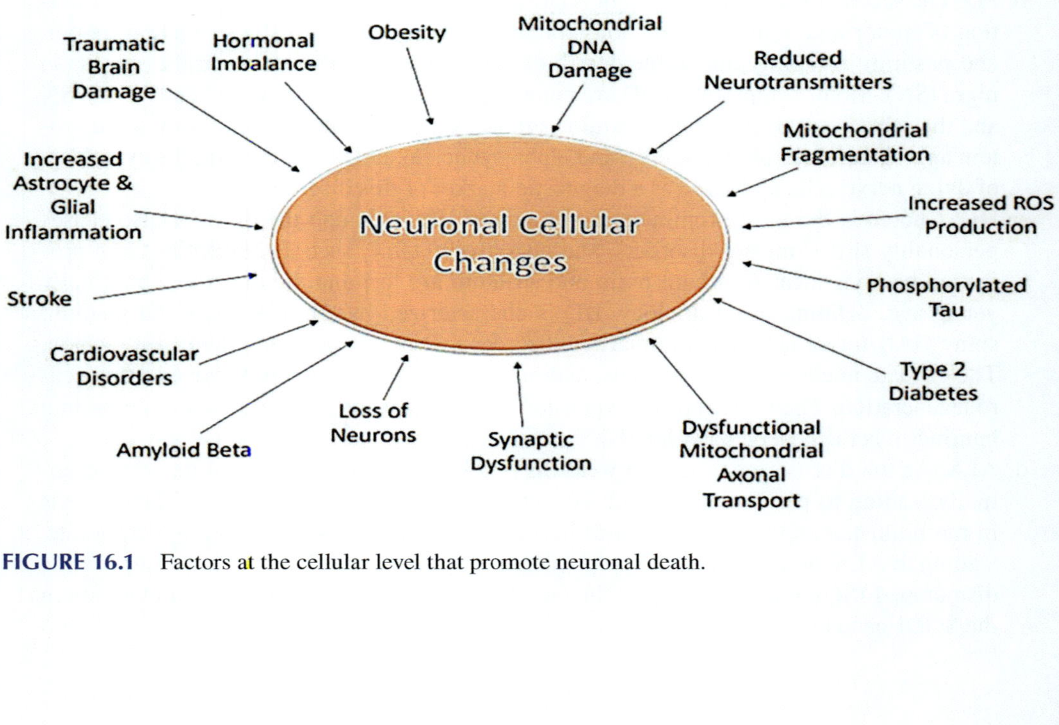

FIGURE 16.1 Factors at the cellular level that promote neuronal death.

including AD, PD, HD, and ALS. Understanding the mechanisms driving neurodegeneration is critical for developing targeted therapies. In the following text, we explore the principal pathways implicated in neuronal injury and death (10).

1. Protein Misfolding and Aggregation
   Protein homeostasis is vital for neuronal health. In many neurodegenerative disorders, abnormal folding and aggregation of specific proteins are key features:
   AD: Characterized by extracellular amyloid-beta plaques and intracellular tau tangles.
   PD: Marked by alpha-synuclein aggregates forming Lewy bodies.
   HD: Caused by expanded polyglutamine tracts in the huntingtin protein.
   Mechanism: Misfolded proteins may evade normal degradation pathways, accumulate, and form toxic aggregates, disrupting cellular processes and inducing oxidative stress.
2. Mitochondrial Dysfunction
   Neurons are highly dependent on mitochondrial energy production, making them particularly vulnerable to mitochondrial defects (11).
   - Impaired oxidative phosphorylation leads to reduced ATP production.
   - Excessive generation of reactive oxygen species (ROS) damages proteins, lipids, and DNA.
   - Mitochondrial dysfunction is a common feature in PD, ALS, and other disorders.

   Mechanism: Mitochondrial damage can initiate apoptosis and exacerbate cellular energy deficits.
3. **Oxidative Stress**
   Oxidative stress arises when the production of ROS exceeds the capacity of antioxidant defenses.
   Sources: Mitochondria, inflammatory cells, and environmental toxins.
   Consequences: ROS react with cellular components, leading to lipid peroxidation, protein modification, and DNA damage.
   Oxidative stress is a hallmark of many NDDs and often acts synergistically with other pathological mechanisms (Figure 16.2).
4. **Impaired Protein Clearance Pathways**
   - The ubiquitin-proteasome system (UPS) and autophagy-lysosome pathway are crucial for removing damaged or misfolded proteins.
   - Dysfunction in these systems leads to protein accumulation and aggregate formation.

   Examples: Impaired UPS function is implicated in PD and HD, while autophagy defects are associated with AD and ALS.
   Mechanism: Reduced proteolysis contributes to cellular toxicity and neurodegeneration.
5. **Neuroinflammation**
   - Chronic activation of microglia and astrocytes can result in harmful neuroinflammation.
   - Activated glial cells release pro-inflammatory cytokines, ROS, and other mediators that exacerbate neuronal damage.
   - Neuroinflammation is a key feature of AD, PD, and multiple sclerosis (MS).

   Mechanism: While acute inflammation may protect against infection or injury, chronic inflammation disrupts neuronal homeostasis.
6. **Excitotoxicity**
   - Excitotoxicity refers to neuronal damage caused by excessive stimulation of glutamate receptors.
   - Overactivation of *N*-methyl-D-aspartate (NMDA) and other glutamate receptors results in calcium influx.

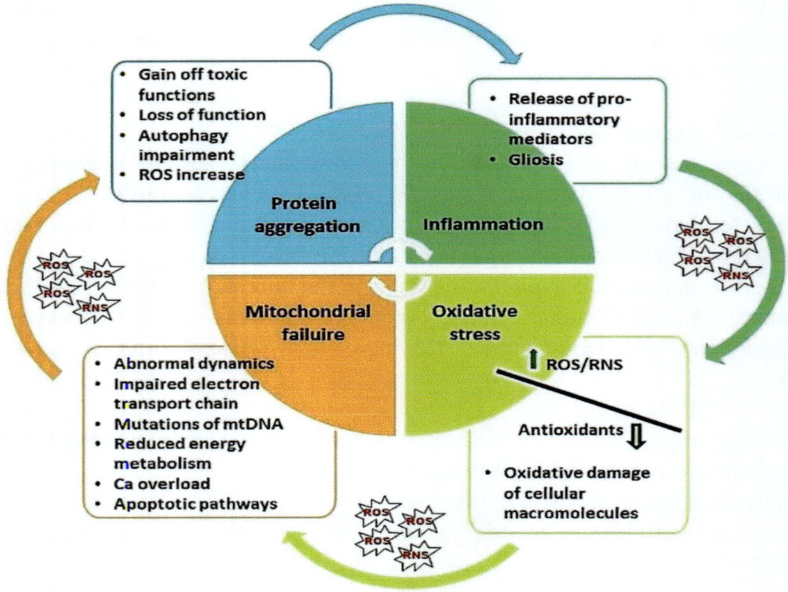

FIGURE 16.2 A vicious loop of oxidative stress, mitochondrial dysfunction, protein aggregation, and inflammation that contributes to neurodegenerative disorders.

- Elevated intracellular calcium triggers harmful processes such as mitochondrial dysfunction and activation of degradative enzymes.
- Excitotoxicity is implicated in stroke, traumatic brain injury, and chronic NDDs.

7. **Axonal Transport Defects**
   - Efficient transport of organelles, proteins, and vesicles along axons is essential for neuronal function.
   - Defects in microtubules or motor proteins disrupt axonal transport, leading to synaptic dysfunction and neuronal degeneration.

   Examples: Tau pathology in AD interferes with microtubules, while similar disruptions are observed in ALS and HD.
8. **Genetic and Epigenetic Factors**

   Mutations in specific genes underlie several NDDs:
   AD: Mutations in APP, PSEN1, and PSEN2.
   PD: Mutations in SNCA, LRRK2, and PINK1.
   HD: Expansion of CAG repeats in the huntingtin gene.
   Epigenetic modifications, including DNA methylation and histone acetylation, also regulate gene expression and may contribute to disease progression.
9. **Vascular Contributions**
   - Cerebrovascular dysfunction often exacerbates neurodegenerative processes.
   - Compromised blood–brain barrier integrity allows neurotoxic substances to enter the brain.
   - Reduced cerebral perfusion leads to hypoxia and energy deficits.
   - Vascular contributions are prominent in vascular dementia and mixed dementia cases.

16.1.3 Importance of Research in Neurodegenerative Disease Prevention and Treatment

The study of neurodegeneration is crucial for developing preventive strategies and therapeutic interventions. NDDs impose significant societal and economic burdens, affecting millions worldwide (12).

Early Detection and Prevention: Research enables the identification of biomarkers for early diagnosis, offering opportunities for preventive measures before significant neuronal loss occurs.

Therapeutic Development: Understanding the underlying mechanisms of neurodegeneration aids in the design of targeted therapies, including drugs, gene editing, and cell-based treatments.

Improved Quality of Life: Advances in research contribute to interventions that enhance the quality of life for affected individuals and their caregivers.

Societal Impact: Effective prevention and treatment reduce healthcare costs and improve productivity, benefiting society as a whole.

16.2 FLAVONOID BIOAVAILABILITY AND MECHANISM OF ACTION

16.2.1 Absorption and Metabolism of Flavonoids

Flavonoids are polyphenolic compounds found in various fruits, vegetables, and beverages. Their health benefits depend on their bioavailability, which involves absorption, metabolism, and distribution in the body.

Absorption: Flavonoids are absorbed in the small intestine and undergo extensive metabolism in the liver.

Metabolism: Phase I and Phase II metabolic processes modify flavonoids into metabolites, which may have distinct biological activities.

Factors Influencing Absorption: Food matrix, gut microbiota composition, and individual genetic differences affect flavonoid bioavailability (13).

16.2.2 Mechanisms by which Flavonoids Affect the Brain

Flavonoids exert neuroprotective effects through multiple mechanisms:

Antioxidant Activity: Scavenging ROS and enhancing endogenous antioxidant defenses.

Anti-inflammatory Effects: Modulating microglial activation and reducing pro-inflammatory cytokines.

Synaptic Plasticity: Enhancing synaptic function and promoting neurogenesis.

Neurovascular Protection: Improving blood–brain barrier integrity and cerebral blood flow.

16.2.3 Bioavailability Challenges in Neurodegenerative Disease Research

Despite their potential, flavonoid-based interventions face challenges:

Low Bioavailability: Rapid metabolism and poor absorption limit their effective concentrations in the brain.

Blood–Brain Barrier: Limited penetration into the brain reduces therapeutic efficacy.

Variability in Efficacy: Differences in individual metabolism and gut microbiota affect flavonoid activity.

Solutions: Research is exploring advanced delivery systems (e.g., nanoparticles) and synthetic derivatives to enhance bioavailability and target specificity (14).

16.3 FLAVONOIDS AND OXIDATIVE STRESS

16.3.1 Role of Oxidative Stress in NDDs

Oxidative stress plays a pivotal role in the pathogenesis of NDDs:

> Oxidative stress plays a crucial role in the development and progression of NDDs, such as AD, PD, HD, and ALS. It results from an imbalance between the production of ROS and reactive nitrogen species (RNS) and the body's ability to detoxify these reactive intermediates or repair the resulting damage (15). The following mechanisms highlight the role of oxidative stress in neurodegeneration:

1. **Lipid Peroxidation**
   Neurons have a high content of polyunsaturated fatty acids, which are susceptible to oxidation. Oxidative stress triggers lipid peroxidation, leading to the formation of toxic aldehydes like malondialdehyde (MDA) and 4-hydroxynonenal (HNE). These products damage cellular membranes, disrupt ion homeostasis, and impair synaptic transmission.
2. **Protein Oxidation**
   ROS and RNS can oxidize proteins, causing structural changes, aggregation, or loss of function. For instance:
   - In AD, oxidative stress promotes the aggregation of amyloid-beta (Aβ) and the hyperphosphorylation of tau, leading to neurofibrillary tangles.
   - In PD, oxidative modifications of alpha-synuclein contribute to Lewy body formation.
3. **DNA Damage**
   ROS can damage nuclear and mitochondrial DNA, leading to mutations, impaired gene expression, and defective mitochondrial function. Mitochondrial DNA damage exacerbates oxidative stress by impairing the electron transport chain, creating a vicious cycle of ROS production.
4. **Mitochondrial Dysfunction**
   Mitochondria are both a source and target of ROS. Oxidative stress impairs mitochondrial function, reducing ATP production and triggering apoptosis. Mitochondrial dysfunction is a common feature in NDDs.
5. **Neuroinflammation**
   Oxidative stress activates microglia and astrocytes, leading to the release of pro-inflammatory cytokines and further ROS/RNS production. Chronic neuroinflammation exacerbates neuronal damage and contributes to disease progression.
6. **Glutamate Excitotoxicity**
   Oxidative stress can impair the regulation of glutamate uptake and metabolism, leading to excitotoxicity. Excessive glutamate activity causes calcium influx and further ROS production, ultimately damaging neurons.

16.3.1.1 Evidence in NDDs

- **AD:** Elevated levels of oxidative markers, such as MDA and oxidized DNA bases, are observed in the brains of AD patients. Aβ plaques induce oxidative stress by interacting with mitochondria and disrupting cellular functions.

- **PD:** Oxidative stress from dopamine metabolism and mitochondrial dysfunction contributes to the degeneration of dopaminergic neurons in the SN.
- **HD:** Mutant huntingtin protein increases oxidative damage to neurons, particularly in the striatum.
- **ALS:** Mutations in superoxide dismutase 1 (SOD1) impair ROS detoxification, leading to motor neuron death.

16.3.1.2 Therapeutic Implications

Targeting oxidative stress is a promising strategy for NDDs. Antioxidants (e.g., vitamin E, coenzyme Q10, and *N*-acetylcysteine) and compounds enhancing endogenous antioxidant defenses (e.g., Nrf2 activators) are under investigation. Strategies to protect mitochondria, reduce neuroinflammation, and modulate glutamate excitotoxicity are also being explored.

Understanding the interplay between oxidative stress and other pathological processes can aid in developing multifaceted therapeutic approaches for NDDs (see Figure 16.3).

16.3.2 Antioxidant Properties of Flavonoids

Flavonoids are a diverse group of naturally occurring polyphenolic compounds found in fruits, vegetables, tea, wine, and other plant-based foods. They exhibit significant antioxidant properties, which play a critical role in protecting cells from oxidative damage and mitigating the effects of ROS and RNS. The antioxidant potential of flavonoids arises from their unique chemical structures, which allow them to act through various mechanisms.

16.3.2.1 Mechanisms of Antioxidant Action

1. **Direct Scavenging of Free Radicals**
   Flavonoids neutralize ROS/RNS (e.g., superoxide anion, hydroxyl radicals, peroxynitrite) by donating hydrogen atoms or electrons, stabilizing these reactive species and preventing cellular damage.

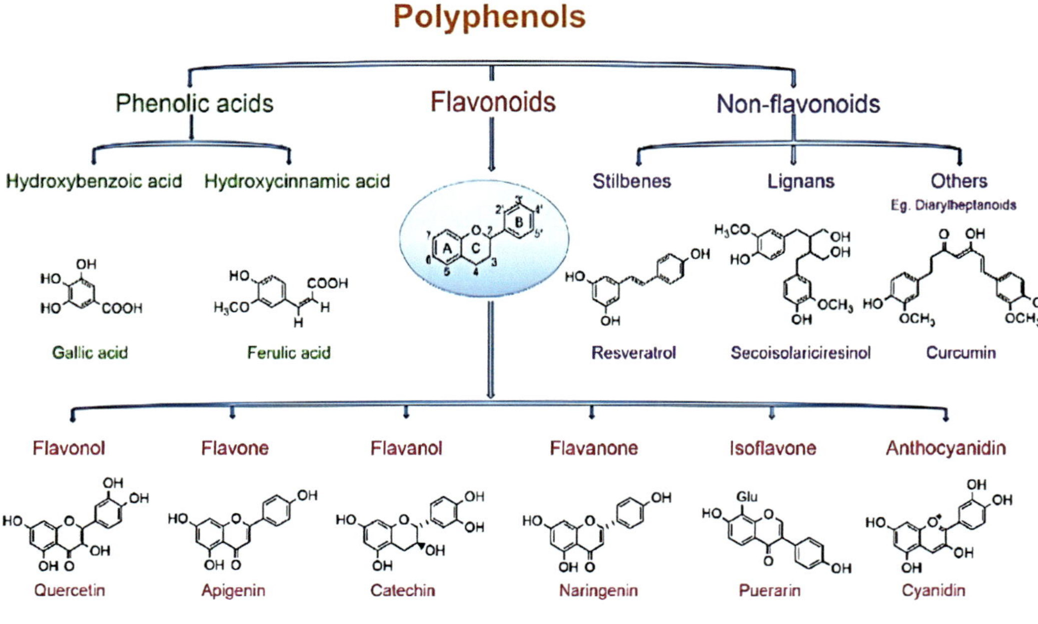

FIGURE 16.3 Chemical classification of polyphenolic compounds (Maher, 2019).

2. **Metal Chelation**
   Transition metals like iron and copper can catalyze the formation of ROS through Fenton and Haber-Weiss reactions. Flavonoids chelate these metals, reducing their availability for ROS generation.
3. **Regeneration of Endogenous Antioxidants**
   Flavonoids support the regeneration of endogenous antioxidants such as glutathione, vitamin C, and vitamin E, enhancing the body's natural defense mechanisms against oxidative stress.

16.3.2.2 Inhibition of Pro-Oxidant Enzymes

Flavonoids inhibit enzymes like xanthine oxidase, NADPH oxidase, and lipoxygenases, which contribute to ROS production.

16.3.2.3 Modulation of Redox Signaling Pathways

Flavonoids influence cellular signaling pathways, such as those involving nuclear factor erythroid 2-related factor 2 (Nrf2), leading to the upregulation of antioxidant enzymes like superoxide dismutase (SOD), catalase, and glutathione peroxidase.

16.3.2.4 Structural Features Contributing to Antioxidant Activity

The antioxidant potency of flavonoids is determined by their molecular structure:

- **Hydroxyl Groups**: The presence and position of hydroxyl groups enhance the ability of flavonoids to donate hydrogen and scavenge radicals.
- **Double Bonds and Carbonyl Groups**: A C2=C3 double bond in conjugation with a 4-oxo group increases the stability of radical forms and enhances metal-chelating properties.
- **B-Ring Catechol Structure**: Flavonoids with hydroxyl groups at positions 3' and 4' on the B-ring (e.g., quercetin) exhibit strong free radical-scavenging activity (16).

16.3.2.5 Examples of Antioxidant Flavonoids

1. **Quercetin:** Found in onions, apples, and tea, quercetin has potent free radical-scavenging activity and can chelate metal ions effectively.
2. **Epicatechin**
   Present in green tea and cocoa, epicatechin exhibits strong antioxidant properties and contributes to cardiovascular and neuroprotective effects.
3. **Kaempferol**
   Found in broccoli, kale, and beans, kaempferol scavenges ROS and induces endogenous antioxidant enzyme expression.
4. **Anthocyanins**
   Responsible for the red, blue, and purple colors in berries and grapes, anthocyanins combat oxidative stress and reduce inflammation.

16.3.2.6 Health Benefits of Flavonoid Antioxidants (14, 16)

Neuroprotection

- Flavonoids protect neurons from oxidative damage, reduce neuroinflammation, and may slow the progression of NDDs like Alzheimer's and Parkinson's.

Cardiovascular Health

- By reducing oxidative stress, flavonoids prevent lipid peroxidation, improve endothelial function, and reduce the risk of atherosclerosis.

Anti-Aging Effects

- Flavonoids protect cellular components (DNA, proteins, lipids) from oxidative damage, delaying aging and related diseases.

Cancer Prevention

- Flavonoids can prevent oxidative DNA damage, inhibit tumor growth, and induce apoptosis in cancerous cells.

Anti-Inflammatory Effects

- Flavonoids modulate inflammatory pathways, partly through their antioxidant actions, reducing chronic inflammation linked to various diseases.

Flavonoids serve as a vital component of dietary antioxidants, contributing to the prevention and management of oxidative stress-related diseases. Their dual action as direct ROS scavengers and modulators of endogenous antioxidant defenses makes them promising candidates for therapeutic interventions aimed at oxidative stress and its pathological consequences.

16.3.3 Evidence of Flavonoid Efficacy in Reducing Oxidative Damage

Numerous studies, both in vitro and in vivo, have demonstrated the efficacy of flavonoids in reducing oxidative damage and mitigating the associated cellular and systemic consequences. The antioxidant properties of flavonoids, coupled with their ability to modulate cellular signaling pathways, have shown promise in protecting against oxidative stress-induced damage in various disease models (12, 17).

1. In Vitro Evidence

a. Free Radical-Scavenging Assays

Studies using assays like DPPH (2,2-diphenyl-1-picrylhydrazyl) and ABTS (2,2'-azino-bis(3-ethylbenzothiazoline-6-sulfonic acid)) consistently demonstrate the strong radical-scavenging activity of flavonoids like quercetin, epicatechin, and rutin.

Quercetin exhibits higher antioxidant activity than common antioxidants like vitamin C in scavenging superoxide anions and hydroxyl radicals.

i. **Lipid Peroxidation Inhibition**

Flavonoids have been shown to inhibit lipid peroxidation in cell membranes. For instance, catechins from green tea protect lipid-rich environments from oxidative degradation.

b. Protection against Oxidative DNA Damage

In studies using human lymphocytes exposed to oxidative stress, flavonoids such as kaempferol and luteolin significantly reduced DNA strand breaks and base modifications caused by ROS.

2. In Vivo Evidence

a. Neurodegenerative Disease Models

i. **AD:**

Flavonoids like epigallocatechin gallate (EGCG) from green tea reduce amyloid-beta (Aβ)-induced oxidative damage in neuronal cells, as evidenced in animal models of AD.

   ii. **PD:**
   Naringenin, a citrus flavonoid, attenuates dopaminergic neuronal loss and oxidative stress in rodent models of PD.

b. **Cardiovascular Protection**
   i. Clinical studies show that flavonoid-rich diets reduce biomarkers of oxidative stress, such as malondialdehyde (MDA) and oxidized low-density lipoprotein.
   ii. A study on cocoa flavonoids demonstrated a reduction in oxidative stress markers and improved endothelial function in individuals with cardiovascular risk factors.

c. **Cancer Models**
   i. In rodent models of chemically induced carcinogenesis, flavonoids such as genistein and hesperidin reduce oxidative damage to DNA and inhibit tumor growth.

d. **Diabetes and Metabolic Disorders**
   Flavonoids like hesperidin and rutin lower ROS levels and improve antioxidant enzyme activity in diabetic rats, mitigating complications like neuropathy and nephropathy.

16.4 CLINICAL EVIDENCE

16.4.1 Dietary Flavonoid Intake and Biomarkers

Epidemiological studies reveal that individuals consuming diets rich in flavonoids exhibit lower levels of oxidative stress biomarkers, such as F2-isoprostanes and MDA.

16.4.1.1 Interventional Studies

A clinical trial involving green tea extract supplementation showed decreased oxidative DNA damage and increased activity of endogenous antioxidants, such as SOD and glutathione peroxidase.

Anthocyanin-rich berry supplementation improved oxidative stress markers and lipid profiles in patients with metabolic syndrome.

16.4.1.2 Flavonoid Supplementation

Quercetin supplementation has been shown to reduce oxidative damage in endurance athletes by decreasing markers of lipid peroxidation and protein oxidation.

1. **Molecular Evidence**

Upregulation of Antioxidant Pathways

Flavonoids activate the Nrf2 signaling pathway, leading to the upregulation of antioxidant enzymes like SOD, catalase, and glutathione peroxidase.

Inhibition of Oxidative Stress-Induced Apoptosis

a. In models of neuronal injury, flavonoids like luteolin prevent apoptosis by inhibiting ROS-mediated mitochondrial dysfunction and caspase activation.

Reduction of Inflammatory Oxidative Stress

a. Flavonoids suppress the activity of NADPH oxidase and pro-inflammatory enzymes, such as cyclooxygenase-2 (COX-2), reducing ROS production in inflamed tissues.

The antioxidant efficacy of flavonoids is supported by extensive evidence from in vitro studies, animal models, and clinical trials. These compounds exhibit a broad spectrum of protective effects against oxidative stress and associated pathologies. Their integration into dietary and therapeutic interventions offers a promising strategy for reducing oxidative damage and improving health outcomes (17, 18).

16.5 FLAVONOID MECHANISMS IN NEUROPROTECTION

Flavonoids play a significant role in protecting the nervous system against damage and dysfunction. Through their antioxidant, anti-inflammatory, and signaling-modulatory properties, they offer protection against various neurological disorders. This section explores the mechanisms by which flavonoids exert their neuroprotective effects.

16.5.1 Modulation of Inflammation and Immune Response

Neuroinflammation is a hallmark of many NDDs, including AD, PD, and MS. Flavonoids modulate the inflammatory response in the brain through the following mechanisms:

1. **Inhibition of Microglial Activation**
   Flavonoids like quercetin and luteolin suppress the overactivation of microglia, the resident immune cells of the CNS, thereby reducing the release of pro-inflammatory cytokines (e.g., TNF-α, IL-6) and ROS.
2. **Downregulation of Pro-inflammatory Pathways**
   Flavonoids inhibit key inflammatory signaling pathways, including nuclear factor-kappa B (NF-κB) and mitogen-activated protein kinases, thereby reducing the production of inflammatory mediators like prostaglandins and nitric oxide.
3. **Regulation of Peripheral Immune Cells**
   Some flavonoids modulate the infiltration of peripheral immune cells into the CNS by regulating the blood–brain barrier (BBB) integrity and chemokine expression, preventing excessive neuroinflammation.
4. **Promotion of Anti-inflammatory Cytokines**
   Flavonoids enhance the production of anti-inflammatory cytokines such as IL-10, which counteracts the effects of pro-inflammatory molecules and promotes resolution of inflammation (19).

Example:

- Epigallocatechin gallate (EGCG) from green tea attenuates neuroinflammation in rodent models of NDDs by reducing microglial activation and suppressing NF-κB signaling.

16.5.2 Flavonoids and Mitochondrial Function

Mitochondria are central to cellular energy production and play a critical role in neuronal health. Dysfunctional mitochondria contribute to oxidative stress, energy deficits, and apoptosis in NDDs. Flavonoids help maintain mitochondrial function in the following ways:

1. **Antioxidant Protection**
   - Flavonoids scavenge mitochondrial ROS, reducing oxidative damage to mitochondrial DNA, lipids, and proteins.
2. **Stabilization of Mitochondrial Membranes**
   - Flavonoids prevent the loss of mitochondrial membrane potential, preserving mitochondrial integrity and function.

3. **Enhancement of Mitochondrial Biogenesis**
   - Certain flavonoids activate the peroxisome proliferator-activated receptor-gamma coactivator 1-alpha (PGC-1α) pathway, promoting mitochondrial biogenesis and improving energy metabolism.
4. **Regulation of Apoptotic Pathways**
   - Flavonoids modulate mitochondrial-mediated apoptosis by inhibiting cytochrome c release and caspase activation, protecting neurons from programmed cell death.
5. **Support of Mitochondrial Dynamics**
   - Flavonoids like resveratrol improve mitochondrial dynamics, balancing fission and fusion processes, which are critical for mitochondrial quality control.

Example:

- Resveratrol has been shown to improve mitochondrial function in animal models of Alzheimer's and Parkinson's diseases by enhancing antioxidant defenses and promoting mitochondrial biogenesis.

16.5.3 Effects on Neurogenesis and Synaptic Plasticity

Neurogenesis and synaptic plasticity are essential processes for learning, memory, and brain repair. Flavonoids positively influence these processes through the following mechanisms:

1. **Stimulation of Brain-Derived Neurotrophic Factor (BDNF)**
   - Flavonoids enhance the expression of BDNF, a key neurotrophin involved in neuronal survival, growth, and synaptic plasticity.
   - BDNF activation promotes the TrkB signaling pathway, enhancing neurogenesis in the hippocampus.
2. **Promotion of Neurogenesis**
   a Dietary flavonoids like epicatechin stimulate the proliferation and differentiation of neural stem cells in the hippocampus, a brain region critical for memory.
3. **Enhancement of Synaptic Plasticity**
   - Flavonoids modulate synaptic strength by influencing long-term potentiation and long-term depression, mechanisms underlying learning and memory.
4. **Reduction of Synaptic Loss**
   - By mitigating oxidative stress and neuroinflammation, flavonoids protect synapses from damage in NDDs.
5. **Regulation of Ion Channels and Receptors**
   - Flavonoids interact with ion channels (e.g., calcium channels) and neurotransmitter receptors (e.g., NMDA and AMPA receptors), facilitating synaptic signaling and plasticity.

Example:

- Blueberry anthocyanins improve spatial memory in aged animals by increasing BDNF levels and enhancing hippocampal neurogenesis and synaptic plasticity.

Flavonoids exhibit neuroprotective effects by modulating neuroinflammation, preserving mitochondrial function, and promoting neurogenesis and synaptic plasticity. These mechanisms highlight their potential as therapeutic agents for the prevention and management of NDDs and cognitive decline. Further studies and clinical trials are needed to translate these findings into effective interventions for human health.

Despite the promising therapeutic potential of flavonoids, there are significant challenges and limitations in the research and application of these compounds. These issues arise from variability in experimental design, limitations in bioavailability, and challenges in translating preclinical findings into clinical success.

16.5.4 Variability in Study Design and Outcomes

1. **Differences in Experimental Models**
   a. **In vitro vs. In vivo Studies**: Many flavonoid studies rely on in vitro models, which do not fully replicate the complexity of living organisms. In vivo studies often yield different outcomes due to factors like metabolism, absorption, and interactions with other biological systems.
   b. **Animal Models**: Preclinical studies frequently use rodent models, which may not accurately predict human responses due to differences in physiology and metabolism.
2. **Heterogeneity of Flavonoid Sources**
   a. Flavonoids are derived from diverse dietary sources and supplements, often with varying purity, extraction methods, and concentrations, leading to inconsistencies in study outcomes.
3. **Dosing and Duration**
   a. Studies employ different doses and treatment durations, making it challenging to establish optimal therapeutic ranges for flavonoids.
   b. High concentrations used in in vitro studies may not be achievable in vivo due to limitations in bioavailability.
4. **Lack of Standardized Outcomes**
   a. Variability in biomarkers and endpoints used to assess flavonoid efficacy (e.g., antioxidant capacity, neuroprotection, cognitive improvement) complicates comparisons across studies.
5. **Interindividual Variability**
   b. Human studies show significant variability in flavonoid responses due to genetic differences, microbiome composition, diet, and lifestyle.

16.5.5 Issues with Flavonoid Bioavailability

Flavonoids often have poor bioavailability, which limits their therapeutic potential. Key factors contributing to this issue include:

1. **Low Absorption**
   a. Flavonoids are typically present as glycosides in foods, requiring enzymatic hydrolysis for absorption. The efficiency of this process varies between individuals.
   b. The absorption of flavonoids across the intestinal epithelium is limited due to their size and polarity.
2. **Rapid Metabolism**
   a. Flavonoids are extensively metabolized in the liver and intestines, forming glucuronides, sulfates, and methylated derivatives, which may have reduced biological activity.
3. **Poor Stability**
   a. Flavonoids are susceptible to degradation under physiological conditions, such as changes in pH and exposure to digestive enzymes.
4. **Short Half-life**
   a. Rapid clearance of flavonoids from the bloodstream limits their sustained effects, necessitating frequent dosing or alternative delivery methods.

5. **Impact of Gut Microbiota**
   a. Gut bacteria play a critical role in metabolizing flavonoids, and variations in microbiota composition between individuals can lead to inconsistent bioavailability and efficacy.

Potential Solutions:

- Development of flavonoid analogs with improved bioavailability.
- Encapsulation technologies, such as liposomes or nanoparticles, to enhance stability and absorption.
- Co-administration with other compounds (e.g., piperine) to inhibit flavonoid metabolism and improve systemic exposure.

16.5.6 Translational Challenges in Clinical Applications

1. **Limited Clinical Evidence**
   - While preclinical studies show promising results, there is a paucity of high-quality clinical trials validating flavonoid efficacy in humans.
   - Many clinical studies lack robust design features such as randomized controls, adequate sample sizes, and long-term follow-up.
2. **Complexity of Flavonoid Mechanisms**
   - The multifunctional nature of flavonoids complicates the identification of specific molecular targets and pathways for therapeutic interventions.
3. **Dietary versus Pharmaceutical Use**
   - The therapeutic potential of dietary flavonoids is limited by their concentrations in foods, which are often insufficient for clinical efficacy.
     Isolated flavonoid supplements may not replicate the synergistic effects observed in whole foods.
4. **Regulatory Hurdles**
   Flavonoids are often classified as dietary supplements, which are subject to less stringent regulations than pharmaceuticals. This can lead to variability in product quality and hinder their acceptance as therapeutic agents.
5. **Safety Concerns**
   - High doses of flavonoids may have pro-oxidant effects, interfere with drug metabolism, or cause adverse effects in specific populations (e.g., individuals with compromised liver or kidney function).
6. **Economic and Practical Constraints**
   - The cost and scalability of flavonoid extraction, synthesis, and delivery systems can limit their widespread adoption in clinical settings (18, 19).

The challenges in flavonoid research, including variability in study outcomes, issues with bioavailability, and translational hurdles, highlight the need for rigorous and standardized approaches to studying these compounds. Advancing flavonoid research will require:

- Improved experimental designs.
- Development of novel delivery systems.
- Comprehensive clinical trials.

Overcoming these obstacles can pave the way for the effective integration of flavonoids into therapeutic strategies for managing oxidative stress-related and NDDs.

16.6 FUTURE DIRECTIONS AND THERAPEUTIC POTENTIAL

Flavonoids hold significant promise for the prevention and treatment of NDDs, and ongoing research continues to explore their therapeutic potential. Future directions include identifying promising flavonoid compounds, exploring combination therapies, and integrating personalized medicine to optimize therapeutic outcomes.

16.6.1 Promising Flavonoid Compounds for Clinical Use

These are some promising flavonoid compounds for clinical use, along with their natural sources, therapeutic properties, and current clinical applications (Table 16.1).

16.6.2 Potential for Combination Therapies

Flavonoids can be used in combination with other therapeutic agents to enhance their effectiveness, overcome limitations, and reduce side effects. Some potential combination therapies include:

1. **Flavonoid and Pharmacological Drugs**
   a. Flavonoids may enhance the efficacy of conventional drugs used to treat NDDs. For example, combining flavonoids like quercetin with cholinesterase inhibitors (e.g.,

TABLE 16.1
Summarizing Some Promising Flavonoid Compounds for Clinical Use (19)

S. No	Flavonoid Compound	Natural Sources	Therapeutic Properties	Clinical Applications
1.	**Quercetin**	Onions, apples, berries, and tea	Antioxidant, anti-inflammatory, antiviral	Cardiovascular health, COVID-19 adjunct therapy
2.	**Luteolin**	Celery, parsley, green peppers	Anti-inflammatory, anticancer, neuroprotective	Cancer therapy, NDDs
3.	**Apigenin**	Parsley, chamomile, celery	Anti-inflammatory, anticancer, anxiolytic	Cancer prevention, anxiety relief
4.	**Kaempferol**	Kale, spinach, tea, broccoli	Antioxidant, anticancer, cardioprotective	Cardiovascular disease prevention
5.	**Kaempferol**	Kale, spinach, tea, broccoli	Antioxidant, anticancer, cardioprotective	Cardiovascular disease prevention
6.	**Hesperidin**	Citrus fruits (oranges, lemons)	Antioxidant, anti-inflammatory, vascular protection	Venous insufficiency, vascular disorders
7.	**Naringenin**	Grapefruit, oranges, tomatoes	Antioxidant, anti-inflammatory, lipid-lowering	Metabolic syndrome, obesity management
8.	**Epigallocatechin gallate (EGCG)**	Green tea	Antioxidant, anticancer, neuroprotective	Cancer prevention, NDDs
9.	**Baicalein**	Scutellaria baicalensis (Chinese skullcap)	Antioxidant, anti-inflammatory, neuroprotective	NDDs, cancer
10.	**Fisetin**	Strawberries, apples, persimmons	Antioxidant, senolytic, neuroprotective	Anti-aging, Alzheimer's disease
11.	**Rutin**	Buckwheat, apples, citrus fruits	Antioxidant, anti-inflammatory, vascular protection	Chronic venous insufficiency, hemorrhoids

donepezil) may provide synergistic benefits for Alzheimer's patients by reducing oxidative stress while enhancing neurotransmission.
   b. Combination therapies with anti-inflammatory agents or mitochondrial-targeted drugs may also increase therapeutic efficacy by targeting multiple pathological mechanisms simultaneously
2. **Flavonoid and Stem Cell Therapy**
   a. Combining flavonoids with stem cell therapies could promote neuroprotection and neurogenesis in conditions like PD and AD. Flavonoids like EGCG may improve the survival and differentiation of transplanted stem cells in the brain.
3. **Flavonoids and Natural Compounds**
   a. The combination of flavonoids with other bioactive compounds, such as omega-3 fatty acids or vitamin D, may synergistically reduce inflammation and oxidative damage. These natural compounds together could provide comprehensive neuroprotective effects, promoting brain health.
4. **Flavonoids and Dietary Interventions**
   a. Combining flavonoid-rich diets with other neuroprotective foods, such as those rich in omega-3 fatty acids (e.g., fish, walnuts), may create synergistic effects that enhance cognitive function and reduce neuroinflammation.
5. **Polyphenol Synergy**
   a. Combining multiple polyphenolic compounds with overlapping but distinct mechanisms of action (e.g., quercetin, EGCG, and curcumin) could enhance overall neuroprotective effects and provide a more comprehensive strategy for treating NDDs.

16.6.3 Prospects for Personalized Medicine in Neurodegenerative Disease

Personalized medicine aims to tailor treatments based on individual genetic, environmental, and lifestyle factors. The therapeutic potential of flavonoids can be optimized through personalized approaches, which consider the following factors:

1. **Genetic Variability and Metabolism**
   a. Genetic differences affect how individuals metabolize and respond to flavonoids. For instance, polymorphisms in genes encoding enzymes involved in flavonoid metabolism (e.g., CYP450 enzymes) may influence bioavailability and efficacy. Personalized treatment plans could account for these differences by selecting specific flavonoids or adjusting doses to optimize therapeutic effects.
2. **Microbiome-Driven Variability**
   a. The gut microbiome plays a key role in flavonoid metabolism and bioavailability. Personalized medicine could incorporate microbiome profiling to predict an individual's ability to metabolize flavonoids, improving the efficacy of flavonoid-based therapies.
3. **Disease Subtypes**
   a. NDDs like Alzheimer's and PD are heterogeneous, with varying disease subtypes and progression rates. Personalized medicine could tailor flavonoid therapies to specific subtypes based on biomarkers and clinical characteristics, optimizing treatment outcomes.
4. **Precision Nutritional Interventions**
   a. Diets rich in flavonoids may have different effects based on an individual's genetic predisposition and health status. Personalized dietary recommendations could maximize the neuroprotective benefits of flavonoid-rich foods and supplements, enhancing brain health in at-risk populations.

5. **Biomarkers for Monitoring Efficacy**
   a. The identification of biomarkers for flavonoid response could help monitor treatment progress and adjust interventions as needed. Biomarkers of oxidative stress, neuro-inflammation, and neurogenesis could be used to track the effectiveness of flavonoid therapies in individual patients (20).

The future of flavonoid-based therapies in NDDs is promising, with several flavonoid compounds showing potential for clinical use. Combination therapies and personalized medicine approaches will likely play key roles in enhancing the therapeutic effects of flavonoids. By considering individual differences in metabolism, genetic predisposition, and disease characteristics, personalized strategies can optimize flavonoid efficacy, paving the way for targeted treatments that offer real benefits to patients suffering from NDDs. Continued research, clinical trials, and the development of novel drug delivery systems will be critical in realizing the full potential of flavonoids in clinical practice.

REFERENCES

1. Agnati, L.F.; Zoli, M.; Biagini, G.; Fuxe, K. NeuronalPlasticity and Ageing Processes in the Frame of the "Red Queen Theory". *Acta Physiol Scand.* **1992**, *145*(4), 301–309.
2. Akterin, S., et al. Involvement of Glutaredoxin-1 and Thioredoxin-1 in Betaamyloid Toxicity and Alzheimer's Disease. *Cell Death Differ.* **2006**, *13*(9), 1454–1465.
3. Bamford, K.A., et al. A Prospective Evaluation of Cognitive Decline in Early Huntington's D isease: Functional and Radiographic Correlates. *Neurology.* **1995**, *45*(10), 1867–1873.
4. Bastianetto, S., et al. The Ginkgo Biloba Extract (EGb 761) Protects Hippocampal Neurons Against Cell Death Induced by Beta-Amyloid. *Eur J Neurosci.* **2000**, *12*(6), 1882–1890.
5. Baum, L., et al. Curcumin Interaction with Copper and Iron Suggests One Possible M echanism of Action in Alzheimer's Disease Animal Models. *J Alzheimers Dis.* **2004**, *6*(4), 367–377.
6. Butterfield, D.A., et al. Brain Protein Oxidation in Age-Related Neurodegenerative Disorders that Are Associated with Aggregated Proteins. *Mech Ageing Dev.* **2001**, *122*(9), 945–962.
7. Cadet, J., et al. Oxidative Damage to DNA: Formation, Measurement and Biochemical Features. *Mutat Res.* **2003**, *531*(1–2), 5–23.
8. Tahir, M.S.; Almezgagi, M.; Zhang, Y.; Bashir, A.; Abdullah, H.M.; Gamah, M.; Wang, X.; Zhu, Q.; Shen, X.; Ma, Q.; et al. Mechanistic New Insights of Flavonols on Neurodegenerative Diseases. *Biomed Pharmacother.* **2021**, *137*, 111253.
9. Wilson, D.M.; Cookson, M.R.; Van Den Bosch, L.; Zetterberg, H.; Holtzman, D.M.; Dewachter, I. Hallmarks of Neurodegenerative Diseases. *Cell.* **2023**, *186*, 693–714.
10. Maher, P. The Potential of Flavonoids for the Treatment of Neurodegenerative Diseases. *Int J Mol Sci.* **2019**, *20*, 3056.
11. Rizzo, G.; Baroni, L. Soy, Soy Foods and Their Role in Vegetarian Diets. *Nutrients.* **2018**, *10*, 43.
12. Messina, M.; Redmond, G. Effects of Soy Protein and Soybean Isoflavones on Thyroid Function in Healthy Adults and Hypothyroid Patients: A Review of the Relevant Literature. *Thyroid.* **2006**, *16*, 249–258.
13. Mas-Bargues, C.; Borrás, C.; Viña, J. The Multimodal Action of Genistein in Alzheimer's and Other Age-Related Diseases. *Free Radic Biol Med.* **2022**, *183*, 127–137.
14. Gao, H.; Lei, X.; Ye, S.; Ye, T.; Hua, R.; Wang, G.; Song, H.; Zhou, P.; Wang, Y.; Cai, B. Genistein Attenuates Memory Impairment in Alzheimer's Disease via ERS-Mediated Apoptotic Pathway In Vivo and In Vitro. *J Nutr Biochem.* **2022**, *109*, 109118.
15. Irfan, Z.; Khanam, S.; Karmakar, V.; Firdous, S.M.; El Khier, B.S.I.A.; Khan, I.; Rehman, M.U.; Khan, A. Pathogenesis of Huntington's Disease: An Emphasis on Molecular Pathways and Prevention by Natural Remedies. *Brain Sci.* **2022**, *12*, 1389.
16. Siddique, Y.H.; Varshney, H.; Mantasha, I.; Shahid, M. Effect of Luteolin on the Transgenic Drosophila Model of Huntington's Disease. *Comput Toxicol.* **2021**, *17*, 100148.
17. Bouayed, J., Polyphenols: A potential new strategy for the prevention and treatment of anxiety and depression. *Curr Nutr Food Sci.* **2010**, *6*(1), 13–18.
18. Rawlings-Mortimer, F.; Lazari, A.; Tisca, C.; Tachrount, M.; Martins-Bach, A.B.; Miller, K.L.; Lerch, J.P.; Johansen-Berg, H. 7,8-Dihydroxyflavone Enhances Long-Term Spatial Memory and Alters Brain Volume in Wildtype Mice. *Front Syst Neurosci.* **2023**, *17*, 1134594.

19. Agnati, L.F., et al. On the Key Role Played by Altered Protein Conformation in Parkinson's Disease. *J Neural Transm*. **2008**, *115*(9), 1285–1299.
20. Bowling, A.C., et al. Superoxide Dismutase Activity, Oxidative Damage, and Mitochondrial Energy Metabolism in Familial and Sporadic Amyotrophic Lateral Sclerosis. *J Neurochem*. **1993**, *61*(6), 2322–2325.

17 Side Effects during Treatment of Neurodegenerative Disorder

Swathi Nalla,[1] *Nagaraju Bandaru,*[2] *Nayudu Teja,*[3] *and Vadabam Valasa Reddy*[4]

[1]Department of Pharmacology, Malla Reddy Pharmacy College, Maisammaguda, Hyderabad

[2]Department of Pharmacology, School of Pharmaceutical Sciences (SOPS), Sandip University, Nasik, Maharashtra

[3]Department of Pharmaceutics, JSS College of Pharmacy, JSSAHER, The Nilgiris, Tamil Nadu, India

[4]Department of Pharmaceutical Chemistry, Adarsa College of Pharmacy, G. Kothapalli, Andhra Pradesh, India

17.1 INTRODUCTION

Neurodegenerative disorders, such as Parkinson's disease, Alzheimer's disease, amyotrophic lateral sclerosis (ALS), and Huntington's disease, are characterized by progressive deterioration of neurons, leading to debilitating physical, cognitive, and emotional symptoms. Treatment strategies for these disorders aim to alleviate symptoms, slow disease progression, and enhance patients' quality of life. However, the complexity of the nervous system and the multifaceted nature of these conditions often result in treatments with significant side effects.

Understanding the side effects of treatments is critical for optimizing patient care. Side effects can vary widely, ranging from mild discomfort to severe complications that impact treatment adherence and overall outcomes. For instance, pharmacological interventions may alleviate primary symptoms but introduce challenges such as nausea, fatigue, or psychological disturbances. Non-pharmacological approaches, such as deep brain stimulation or physical therapy, also carry risks that require careful management (1).

This chapter delves into the spectrum of side effects associated with current and emerging treatments for neurodegenerative disorders. It examines pharmacological therapies, non-pharmacological interventions, and experimental modalities, highlighting their benefits and limitations. By understanding these side effects, clinicians and researchers can work toward personalized treatment plans that minimize risks and improve the quality of life for individuals affected by these devastating conditions (2).

17.1.1 Brief Overview of Neurodegenerative Disorders

Neurodegenerative disorders are a group of conditions characterized by the progressive loss of structure and function of neurons, often leading to irreversible damage in the central and peripheral nervous systems. These disorders include Parkinson's disease (PD), Alzheimer's disease (AD), ALS, Huntington's disease (HD), and multiple sclerosis (MS), among others.

Each disorder presents unique clinical manifestations, ranging from motor impairments and cognitive decline to behavioral changes and sensory deficits. The underlying pathology often involves mechanisms such as protein misfolding, oxidative stress, mitochondrial dysfunction, and neuroinflammation. Despite differences in clinical presentation, these disorders share a common trajectory

DOI: 10.1201/9781003478652-19

of progressive neurological decline, posing significant challenges for patients, caregivers, and healthcare systems worldwide (3).

17.1.2 Common Treatments and Medications

The management of neurodegenerative disorders primarily focuses on symptom control, delaying disease progression, and improving quality of life. Treatments include:

Pharmacological Interventions:

- **PD:** Levodopa, dopamine agonists, MAO-B inhibitors, and anticholinergics.
- **Alzheimer's Disease:** Cholinesterase inhibitors (donepezil, rivastigmine) and NMDA receptor antagonists (memantine).
- **ALS:** Riluzole and edaravone to modestly slow disease progression.
- **HD:** Tetrabenazine and antipsychotics to manage chorea and psychiatric symptoms.

Non-Pharmacological Therapies:

- Deep brain stimulation for PD and tremor disorders.
- Physical and occupational therapy to manage mobility and functional impairments.
- Cognitive therapy and behavioral interventions for Alzheimer's disease and related dementias.

Emerging Therapies:

- Monoclonal antibodies (e.g., aducanumab for Alzheimer's disease).
- Gene therapy and RNA-based approaches for ALS and HD (4).

These treatments have shown varying degrees of efficacy but are often associated with side effects that can limit their use or effectiveness.

17.1.3 Importance of Understanding Side Effects

The benefits of treatment for neurodegenerative disorders are frequently accompanied by unintended side effects that can profoundly affect patients' health, comfort, and treatment adherence. For example, while levodopa significantly alleviates motor symptoms in PD, long-term use may result in dyskinesias and motor fluctuations. Similarly, cholinesterase inhibitors for Alzheimer's disease can lead to gastrointestinal (GI) upset, which can deter patients from continuing therapy.

Understanding these side effects is crucial for several reasons:

1. **Optimizing Treatment Outcomes:** By anticipating and managing side effects, clinicians can tailor treatment plans to each patient's needs, enhancing therapeutic efficacy.
2. **Improving Quality of Life:** Mitigating side effects helps maintain patients' comfort and functionality, which is vital for their well-being.
3. **Supporting Treatment Adherence:** Minimizing adverse effects encourages patients to stay on prescribed therapies, ensuring sustained benefits.
4. **Informing Future Research:** Studying treatment side effects provides insights into the mechanisms of neurodegeneration and opportunities for safer therapeutic development (5).

This chapter explores the multifaceted side effects associated with treatments for neurodegenerative disorders, offering a comprehensive understanding to guide clinical practice and future innovations.

17.2 COMMON SIDE EFFECTS OF NEURODEGENERATIVE DISORDER TREATMENTS

Treatment of neurodegenerative disorders often comes with a spectrum of side effects that can impact the patient's overall well-being and adherence to prescribed therapies. Among the most commonly observed are GI issues, which are frequently reported with pharmacological interventions. This section delves into the GI side effects, focusing on their prevalence, underlying mechanisms, and management strategies (6, 7).

17.2.1 GI Issues

GI side effects are a significant concern in the treatment of neurodegenerative disorders. They arise due to the interaction of medications with the enteric nervous system, systemic effects of drugs, or changes in neurotransmitter levels. These side effects can range from mild discomfort to severe disturbances that necessitate treatment adjustments or discontinuation.

17.2.1.1 Nausea and Vomiting

Prevalence:

Nausea and vomiting are common side effects, particularly with medications such as:

- **Levodopa-carbidopa (PD):** Caused by dopamine stimulation in the chemoreceptor trigger zone (CTZ).
- **Cholinesterase inhibitors (Alzheimer's disease):** Result from increased acetylcholine activity in the gastrointestinal tract.

Mechanism:

- Dopaminergic drugs stimulate the CTZ in the brainstem, leading to nausea.
- Increased neurotransmitter activity in the GI tract can result in excessive peristalsis, triggering nausea and vomiting.

Management Strategies:

- Co-administration of antiemetics (e.g., domperidone, ondansetron).
- Gradual dose titration to allow the body to adapt.
- Taking medications with food to minimize stomach irritation.

17.2.1.2 Diarrhea or Constipation

Prevalence:

- **Diarrhea:** Often associated with cholinesterase inhibitors like donepezil and rivastigmine.
- **Constipation:** Frequently observed in PD due to dopaminergic dysfunction and medications such as anticholinergics.

Mechanism:

- Diarrhea: Increased cholinergic activity enhances GI motility, leading to loose stools.
- Constipation: Reduced dopamine levels in the enteric nervous system and the anticholinergic effects of certain drugs slow down bowel motility.

Management Strategies:

- For diarrhea: Adjusting medication dosages, hydration, and using antidiarrheal agents if necessary.
- For constipation: High-fiber diets, increased fluid intake, regular exercise, and laxatives (e.g., polyethylene glycol) as needed.

17.2.1.3 Loss of Appetite

Prevalence:

Loss of appetite is a common complaint among patients using medications like cholinesterase inhibitors and dopamine agonists.

Mechanism:

- Drug-induced nausea or dysregulation of appetite-controlling neurotransmitters may reduce food intake.
- In advanced neurodegenerative disorders, systemic factors like disease progression and depression can also contribute.

Management Strategies:

- Encouraging small, frequent meals to reduce the sensation of fullness.
- Addressing underlying nausea or other contributing factors.
- Considering appetite stimulants (e.g., megestrol acetate) in severe cases.

17.2.2 Neurological Side Effects

Neurological side effects are commonly observed during the treatment of neurodegenerative disorders. These side effects often stem from alterations in the nervous system's chemical balance or direct interactions with the central or peripheral nervous systems. While some of these effects are mild, others can significantly impact patients' quality of life and necessitate clinical intervention. This section discusses the key neurological side effects associated with treatment (8, 9).

17.2.2.1 Headache

Prevalence:

Headaches are frequently reported with certain medications, particularly those that modulate neurotransmitters, such as dopamine agonists and cholinesterase inhibitors.

Mechanism:

- Vasodilation induced by some medications (e.g., dopamine agonists).
- Changes in cerebral blood flow or pressure due to neurological modulation.

Management Strategies:

- Over-the-counter (OTC) analgesics (e.g., acetaminophen, ibuprofen) for mild cases.
- Adjusting medication dosage or timing.
- Ensuring adequate hydration and monitoring for signs of more serious complications, such as intracranial pressure changes.

17.2.2.2 Dizziness

Prevalence:

Dizziness is a common side effect of medications like levodopa, dopamine agonists, and some NMDA receptor antagonists.

Mechanism:

- Orthostatic hypotension due to dopaminergic medications.
- Central nervous system effects, such as impaired vestibular function or neurotransmitter imbalance.

Management Strategies:

- Gradual position changes to prevent orthostatic hypotension.
- Adjusting medication timing to avoid peaks that induce dizziness.
- Medications like fludrocortisone or midodrine for persistent orthostatic symptoms.

17.2.2.3 Tremors

Prevalence:

Tremors can occur as a paradoxical response to medications intended to manage motor symptoms, such as levodopa. They may also result from disease progression or inadequate symptom control.

Mechanism:

- Dysregulated dopamine levels in the basal ganglia.
- Off-periods in PD treatment when medication effects wear off.

Management Strategies:

- Adjusting levodopa dosing schedules or using adjunctive therapies like MAO-B inhibitors or COMT inhibitors.
- Non-pharmacological interventions such as physical therapy to manage residual motor symptoms.

17.2.2.4 Insomnia

Prevalence:

Insomnia is a frequent complaint among patients using medications like dopamine agonists, cholinesterase inhibitors, or stimulants for fatigue.

Mechanism:

- Increased central nervous system activity or disrupted circadian rhythms.
- Side effects like vivid dreams or nightmares, particularly with cholinesterase inhibitors.

Management Strategies:

- Encouraging good sleep hygiene practices (e.g., consistent sleep schedules, avoiding caffeine).

- Switching medication administration to the morning if nighttime doses are stimulating.
- Considering sedatives or hypnotics (e.g., melatonin, low-dose benzodiazepines) in refractory cases.

17.2.2.5 Anxiety or Depression

Prevalence:

Mood disturbances, including anxiety and depression, are both symptoms of neurodegenerative disorders and potential side effects of their treatments. For example:

- Dopaminergic therapies may cause anxiety or mood swings.
- Long-term use of anticholinergics or NMDA antagonists can exacerbate depressive symptoms.

Mechanism:

- Dysregulation of neurotransmitters such as dopamine, serotonin, and acetylcholine.
- Psychological effects of chronic illness and treatment burdens.

Management Strategies:

- Screening for baseline mood disorders before initiating therapy.
- Using antidepressants or anxiolytics as needed (e.g., SSRIs or SNRIs).
- Non-pharmacological approaches, such as psychotherapy or mindfulness-based stress reduction.

Neurological side effects, including headaches, dizziness, tremors, insomnia, and mood disturbances, are prevalent during treatment for neurodegenerative disorders. These complications can arise from direct medication effects or interactions with preexisting disease pathology. Tailored management strategies and a multidisciplinary approach are essential to address these side effects, ensuring better patient outcomes and quality of life. Subsequent sections will explore additional categories of treatment-related side effects.

17.2.3 Cardiovascular Side Effects (8, 10)

Cardiovascular side effects are significant concerns in the treatment of neurodegenerative disorders, as many therapies can influence the autonomic nervous system or interact with cardiovascular physiology. These effects, ranging from mild to severe, require close monitoring and management to prevent complications. This section examines two common cardiovascular side effects: high blood pressure and irregular heart rhythm.

17.2.3.1 High Blood Pressure

Prevalence:

Elevated blood pressure is observed with certain medications, particularly those affecting the autonomic nervous system, such as dopamine agonists and NMDA receptor antagonists.

Mechanism:

- Dopaminergic drugs can enhance sympathetic activity, leading to vasoconstriction and increased blood pressure.
- Interactions between medications and preexisting hypertension may exacerbate the condition.

Management Strategies:

- **Monitoring:** Regular blood pressure checks to detect and address changes early.
- **Medication Adjustments:** Modifying dosages or switching to alternative treatments with lower hypertensive potential.
- **Antihypertensive Therapy:** Adding medications like ACE inhibitors, beta-blockers, or calcium channel blockers as needed.
- **Lifestyle Modifications:** Encouraging a low-sodium diet, regular exercise, and stress management to control blood pressure.

17.2.3.2 Irregular Heart Rhythm (Arrhythmias)

Prevalence:

Irregular heart rhythms, including bradycardia, tachycardia, and atrial fibrillation, can occur with treatments such as cholinesterase inhibitors and certain dopamine-related therapies.

Mechanism:

- **Cholinesterase Inhibitors:** Increase parasympathetic tone, potentially causing bradycardia.
- **Dopaminergic Drugs:** May induce tachycardia or arrhythmias by stimulating beta-adrenergic receptors.
- **QT Prolongation:** Some medications can prolong the QT interval, increasing the risk of torsades de pointes or other arrhythmias.

Management Strategies:

- **Cardiac Monitoring:** Electrocardiograms (ECGs) to detect and monitor arrhythmias.
- **Medication Review:** Adjusting or discontinuing medications contributing to cardiac rhythm abnormalities.
- **Treatment of Underlying Conditions:** Managing electrolyte imbalances or preexisting heart conditions that may exacerbate arrhythmias.
- **Pharmacological Intervention:** Using antiarrhythmic drugs or beta-blockers when indicated.
- **Device-Based Therapies:** In severe cases, implantation of a pacemaker or defibrillator may be required.

17.2.4 Psychiatric Side Effects

Psychiatric side effects are frequently encountered during the treatment of neurodegenerative disorders, often complicating management and impacting patient safety and quality of life. These effects can range from mild disorientation to severe symptoms, such as hallucinations or suicidal ideation, which require urgent intervention. This section addresses four common psychiatric side effects: confusion, hallucinations, agitation, and suicidal thoughts (11, 12, 13).

17.2.4.1 Confusion

Prevalence:

Confusion is a common side effect of medications used in PD, Alzheimer's disease, and other neurodegenerative conditions.

Mechanism:

- Cholinesterase inhibitors and NMDA receptor antagonists may disrupt cognitive processing by altering neurotransmitter activity.
- Dopaminergic therapies can induce confusion by overstimulating the central nervous system.

Management Strategies:

- **Medication Review:** Identifying and discontinuing or adjusting medications contributing to confusion.
- **Non-Pharmacological Interventions:** Simplifying routines and providing cognitive support to reduce disorientation.
- **Addressing Underlying Causes:** Monitoring for infections, electrolyte imbalances, or other systemic factors that may exacerbate confusion.

17.2.4.2 Hallucinations

Prevalence:

Hallucinations, particularly visual, are common in patients with PD and dementia-related disorders, especially when treated with dopaminergic drugs or anticholinergic agents.

Mechanism:

- Overactivation of dopaminergic pathways or imbalances in acetylcholine and serotonin can lead to false sensory perceptions.
- Neurodegeneration itself may predispose patients to hallucinations.

Management Strategies:

- **Dose Adjustments:** Lowering the dose of dopaminergic medications or switching to alternatives with fewer psychotropic effects.
- **Medications:** Adding atypical antipsychotics (e.g., quetiapine, clozapine) specifically for PD psychosis.
- **Reassurance:** Offering reassurance and creating a safe, low-stimulation environment to help patients cope with hallucinations.

17.2.4.3 Agitation

Prevalence:

Agitation is frequently seen in Alzheimer's disease and other dementias, often exacerbated by medications or environmental factors.

Mechanism:

- Neurochemical imbalances, such as increased glutamate or dopamine activity, can heighten agitation.
- Disease-related changes in brain regions involved in emotional regulation may contribute.

Management Strategies:

- **Behavioral Approaches:** Identifying and minimizing triggers (e.g., noise, discomfort, unmet needs).

- **Medications:** Using anxiolytics or mood stabilizers (e.g., benzodiazepines, valproate) cautiously, given the risk of side effects.
- **Caregiver Support:** Training caregivers to manage agitation effectively through non-confrontational communication and de-escalation techniques.

17.2.4.4 Suicidal Thoughts

Prevalence:

While less common, suicidal thoughts or behaviors may occur in patients with neurodegenerative disorders, particularly those experiencing depression or receiving medications such as dopamine agonists.

Mechanism:

- Dopaminergic drugs can cause impulsivity and mood dysregulation, increasing the risk of suicidal ideation.
- The psychological burden of chronic illness and cognitive decline can exacerbate feelings of hopelessness.

Management Strategies:

- **Mental Health Assessment:** Regular screening for depression and suicidal thoughts during treatment.
- **Emergency Intervention:** Immediate psychiatric care for patients expressing active suicidal ideation.
- **Medications:** Prescribing antidepressants (e.g., SSRIs, SNRIs) to manage underlying mood disorders.
- **Psychological Support:** Providing access to psychotherapy, support groups, and resources to help patients and caregivers cope.

These complications necessitate a combination of pharmacological and non-pharmacological strategies tailored to each patient's needs. Addressing psychiatric side effects is crucial for ensuring treatment success, patient safety, and improved quality of life.

17.2.5 Other Side Effects

In addition to GI, neurological, cardiovascular, and psychiatric side effects, treatments for neurodegenerative disorders can cause various other adverse effects. These include changes in weight, fatigue, sexual dysfunction, and allergic reactions. While these side effects may not always be life-threatening, they can significantly affect patients' quality of life and adherence to treatment plans. This section explores these side effects, their mechanisms, and strategies for management (14).

17.2.5.1 Weight Gain or Loss

Prevalence:

- Weight gain: Associated with medications such as dopamine agonists, antipsychotics, or certain mood stabilizers.
- Weight loss: Frequently observed with cholinesterase inhibitors or as a consequence of disease progression (e.g., ALS, HD).

Mechanism:

- Weight gain: Increased appetite due to medication effects on dopamine and serotonin pathways.
- Weight loss: Reduced appetite, GI side effects (e.g., nausea), or increased metabolic demand from the disease itself.

Management Strategies:

- **Monitoring:** Regular weight assessments to detect significant changes.
- **Nutritional Support:** Dietitian consultations for personalized meal plans to address weight fluctuations.
- **Medication Adjustments:** Switching to alternatives with less impact on weight when possible.

17.2.5.2 Fatigue

Prevalence:

Fatigue is a common and debilitating side effect, often reported with medications like dopamine agonists, antidepressants, or antiepileptics.

Mechanism:

- Central nervous system depression caused by sedative medications.
- Energy depletion from disease-related muscle weakness or cognitive effort.

Management Strategies:

- **Lifestyle Modifications:** Incorporating regular light exercise, structured rest periods, and proper sleep hygiene.
- **Medications:** Using stimulants (e.g., modafinil) or adjusting dosages of sedative medications.
- **Energy Conservation Techniques:** Teaching patients to prioritize activities and manage energy effectively.

17.2.5.3 Sexual Dysfunction

Prevalence:

Sexual dysfunction, including reduced libido, erectile dysfunction, or difficulty achieving orgasm, is common in patients with neurodegenerative disorders and can be exacerbated by medications such as SSRIs, anticholinergics, and dopamine agonists.

Mechanism:

- Neurochemical changes affecting dopamine and serotonin levels, which are crucial for sexual function.
- Psychological factors like depression or anxiety related to the disorder.

Management Strategies:

- **Open Communication:** Encouraging discussions about sexual health between patients and healthcare providers.

- **Medication Review:** Switching to alternatives with fewer sexual side effects when possible.
- **Targeted Therapies:** Prescribing phosphodiesterase inhibitors (e.g., sildenafil) or testosterone replacement therapy if indicated.

17.2.5.4 Allergic Reactions

Prevalence:

Allergic reactions, ranging from mild skin rashes to severe anaphylaxis, can occur with various medications used to treat neurodegenerative disorders.

Mechanism:

- Immune system hypersensitivity to active ingredients or excipients in medications.

Management Strategies:

- **Monitoring:** Observing for early signs of allergic reactions, such as rashes, itching, or swelling.
- **Immediate Intervention:** Discontinuing the offending medication and administering antihistamines, corticosteroids, or epinephrine for severe reactions.
- **Allergy Testing:** Identifying specific allergens to guide future medication choices.
- **Patient Education:** Instructing patients to recognize symptoms of allergic reactions and seek timely medical help.

These effects can interfere with patients' physical and emotional well-being, requiring proactive management and a multidisciplinary approach. By addressing these side effects effectively, healthcare providers can improve treatment adherence and overall patient satisfaction.

17.3 MANAGING SIDE EFFECTS

Effectively managing side effects is essential for ensuring that treatments for neurodegenerative disorders remain both safe and effective. Side effects can range in severity, and their impact on patients' quality of life may vary. A critical component of managing these challenges is open communication with healthcare providers, allowing for timely interventions such as dosage adjustments or alternative therapies (15, 16).

17.3.1 Communication with Healthcare Provider

Open and consistent communication between patients and their healthcare providers is vital for addressing side effects. Many adverse effects can be mitigated or eliminated through collaborative efforts, improving treatment adherence and overall outcomes.

17.3.1.1 Importance of Reporting Side Effects

Understanding the Importance:

- Reporting side effects helps healthcare providers identify adverse reactions early, potentially preventing complications.
- It allows clinicians to differentiate between medication-related issues and symptoms of the underlying disorder.

Challenges in Reporting:

- Patients may hesitate to report side effects due to embarrassment, fear of being perceived as noncompliant, or misunderstanding what constitutes a side effect.
- Cognitive impairments common in neurodegenerative disorders may make it difficult for patients to articulate their experiences, necessitating caregiver involvement.

Encouraging Reporting:

- Providers should educate patients and caregivers about the importance of reporting even minor or seemingly unrelated symptoms.
- Clear communication channels, such as scheduled follow-ups or telemedicine appointments, can facilitate easier reporting.

17.3.1.2 Adjusting Medication Dosage or Switching Medications

Medication Dosage Adjustments:

- **Titration:** Gradually increasing or decreasing medication dosages can help minimize side effects by allowing the body to adapt.
- **Timing Modifications:** Adjusting the time of day a medication is taken may reduce side effects like nausea or insomnia.

Switching Medications:

- In cases where side effects are severe or persistent, switching to an alternative medication within the same class may provide relief.
- For example:
  - Patients experiencing hallucinations with certain dopamine agonists may benefit from switching to drugs with lower psychotropic potential.
  - Those with GI side effects from cholinesterase inhibitors could try transdermal formulations, such as rivastigmine patches, to bypass the GI tract.

Considerations for Switching:

- Cross-tapering strategies are often necessary to avoid withdrawal symptoms or exacerbation of the underlying condition.
- Careful monitoring during the transition period ensures the effectiveness of the new therapy and minimizes risks.

Effective management of side effects begins with open communication between patients, caregivers, and healthcare providers. Reporting side effects promptly enables tailored interventions, such as dosage adjustments or medication changes, ensuring that treatments remain effective while minimizing discomfort. A collaborative approach empowers patients to take an active role in their care, improving adherence and long-term outcomes.

17.3.2 Lifestyle Modifications

Lifestyle modifications play a crucial role in managing the side effects of treatments for neurodegenerative disorders. While medications address the primary symptoms, healthy lifestyle habits can enhance overall well-being, mitigate side effects, and improve patients' quality of life. This section focuses on two key aspects of lifestyle management: diet and exercise, and stress management techniques (17, 18).

17.3.2.1 Diet and Exercise

Dietary Adjustments:

- **Balanced Nutrition:** A diet rich in fruits, vegetables, whole grains, lean proteins, and healthy fats supports overall health and can reduce side effects such as fatigue, GI discomfort, or weight fluctuations.
- **Hydration:** Drinking adequate water helps alleviate constipation and prevent dehydration-related side effects like dizziness and fatigue.
- **Targeted Nutrition:**
  - High-fiber foods for patients experiencing constipation.
  - Easily digestible meals for those with nausea or reduced appetite.
  - Avoidance of foods that interact with medications, such as high-protein meals that may hinder the absorption of levodopa.

Exercise:

- **Regular Physical Activity:** Exercise improves mobility, reduces fatigue, and can alleviate depression or anxiety associated with neurodegenerative disorders.
- **Tailored Programs:** Physical therapy or low-impact exercises (e.g., walking, swimming, yoga) help address specific needs without overexertion.
- **Motor Symptom Management:** Strength and balance training can help minimize tremors and improve coordination.

17.3.2.2 Stress Management Techniques

Stress as a Trigger:

Chronic stress can exacerbate the symptoms of neurodegenerative disorders and worsen treatment side effects, such as insomnia or anxiety. Implementing stress management techniques is essential for overall health.

Mindfulness and Relaxation Practices:

- **Meditation:** Practices like mindfulness meditation or guided imagery can reduce anxiety and improve focus.
- **Deep Breathing Exercises:** Controlled breathing techniques help calm the nervous system, relieving agitation or emotional distress.

Structured Routines:

- Maintaining a consistent daily schedule can provide stability, particularly for patients with cognitive impairments or confusion.

Social Support:

- **Community Engagement:** Participating in support groups for individuals with neurodegenerative disorders fosters a sense of belonging and reduces isolation.
- **Caregiver Involvement:** Educating caregivers on stress-reduction methods ensures a supportive environment for patients.

Professional Support:

- Seeking counseling or therapy for patients and caregivers can address emotional challenges and build resilience in coping with the disease.

Lifestyle modifications, including a balanced diet, regular exercise, and effective stress management techniques, are essential complements to medical treatments for neurodegenerative disorders. These adjustments not only mitigate side effects but also enhance patients' physical and emotional well-being. Integrating such practices into daily routines empowers individuals to take an active role in their care and achieve better overall outcomes.

17.3.3 OTC Medications

OTC medications can play a significant role in managing mild side effects associated with the treatment of neurodegenerative disorders. These medications are often used to alleviate symptoms like pain, nausea, or minor GI discomfort. However, it is important for patients and caregivers to use OTC medications cautiously, as they may interact with prescribed treatments or worsen certain conditions. This section explores the use of OTC medications for common mild symptoms.

17.3.3.1 Treating Mild Symptoms Like Pain or Nausea

Pain Management:

- **Acetaminophen (Tylenol):** A common OTC option for mild to moderate pain, acetaminophen can be used to alleviate discomfort from headaches, muscle aches, or minor joint pain. However, it should be used sparingly to avoid liver toxicity, especially in patients with preexisting liver conditions.
- **Non-Steroidal Anti-Inflammatory Drugs:** Medications like ibuprofen or naproxen may be used for pain relief but should be used cautiously due to potential GI side effects, especially in patients already at risk for ulcers or bleeding.
- **Topical Analgesics:** Creams or gels containing menthol, capsaicin, or lidocaine may provide localized pain relief, such as for muscle stiffness or neuropathic pain. These are generally safer alternatives with minimal systemic side effects.

Nausea Relief:

- **Antihistamines (e.g., meclizine, dimenhydrinate):** These medications are commonly used to treat mild nausea or dizziness and can be helpful for patients who experience nausea as a result of medications like dopamine agonists or cholinesterase inhibitors.
- **Ginger Supplements:** Ginger has been shown to alleviate nausea and can be used as a natural remedy, often found in teas, capsules, or lozenges. It is generally considered safe but should be used cautiously in patients taking blood thinners, as it may have mild anticoagulant effects.
- **Prokinetic Agents (e.g., metoclopramide):** Although typically prescription-only, low-dose prokinetic agents may sometimes be used off-label for nausea. These medications improve gastric motility and help reduce nausea and vomiting, particularly related to GI disturbances caused by medications (19, 20).

General Considerations:

- **Consultation with Healthcare Providers:** Before using OTC medications, patients should consult their healthcare provider to ensure there are no contraindications with their prescribed medications or underlying conditions.
- **Avoiding Overuse:** OTC medications, while useful for managing mild symptoms, should not be overused. Chronic reliance on OTC pain relievers or anti-nausea medications can mask underlying issues or lead to additional health concerns.

- **Monitoring for Side Effects:** Patients using OTC medications should be aware of potential side effects, such as drowsiness, dizziness, or stomach irritation, and discontinue use if these side effects occur.

Careful monitoring and consideration of potential drug interactions are essential to ensure patient safety and well-being.

17.3.4 Alternative Therapies

Alternative therapies, including complementary treatments such as herbal supplements, acupuncture, or mindfulness practices, are increasingly sought after by patients with neurodegenerative disorders. These therapies may offer additional relief for managing side effects or symptoms, but they should always be considered as part of a holistic treatment plan. Consulting with a healthcare professional before trying alternative treatments is crucial to ensure their safety, effectiveness, and compatibility with conventional treatments.

17.3.4.1 Consulting with a Healthcare Professional before Trying Alternative Treatments

Why Consultation Is Crucial:

- **Safety Concerns:** Many alternative therapies, such as herbal supplements or certain vitamins, can interact with prescribed medications, potentially diminishing their effectiveness or causing harmful side effects. For example, herbal supplements like St. John's wort can interfere with medications used to treat neurodegenerative disorders, including those for PD or Alzheimer's disease.
- **Evidence-Based Practices:** Healthcare providers can guide patients toward alternative treatments that have scientific backing or provide evidence of efficacy. This ensures that the chosen therapies are not just popular but also grounded in research, maximizing their potential benefits.
- **Risk of Worsening Symptoms:** Some alternative therapies, such as high-dose supplements or certain dietary changes, may inadvertently worsen symptoms. For example, a supplement that increases dopamine levels could interact dangerously with medications like Levodopa.
- **Customized Recommendations:** Every patient's needs are unique, especially in the context of neurodegenerative disorders. Healthcare providers can help determine which alternative therapies, if any, are suitable based on the individual's medical history, current medications, and overall health condition.

17.3.4.2 Examples of Alternative Therapies and Their Considerations

Table 17.1 summarizes examples of alternative therapies for neurodegenerative disorders along with their mechanisms, potential benefits, and key considerations.

1. **Herbal Supplements:**
   - **Ginkgo biloba:** Often used to improve cognitive function, particularly in Alzheimer's disease, but it can interact with blood-thinning medications, increasing the risk of bleeding.
   - **Curcumin (turmeric):** Known for its anti-inflammatory properties, curcumin may help with cognitive decline but can cause GI issues and interact with certain medications.

TABLE 17.1
Alternative Therapies for Neurodegenerative Disorders (21, 22)

S. No	Alternative Therapy	Mechanism of Action	Potential Benefits	Considerations/ Limitations
1.	**Acupuncture**	Modulates neurotransmitter release, improves blood flow	Reduces pain, improves motor function, relieves anxiety	Requires skilled practitioner; limited large-scale evidence
2.	**Herbal Medicine (e.g., Ginkgo biloba, Bacopa monnieri)**	Antioxidant, anti-inflammatory, neuroprotective	Enhances memory, cognitive function	Possible herb-drug interactions; dosage standardization needed
3.	**Omega-3 fatty acids**	Anti-inflammatory, supports neuronal membrane integrity	Slows cognitive decline, improves mood	May interact with anticoagulants; inconsistent study results
4.	**Meditation and mindfulness**	Reduces stress, regulates cortisol, enhances brain plasticity	Reduces anxiety, depression, and cognitive decline	Requires regular practice; varying individual response
5.	**Music therapy**	Stimulates cognitive and emotional centers in the brain	Improves mood, memory recall, and communication	Best as adjunct therapy; tailored approaches needed
6.	**Physical exercise (yoga, tai chi)**	Enhances neuroplasticity, increases BDNF levels	Improves balance, motor coordination, cognitive health	Risk of injury if unsupervised; adaptation for patient ability
7.	**Aromatherapy (essential oils)**	Influences limbic system, reduces oxidative stress	Reduces anxiety, improves sleep, mild cognitive improvement	May trigger allergies; lacks strong clinical evidence
8.	**Nutritional supplements (Vitamin D, B12, CoQ10)**	Supports mitochondrial function, reduces oxidative stress	Enhances energy, slows progression of cognitive symptoms	Over-supplementation risks; evidence varies by supplement

- **Cannabis (CBD/THC):** Some evidence suggests that cannabis may help with symptoms such as pain or spasticity in conditions like MS or PD, but it can cause dizziness, sedation, and cognitive changes.

2. **Acupuncture and Acupressure:**
   - **Potential Benefits:** Acupuncture may help manage symptoms like pain, muscle stiffness, and insomnia, commonly seen in PD or other neurodegenerative conditions.
   - **Considerations:** While acupuncture is generally considered safe when performed by a qualified practitioner, it's important to discuss it with a healthcare provider to ensure it does not interfere with medical treatments or cause unintended side effects.
3. **Mindfulness Practices (e.g., Meditation, Yoga):**
   - **Potential Benefits:** Mindfulness techniques, such as meditation and yoga, can significantly reduce stress, anxiety, and depression, which are common in patients with neurodegenerative disorders. These practices may also improve cognitive function and enhance quality of life.
   - **Considerations:** These therapies are typically safe and well-tolerated, but it's still important to tailor practices to the individual's ability. For example, certain types of yoga or meditation may require modifications for those with physical or cognitive limitations.

4. **Diet and Nutrition (e.g., Ketogenic Diet, Omega-3 Fatty Acids):**
   - **Potential Benefits:** Some patients explore the ketogenic diet as a way to manage symptoms of PD or Alzheimer's disease, as it may help reduce brain inflammation and improve energy metabolism. Omega-3 fatty acids, often found in fish oil, are also considered beneficial for brain health.
   - **Considerations:** Any major dietary change should be discussed with a healthcare provider to avoid potential nutrient imbalances or interactions with medications (23, 24).

Consulting a healthcare provider before trying any alternative treatment ensures that patients make informed decisions, minimize risks, and optimize their overall treatment plan. By integrating alternative therapies in collaboration with conventional medical care, patients can achieve a balanced, holistic approach to managing their condition.

17.4 SUMMARY

The management of side effects during the treatment of neurodegenerative disorders is a critical aspect of ensuring that patients can maintain their quality of life while receiving effective therapy. This chapter has provided a comprehensive overview of the various side effects that may occur with common treatments, ranging from GI issues and neurological side effects to psychiatric and cardiovascular complications. Addressing these side effects is essential not only to alleviate discomfort but also to optimize the effectiveness of treatments and improve overall patient outcomes.

Key strategies for managing side effects include open communication with healthcare providers, making necessary medication adjustments, and adopting lifestyle modifications such as diet, exercise, and stress management techniques. The careful use of OTC medications, along with consulting healthcare professionals before pursuing alternative therapies, further enhances the ability to manage these challenges effectively.

It is vital for patients to remember that they are not alone in this process. Collaboration with healthcare providers, including regular consultations to report side effects and receive guidance on treatment adjustments, ensures that side effects are identified early and managed appropriately. Patients should feel empowered to seek professional advice and actively participate in their treatment journey to minimize the impact of side effects.

Moreover, the importance of ongoing research and the development of new treatments cannot be overstated. Continued scientific advancements are essential to discovering safer and more effective therapies, as well as uncovering novel approaches to managing the side effects of existing treatments. By staying informed and participating in research opportunities, patients contribute to the evolution of care and the potential for better outcomes in the future.

Managing side effects is an integral part of the treatment of neurodegenerative disorders, and a proactive, collaborative approach is key to achieving the best possible results. As research progresses, new treatments will offer hope for improved symptom management and enhanced quality of life for those living with these challenging conditions.

REFERENCES

1. Mattos MK, Chang A, Pitcher K, Whitt C, Ritterband LM, Quigg MS. A review of insomnia treatments for patients with mild cognitive impairment. *Aging Dis*. 2021;4:1036–1042.
2. Ahmad MA, Kareem O, Khushtar M, Akbar M, Haque MR, Iqubal A, Haider MF, Pottoo FH, Abdulla FS, Al-Haidar MB, Alhajri N. Neuroinflammation: A potential risk for dementia. *Int J Mol Sci*. 2022:23. https://doi.org/10.3390/IJMS23020616.
3. Turner DA. Contrasting metabolic insufficiency in aging and dementia. *Aging Dis*. 2021;4:1081–1096.
4. Kirmani BF, Au K, Ayari L, John M, Shetty P, Delorenzo RJ. Super-refractory status epilepticus: Prognosis and recent advances in management. *Aging Dis*. 2021;4:1097–1119.
5. Ajmal MR. Protein misfolding and aggregation in proteinopathies: Causes, mechanism and cellular response. *Diseases*. 2023;11:30. https://doi.org/10.3390/DISEASES11010030.

6. Rincon N, Barr D, Velez-Ruiz N. Neuromodulation in drug resistant epilepsy. *Aging Dis.* 2021;4:1070–1080.
7. O'Brien RJ, Wong PC. Amyloid precursor protein processing and Alzheimer's disease. *Annu Rev Neurosci.* 2011;34:185–204. https://doi.org/10.1146/annurev-neuro-061010-113613.
8. Alsina D, Purroy R, Ros J, Tamarit J. Iron in Friedreich ataxia: A central role in the pathophysiology or an epiphenomenon? *Pharmaceuticals (Basel).* 2018;11(3):89. https://doi.org/10.3390/ph11030089.
9. Wilson DM 3rd, Cookson MR, Van Den Bosch L, Zetterberg H, Holtzman DM, Dewachter I. Hallmarks of neurodegenerative diseases. *Cell.* 2023 Feb 16;186(4):693–714. https://doi.org/10.1016/j.cell.2022.12.032
10. Merelli A, Czornyj L, Lazarowski A. Erythropoietin: A neuroprotective agent in cerebral hypoxia, neurodegeneration, and epilepsy. *Curr Pharm Des.* 2013;19(38):6791–6801. https://doi.org/10.2174/1381612811319380011
11. Tomlinson CL, Herd CP, Clarke CE, Meek C, Patel S, Stowe R, Deane KH, Shah L, Sackley CM, Wheatley K, Ives N. Physiotherapy for Parkinson's disease: A comparison of techniques. *Cochrane Database Syst Rev.* 2014 Jun 17;2014(6):CD002815. https://doi.org/10.1002/14651858.CD002815.pub2
12. Melo A, Monteiro L, Lima RM, Oliveira DM, Cerqueira MD, El-Bachá RS. Oxidative stress in neurodegenerative diseases: Mechanisms and therapeutic perspectives. *Oxid Med Cell Longev.* 2011;2011:467180. https://doi.org/10.1155/2011/467180
13. MA Mora-Navarro, AR, Rincón-Sánchez, FP, Pacheco-Moisés, MA, Mireles-Ramírez, D. Nutritional Status in Patients with Probable Alzheimer's Disease: Effect of Rivastigmine, *The Journal of nutrition, health and aging*, 2021;25(3):340–346, ISSN 1279-7707, https://doi.org/10.1007/s12603-020-1519-9.
14. Morén C, deSouza RM, Giraldo DM, Uff C. Antioxidant therapeutic strategies in neurodegenerative diseases. *Int J Mol Sci.* 2022;23:9328.
15. Nabi M, Tabassum N. Role of environmental toxicants on neurodegenerative disorders. *Front Toxicol.* 2022;4:837579.
16. Pardo-Moreno T, González-Acedo A, Rivas-Domínguez A, García-Morales V, García-Cozar FJ, Ramos-Rodríguez JJ, Melguizo-Rodríguez L. Therapeutic approach to Alzheimer's disease: Current treatments and new perspectives. *Pharmaceutics.* 2022;14:1117.
17. Ma C, Hong F, Yang S. Amyloidosis in Alzheimer's disease: Pathogeny, etiology, and related therapeutic directions. *Molecules.* 2022;27:1210.
18. Hobson DE, Lang AE, Martin WR, Razmy A, Rivest J, Fleming J. Excessive daytime sleepiness and sudden-onset sleep in Parkinson disease: a survey by the Canadian Movement Disorders Group. *JAMA.* 2002 Jan 23-30;287(4):455–63. doi: 10.1001/jama.287.4.455. PMID: 11798367.
19. Camargos EF, Louzada LL, Quintas JL, Naves JO, Louzada FM, Nóbrega OT. Trazodone improves sleep parameters in Alzheimer disease patients: a randomized, double-blind, and placebo-controlled study. *Am J Geriatr Psychiatry.* 2014 Dec;22(12):1565–74. doi: 10.1016/j.jagp.2013.12.174. Epub 2014 Jan 4. PMID: 24495406.
20. Marks Jr., WJ, Ostrem, JL, Verhagen, L, et al. Safety and Tolerability of Intraputaminal Delivery of CERE-120 (Adeno-Associated Virus Serotype 2-Neurturin) to Patients with Idiopathic Parkinson's Disease: An Open-Label, Phase 1 Trial. *The Lancet Neurology*, 2008;5:400–408. http://dx.doi.org/10.1016/S1474-4422(08)70065-6.
21. Kum WF, Durairajan SS, Bian ZX, Man SC, Lam YC, Xie LX, Lu JH, Wang Y, Huang XZ, Li M. Treatment of idiopathic Parkinson's disease with traditional Chinese herbal medicine: a randomized placebo-controlled pilot clinical study. *Evid Based Complement Alternat Med* 2011: 724353.
22. Jason Gandhi, Min Yea Lee, Gunjan Joshi & Sardar Ali Khan. Surfer'smyelopathy: A review of etiology, pathogenesis, evaluation, and management, *The Journal of Spinal Cord Medicine.* 2019.
23. Sharma K. Cholinesterase inhibitors as Alzheimer's therapeutics (review). *Mol Med Rep.* 2019;20:1479–1487. https://doi.org/10.3892/mmr.2019.10374
24. Arici Duz O, Helvaci Yilmaz N. Nocturnal blood pressure changes in Parkinson's disease: Correlation with autonomic dysfunction and vitamin D levels. *Acta Neurol Belg.* 2019. https://doi.org/10.1007/s13760-019-01113-7

18 Differentiation between Synthetic and Natural Treatment Modalities

Deepak Vasudevan Sajini and Amritha Chakkitukandiyil
Department of Pharmacology, National College of Pharmacy, Kerala University of Health Sciences, Manassery, Kozhikode, Kerala, India

18.1 INTRODUCTION TO NEURODEGENERATIVE DISORDERS

Neurodegenerative disorders (NDs) are widely prevalent and have a considerable financial and medical impact on our society (Kawamata and Manfredi 2011). NDs are primarily characterized by neuron loss. It is established that the primary pathophysiological change in most of the brain-related disorders is neurodegeneration. The most common NDs are Alzheimer's disease (AD), Parkinson's disease (PD), Huntington's disease (HD), spinal muscular atrophy, amyotrophic lateral sclerosis (ALS), prion disease, and spinocerebellar ataxia (Merelli, Czornyj, and Lazarowski 2013; Sajini, Krishnamurthy, and Kandiyil 2022). The blood–brain barrier (BBB) is a significant barrier in successfully managing NDs. As a result, neurotherapeutics that can penetrate the BBB have been suggested as a viable means of delaying or perhaps stopping neurodegeneration (Hinge, Kathuria, and Pandey 2022). NDs impact millions of people globally. While aging remains the primary risk factor for developing these conditions, recent research indicates that a person's genetic predisposition and environmental influences also significantly elevate ND risk. Additionally, even though certain genes linked to NDs are expressed in individuals, their surrounding environment heavily influences the timing and severity of neurodegeneration (Jain and Chen-Plotkin 2018).

Since the brain governs numerous bodily functions, NDs significantly impact various aspects of human performance. They impair fundamental abilities (such as speech, movement, and balance) and more complex functions (including cognitive processes and bladder or bowel control). While many NDs advance progressively without remission, treatments often focus on alleviating symptoms, reducing pain when applicable, and improving mobility and stability (Lamptey et al. 2022).

18.2 ALZHEIMER'S DISEASE

AD is a chronic, age-related, progressive, and unrelenting neurological disorder with a frequency of 10–30% in the population over the age of 65 (Chakkittukandiyil et al. 2023; Masters et al. 2015). It affects a large portion of the cerebral cortex and hippocampus. Abnormalities are commonly found in the frontal and temporal lobes of the brain, and then gradually spread to other sections of the neocortex. AD is associated with the deposition of insoluble forms of amyloid-beta (Aβ) in plaques in extracellular spaces and blood vessel walls, as well as aggregation of the microtubule protein tau in neurons (Masters et al. 2015).

DOI: 10.1201/9781003478652-20

18.2.1 Synthetic Treatment Modalities for AD

The development of pharmaceutical treatments to counteract the cognitive and memory impairments caused by AD has been ongoing for many years. Initially, these drugs were designed based on the cholinergic and glutamatergic hypotheses. The cholinergic hypothesis, one of the earliest theories explaining the neuropsychiatric symptoms of AD, suggests that the loss of acetylcholine (ACh) neurons and related enzymatic activities in the amygdala, basal forebrain, cortex, and hippocampus are key contributors to the disease (Zr et al. 2022). Additionally, the decline in α7 or α4β2 nicotinic ACh receptors (nAChRs) and M2 muscarinic ACh receptors (mAChRs) in pre- and postsynaptic neurons is believed to play a role in AD progression. Standard treatments (Figure 18.1), such as the cholinesterase inhibitors (ChEIs) galantamine, rivastigmine, and donepezil, work by preventing the enzymatic breakdown of ACh by acetylcholinesterase (AChE). This enhances cholinergic signaling, allowing ACh to bind more effectively to nAChRs and mAChRs, thereby maintaining neurotransmission. In contrast, memantine targets *N*-methyl-D-aspartate (NMDA) receptors, reducing excessive calcium influx and preventing excitotoxicity that leads to neuronal degeneration. Several double-blind, randomized controlled trials and systematic reviews have established that memantine and ChEIs are effective in enhancing cholinergic activity. These studies demonstrate that AD patients taking these medications often show improvements in cognitive function (Se Thoe et al. 2021).

18.2.2 Natural Treatment Modalities for AD

Natural medicine has the potential to manage symptoms and prevent or slow down the development of AD. Clinical and preclinical studies have assessed the neuroprotective activities of a variety of natural compounds from various sources in the intricate pathology of AD (Chen et al. 2021). Natural sources such as plants, animals, microbes, and marine organisms offer a wealth of bioactive compounds characterized by complex structures and unique pharmacological properties (Yuan et al. 2016). Plant-based compounds, such as alkaloids, polyphenols, lunasin, and tannins, are potential treatment possibilities for AD (Deshpande, Gogia, and Singh 2019). Curcumin, a diphenol from turmeric, and resveratrol, a phenol found in mulberries, red wine, and tea, are two naturally occurring compounds whose potential for treating AD has been thoroughly studied (Figure 18.1) (Tang and Taghibiglou 2017). Both exhibit strong antioxidant properties, capable of breaking down toxic Aβ plaques, reducing tau protein hyperphosphorylation, and facilitating the clearance of these damaged proteins. Ginkgo biloba, which is well-known for its antioxidant properties, has been

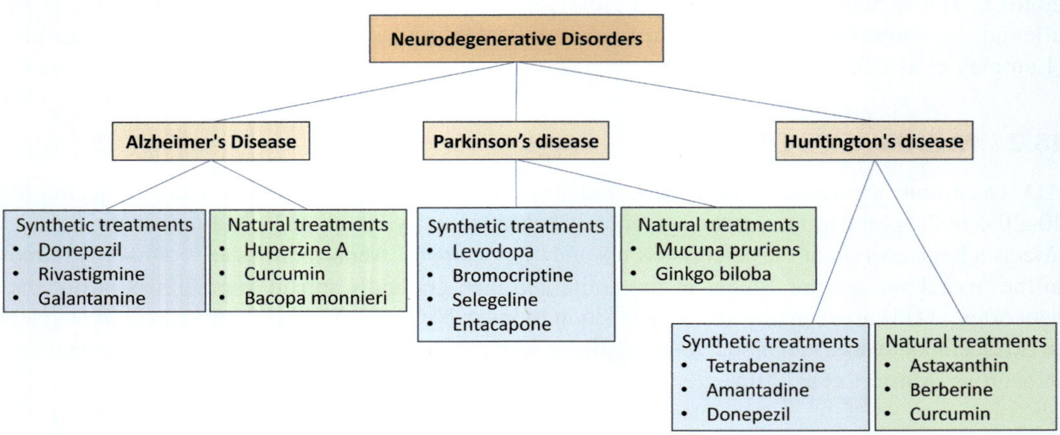

FIGURE 18.1 Synthetic and natural treatment modalities for neurodegenerative disorders.

shown to aid in enhancing cognitive function and lessening the symptoms of dementia. Sage (*Salvia officinalis*) has improved memory and cognitive performance in people with AD. Similarly, lemon balm has been linked to reduced agitation and enhanced cognitive function in AD patients. Ginseng has also been studied for its potential to improve cognitive performance and reduce oxidative stress. Furthermore, saffron has been shown to improve both cognitive function and mood in AD patients (Ahmad et al. 2023). An Ayurvedic herb, Bacopa monnieri, often known as brahmi, is known to increase memory and lower oxidative stress in the brain (Walker and Pellegrini 2024). Huperzine A, derived from the Chinese club moss *Huperzia serrata*, is a natural AChE inhibitor. It enhances cholinergic function and protects against neuronal damage (Gerbarg and Brown 2013).

Omega-3 fatty acids in fish oil, particularly eicosapentaenoic acid (EPA) and docosahexaenoic acid (DHA), help reduce cell damage and possess anti-inflammatory properties, making them beneficial in the treatment of AD (Rahman et al. 2021). Cognitive decline has been associated with deficiencies in calcium, magnesium, and zinc, as well as vitamins B6, B12, and folate. Through the maintenance of homocysteine levels, the reduction of oxidative damage, and the promotion of neurogenesis, these nutrients support the health of neurons (Tardy et al. 2020).

Among the FDA-approved drugs for AD, Galantamine is derived from a natural source, while rivastigmine is a semi-synthetic derivative based on the natural compound, physostigmine (Scott and Goa 2000; V. Kumar 2006). Natural products exert their neuroprotective effects primarily by acting as antioxidants or by reducing neuroinflammation. They help prevent the aggregation of Aβ and tau proteins and boost cholinergic signaling (Venkatesan, Ji, and Kim 2015).

18.3 PARKINSON'S DISEASE

PD is the world's second-most prevalent neurological disorder, which affects about 2–3% of those over 65 (Deepika et al. 2024a, 2004b; Kalia and Lang 2015). One of the main clinical features of PD is the progressive loss of dopamine (DA)-containing neurons in the midbrain's substantia nigra pars compacta (SNpc), which leads to DA insufficiency. The predominant lack of dopaminergic neurons severely impairs motor control (Chakkittukandiyil et al. 2022). Bradykinesia, muscle rigidity, resting tremors, poor postural balance, and later effects like dementia and psychiatric issues are some of its symptoms. Treatment is challenging since these symptoms cannot manifest until 70–80% of dopaminergic neurons have degenerated. It is believed that oxidative stress, mitochondrial failure, and neuroinflammation are the causes of neurodegeneration and the development of PD, even though the precise mechanism behind this disease is unknown (Vasudevan Sajini et al. 2024; Deepika et al. 2024a, 2004b). Environmental (pesticide exposure such as paraquat, rotenone, or traumatic brain injury) and genetic variables (mutation in the SNCA, LRRK2, GBA, MAPT genes) are responsible for oxidative and inflammatory stress in PD (Pang et al. 2019; Chakkittukandiyil et al. 2024).

18.3.1 Synthetic Treatment Modalities for PD

Although PD cannot be cured, several synthetic treatment approaches have been created to control its symptoms and enhance patients' quality of life.

Levodopa (L-DOPA) and Carbidopa: L-DOPA is the cornerstone of PD treatment (Fig.18.1). It is a DA precursor that can pass the BBB and be transformed into DA, correcting for DA shortage in PD patients. Carbidopa is frequently used with levodopa to avoid early conversion of L-DOPA to DA outside the brain, thereby increasing the amount of levodopa that reaches the brain while decreasing adverse effects like nausea and vomiting. They are effective in reducing tremors, stiffness, and bradykinesia. Long-term use can cause motor fluctuations and dyskinesias (uncontrolled movements) (Fahn 2008).

To overcome levodopa's short half-life and enhance motor symptoms during daytime activities, levodopa/carbidopa intestinal gel was created. A levodopa inhalant (CVT-301) has been approved in the United States as a rescue drug during downtime (Murakami et al. 2023).

Dopamine Agonists: Dopamine agonists are synthetic substances that imitate DA's actions by stimulating DA receptors in the brain. Pramipexole, ropinirole, and rotigotine are common DA agonists. Effective in treating motor symptoms, particularly in the early stages of PD. Side effects include impaired impulse control, sleep problems, and hallucinations.

Monoamine Oxidase-B (MAO-B) Inhibitors: MAO-B inhibitors like selegiline and rasagiline function by inhibiting monoamine oxidase-B, an enzyme that is responsible for the degradation of DA in the brain. MAO-B medicines work by blocking this enzyme, increasing DA availability. They are modestly effective in lowering motor symptoms and are frequently taken in conjunction with levodopa. Insomnia, nausea, and drug-food interactions are among the possible side effects.

Catechol-*O*-methyltransferase (COMT) Inhibitors: COMT inhibitors, including entacapone and tolcapone, function by blocking catechol-O-methyltransferase, an enzyme that degrades DA and levodopa. These inhibitors are frequently used in combination with levodopa to extend its effects. They are effective in controlling motor fluctuations and prolonging the benefits of levodopa. Diarrhea, liver damage (when used with tolcapone), and urine staining are also potential side effects.

Amantadine is an antiviral drug that also has dopaminergic and anti-glutamatergic effects. It treats dyskinesias and motor symptoms in PD (Connolly and Lang 2014).

18.3.2 Natural Treatment Modalities for PD

There is currently no conventional medication that can stop or delay the progression of PD. The main treatment for motor impairments in PD is dopaminergic medication; however, this strategy does not address other elements of the disease that affect quality of life. People are looking for a more holistic approach to healthcare because of this. Tai Chi, yoga, acupuncture, natural products, and other traditional treatment systems are often regarded as alternative or complementary therapies for PD (Li et al. 2016). Lots of herbal treatments and nutraceuticals have demonstrated promise in treating PD symptoms. *Mucuna pruriens*, a legume that contains natural levodopa, is a popular alternative to synthetic levodopa for reducing motor symptoms with fewer side effects (Johnson et al. 2018). Resveratrol (found in grapes) and epigallocatechin gallate (EGCG) in green tea have anti-inflammatory and antioxidant properties that make them neuroprotective. These natural compounds help to improve mitochondrial function, a major PD pathogenesis element (Menard, Bastianetto, and Quirion 2013). A number of ginseng constituents, such as Rb1, Rg1, Rd, Re, notoginsenoside R2, and pseudoginsenoside-F11, have garnered significant attention due to their potential as medicinal agents. These compounds demonstrate neuroprotective properties through diverse mechanisms, such as reducing oxidative damage and neural inflammation, diminishing apoptosis and nigral iron levels, and regulating NMDA receptor activity (González-Burgos, Fernandez-Moriano, and Gómez-Serranillos 2015). Ginkgo biloba was clinically reported to show dramatic improvement in PD patients after supplementing with Ginkgo and multivitamin supplements (Conrad 2014). *M. pruriens* and *Vicia faba* are two herbs rich in levodopa. These herbs have been proven to have neuroprotective characteristics, including restoring endogenous DA production and decreasing oxidative stress in the substantia nigra (Yadav et al. 2013). Numerous dietary changes and supplements have demonstrated the potential to reduce symptoms and enhance the quality of life. Foods rich in antioxidants, such as leafy greens, berries, and nuts, help to lower oxidative stress, which is one of the factors that contribute to PD progression (Park and Ellis 2020). Coenzyme Q10 (CoQ10) supplements have the potential to support mitochondrial function and reduce oxidative stress in PD patients (Mischley, Allen, and Bradley 2012). Polyphenols present in tea can slow or stop the progression of disease. Preclinical studies using SH-SY5Y cells and Swiss albino mice demonstrate that Orientin, a flavonoid compound, exerts neuroprotective effects in a rotenone-induced PD model (Vasudevan Sajini et al. 2024). In vitro and in vivo studies have demonstrated that tea and its primary components directly reduce α-synuclein aggregation and modify intracellular signaling pathways (Bieschke et al. 2010; Camilleri et al. 2013).

18.4 HUNTINGTON'S DISEASE

HD is an extremely rare neurological disorder that is caused by a dominantly inherited CAG trinucleotide repeat amplification in the huntingtin gene located on chromosome 4 (Roos 2010). CAG typically has a repeat range length of 10–35. Patients having more than 40 repeats are more likely to manifest the condition, while those with 36–39 repeats have a limited penetrance range (Medina et al. 2022). The average prevalence of HD is about 0.48 occurrences per 100,000 person-years. This means that each year, about 0.48 people out of every 100,000 will develop HD (Kay, Hayden, and Leavitt 2017). HD is a progressive neurological condition characterized by chorea, dystonia, incoordination, cognitive impairment, and behavioral problems. Symptoms typically arise in middle age after a person has had children; however, the condition can emerge at any time between infancy and senescence (Walker 2007). As the disease progresses, it becomes increasingly dependent on daily activities and eventually leads to death. The leading cause of death is pneumonia, followed by suicide. The striatal medium spiny neurons are primarily affected by HD pathogenesis, with the cortex being less affected (P. Kumar, Kalonia, and Kumar 2010). HD alters the quantity of NMDA receptors in the basal ganglia and results in the death of GABA and enkephalin neurons (Ellerby 2002). When a person has an established HD parent, the diagnosis is determined based on clinical symptoms and hints. The huntingtin gene on chromosome 4 has a CAG repeat of 36 or more, according to DNA analysis, which is currently the gold standard (Roos 2010).

18.4.1 Synthetic Treatment Modalities for HD

Currently, there are no known treatments to halt or reduce the progression of HD. However, symptomatic management of chorea, guided by known neurochemical pathways, can benefit certain individuals by improving motor function, enhancing quality of life, and ensuring safety (Frank 2014).

Pharmacological treatment for motor dysfunction in HD includes:

Antidopaminergic Agents: Tetrabenazine (TBZ) is the sole drug licensed by the United States Food and Drug Administration (FDA) for the treatment of HD. While the precise mechanism of action is unknown, the main way that TBZ works is by reversibly inhibiting the presynaptic protein known as vesicular monoamine transporter 2 (VMAT2), which encapsulates monoamines in vesicles. By decreasing VMAT2, TBZ predominantly depletes presynaptic DA, with less influence on norepinephrine and serotonin levels. TBZ is generally well-tolerated and safe, though dosage-dependent side effects might occur (Paleacu 2007). Common side effects include sedation and somnolence, akathisia, Parkinsonism, depression, and suicidal thoughts.

Antipsychotic Agents: Chorea suppression is mediated by antipsychotics blocking postsynaptic DA D2 receptors. Acute dyskinesia, tardive dyskinesia, drug-induced parkinsonism, and neuroleptic malignant syndrome are among the risks connected to the use of neuroleptics. Antipsychotics are used in small doses to lessen chorea, and dosages are adjusted according to patient tolerance (Venuto et al. 2012).

N-Methyl-D-Aspartic Acid Receptor Antagonists: NMDA receptor antagonists are being studied for the treatment of HD because of their ability to diminish excitotoxicity, a process in which excessive NMDA receptor activation by glutamate causes cell death. Antagonists like memantine and amantadine, which inhibit these receptors, may protect neurons from degeneration. Memantine has shown potential in lowering motor and cognitive symptoms (Ondo, Mejia, and Hunter 2007), and amantadine may aid with chorea (O'Suilleabhain and Dewey 2003).

Omega-3 Fatty Acids: There has been interest in omega-3 fatty acids as possible therapies for HD, especially DHA and EPA. These polyunsaturated fatty acids, found predominantly in fish oil and select plant sources, have neuroprotective and anti-inflammatory characteristics that may assist HD patients (Vaddadi et al. 2002).

The pharmacological therapy of cognitive dysfunction in HD is complicated because there are limited choices, particularly licensed for this purpose. Pharmacological treatment for HD-associated cognitive dysfunction in HD includes:

Cholinesterase Inhibitors: These medications, such as donepezil and rivastigmine, are occasionally used off-label to improve cognition by boosting the level of ACh in the brain.
Antipsychotics: Olanzapine and risperidone can be used to control psychiatric symptoms, which may indirectly improve cognitive function.
Cognitive enhancers: Medicines like methylphenidate have been examined for their potential to improve cognitive performance in HD patients (Frank 2014).

18.4.2 Natural Treatment Modalities for HD

Numerous papers discuss the use of natural medications to treat HD. HD can be cured or treated using natural compounds produced from herbal plants that have anti-inflammatory, anti-apoptotic, immune-stimulatory, antioxidant, calcium-antagonistic, and neuroprotective properties. While natural treatment modalities cannot cure the disease, they can help patients manage symptoms and improve their quality of life significantly (Irfan et al. 2022). These approaches take a holistic approach to health, focusing on the body's natural healing processes and overall well-being. Various antioxidants and free radical scavengers, including α-mangostin, lipoic acid, astaxanthin, berberine, curcumin, ferritin, transferring EGCG, hesperidin, kaempferol, lutein, lycopene, melatonin, resveratrol, ceruloplasmin, α-tocopherol, β carotene, lycopene, and ascorbic acid and sesamol, have shown efficacy in HD models (Figure 18.1).

18.5 EMERGING TRENDS AND INNOVATIONS

Promising new developments and trends in the treatment of neurodegenerative diseases have resulted from recent discoveries in medical research. They mainly focus on novel therapeutic targets and advanced technologies.

Gene Therapy: Through gene therapy, the defective or impaired genes are replaced or corrected. This method has demonstrated potential in illnesses such as PD, HD, and spinal muscular atrophy (García-González et al. 2024). Gene-based diagnostics and screening also allow for the early detection of individuals who are at risk of developing a specific disease.

Stem Cell Therapy: Skin cells can be used to develop induced pluripotent stem cells, which can develop into either glial or neuronal cells. These cells have the potential to be transplanted to replace missing neurons and mend damaged brain circuits. To determine their safety and effectiveness in treating PD and ALS, clinical trials are now being conducted (Albert et al. 2021).

Advanced Drug Delivery Systems: Nanoparticles are a viable technique for bypassing the BBB and carrying drugs straight to the brain. These particles can be loaded with therapeutics and functionalized with targeted molecules to improve selectivity. They focus on the localization of intracellular or extracellular materials, like the Aβ plaques in AD (Asil et al. 2020). Intranasal drug delivery is a non-invasive alternative for brain-targeted delivery. Three major proposed routes for nose-to-brain delivery are the olfactory nerve, the trigeminal nerve, and the perivascular transport. A few intranasal delivery systems utilized in clinical trials showing promising outcomes in terms of safety and efficacy are ViaNase™, Precision olfactory delivery®, Aero Pump system, Mistette MK Pump II, GL18, Metered nasal dispenser, OptiMist™, and Sipnose (Trevino et al. 2020).

Improved diagnostic tools: Early identification of neurodegenerative illnesses is critical to effective therapy. New imaging techniques and biomarkers are being developed to detect these diseases at an early stage. This enables more effective symptom control and timely intervention (Akram et al., n.d.).

Neurovascular Interactions: Another area of research understands how the neurological system and blood arteries interact. This study attempts to discover new therapeutic targets that could help prevent or treat neurodegeneration (Wareham et al. 2022).

Bioenergetics and Metabolism: Researchers are looking into how cellular energy production and metabolism changes lead to neurodegenerative disorders. By addressing these pathways, new medicines could be developed to support neuronal health and function (Wareham et al. 2022).

Personalized Medicine: Approaches to individualized treatment are being made possible by developments in genetic and molecular diagnostics. This implies that therapies can be customized to each patient's unique genetic and molecular profile, which could lead to better results (Mathur and Sutton 2017).

18.6 CONCLUSION

NDs, such as AD, PD, and HD, pose substantial problems due to their progressive nature and global incidence. While synthetic medications have long been the foundation of treatment, their symptom-focused approach and risk of undesirable side effects restrict their long-term utility. Natural therapeutic techniques, on the other hand, provide intriguing options, harnessing their multiple mechanisms such as antioxidant and anti-inflammatory properties to address the complex pathophysiology of neurodegeneration while posing fewer risks. Future medicines that combine the capabilities of synthetic and natural remedies could not only relieve symptoms but also reduce or stop disease progression. Continued research into merging these modalities has the potential to transform the care of neurodegenerative illnesses, providing hope for better outcomes and quality of life for those affected.

REFERENCES

Ahmad Saara, Saad Bilal Ahmed, Asra Khan, Muhammad Wasim, Saiqa Tabassum, Saida Haider, Fatima Ahmed, et al. 2023. "Natural Remedies for Alzheimer's Disease: A Systematic Review of Randomized Controlled Trials." *Metabolic Brain Disease* 38 (1): 17–44. https://doi.org/10.1007/s11011-022-01063-9.

Akram Ahmed S., Han Grezenko, Prem Singh, Muhammad Ahmed, Baran Dilshad Hassan, Vibhavari Hagenahalli Anand, Abdelrahman A Elashry, Faran Nazir, and Rehman Khan et al. 2024. "Advancing the Frontier: Neuroimaging Techniques in the Early Detection and Management of Neurodegenerative Diseases." *Cureus* 16 (5): e61335. https://doi.org/10.7759/cureus.61335.

Albert Katrina, Jonna Niskanen, Sara Kälvälä, and Šárka Lehtonen. 2021. "Utilising Induced Pluripotent Stem Cells in Neurodegenerative Disease Research: Focus on Glia." *International Journal of Molecular Sciences* 22 (9): 4334. https://doi.org/10.3390/ijms22094334.

Asil Shima Masoudi, Jyoti Ahlawat, Gileydis Guillama Barroso, and Mahesh Narayan. 2020. "Nanomaterial Based Drug Delivery Systems for the Treatment of Neurodegenerative Diseases." *Biomaterials Science* 8 (15): 4109–28. https://doi.org/10.1039/d0bm00809e.

Bieschke Jan, Jenny Russ, Ralf P. Friedrich, Dagmar E. Ehrnhoefer, Heike Wobst, Katja Neugebauer, and Erich E. Wanker. 2010. "EGCG Remodels Mature α-Synuclein and Amyloid-β Fibrils and Reduces Cellular Toxicity." *Proceedings of the National Academy of Sciences* 107 (17): 7710–15. https://doi.org/10.1073/pnas.0910723107.

Camilleri Angelique, Claire Zarb, Mario Caruana, Ulrike Ostermeier, Stephanie Ghio, Tobias Högen, Felix Schmidt, Armin Giese, and Neville Vassallo. 2013. "Mitochondrial Membrane Permeabilisation by Amyloid Aggregates and Protection by Polyphenols." *Biochimica et Biophysica Acta (BBA) – Biomembranes* 1828 (11): 2532–43. https://doi.org/10.1016/j.bbamem.2013.06.026.

Chakkittukandiyil Amritha, Saurav Chakraborty, Ram Kothandan, Emdormi Rymbai, Santhosh Kumar Muthu, Soumya Vasu, Deepak Vasudevan Sajini, et al. 2023. "Side Effects Based Network Construction and Drug Repositioning of Ropinirole as a Potential Molecule for Alzheimer's Disease: An in-Silico, in-Vitro, and in-Vivo Study." *Journal of Biomolecular Structure & Dynamics* (September): 1–15. https://doi.org/10.1080/07391102.2023.2258968.

Chakkittukandiyil Amritha, Deepak Vasudevan Sajini, Arjunan Karuppaiah, and Divakar Selvaraj. 2022. "The Principal Molecular Mechanisms behind the Activation of Keap1/Nrf2/ARE Pathway Leading to Neuroprotective Action in Parkinson's Disease." *Neurochemistry International* 156:105325.

Chakkittukandiyil Amritha, Deepak Vasudevan Sajini, Emdormi Rymbai, Deepa Sugumar, Jinu Mathew, Suresh Arumugam, Vadivelan Ramachandran, and Divakar Selvaraj. 2024. "Synthesis and Evaluation of Novel Ethyl Ferulate Derivatives as Potent Keap1 Inhibitors to Activate the Nrf2/ARE Pathway in Parkinson's Disease." *Toxicology and Applied Pharmacology*, November: 117172. https://doi.org/10.1016/j.taap.2024.117172.

Chen Xin, Joshua Drew, Wren Berney, and Wei Lei. 2021. "Neuroprotective Natural Products for Alzheimer's Disease." *Cells* 10 (6): 1309. https://doi.org/10.3390/cells10061309.

Connolly Barbara S., and Anthony E. Lang. 2014. "Pharmacological Treatment of Parkinson Disease: A Review." *JAMA* 311 (16): 1670–83. https://doi.org/10.1001/jama.2014.3654.

Conrad Gary D. 2014. "Is Ginkgo Biloba and/or a Multivitamin-Multimineral Supplement a Therapeutic Option for Parkinson's Disease? A Case Report." *Global Advances in Health and Medicine* 3 (4): 43–44. https://doi.org/10.7453/gahmj.2013.096.

Deepika N. P., Mohammed Shameer Kondengadan, Sherouk Hussein Sweilam, Md. Habibur Rahman, K.M Muhasina, Puja Ghosh, Divya Bhargavi, Divya Jyothi Palati, Fathi Maiz, and B. Duraiswamy. 2024a. "Neuroprotective Role of Coconut Oil for the Prevention and Treatment of Parkinson's Disease: Potential Mechanisms of Action." *Biotechnology and Genetic Engineering Reviews* 40 (4): 3346–78. https://doi.org/10.1080/02648725.2022.2122296.

Deepika N. P., Praveen Thaggikuppe Krishnamurthy, Magham Sai Varshini, Mudavath Ravi Naik, Deepak Vasudevan Sajini, Ammu VVV Ravi Kiran, Kusuma Kumari Garikapati, Basavan Duraiswamy, and Rohit Sharma. 2024b. "Ethnopharmacological Validation of Karkataka Taila - an Edible Crab Rasayana in Rotenone-Induced *in Vitro* and *in Vivo* Models of Parkinson's Disease." *Journal of Ethnopharmacology* 335 (December): 118691. https://doi.org/10.1016/j.jep.2024.118691.

Deshpande Prajakta, Neha Gogia, and Amit Singh. 2019. "Exploring the Efficacy of Natural Products in Alleviating Alzheimer's Disease." *Neural Regeneration Research* 14 (8): 1321. https://doi.org/10.4103/1673-5374.253509.

Ellerby Lisa M. 2002. "Hunting for Excitement: NMDA Receptors in Huntington's Disease." *Neuron* 33 (6): 841–42. https://doi.org/10.1016/S0896-6273(02)00631-1.

Fahn Stanley. 2008. "The History of Dopamine and Levodopa in the Treatment of Parkinson's Disease." *Movement Disorders: Official Journal of the Movement Disorder Society* 23 (Suppl 3): S497–508. https://doi.org/10.1002/mds.22028.

Frank Samuel. 2014. "Treatment of Huntington's Disease." *Neurotherapeutics* 11 (1): 153–60. https://doi.org/10.1007/s13311-013-0244-z.

García-González Nerea, Jaime Gonçalves-Sánchez, Ricardo Gómez-Nieto, Jesús M. Gonçalves-Estella, and Dolores E. López. 2024. "Advances and Challenges in Gene Therapy for Neurodegenerative Diseases: A Systematic Review." *International Journal of Molecular Sciences* 25 (23): 12485. https://doi.org/10.3390/ijms252312485.

Gerbarg Patricia L., and Richard P. Brown. 2013. "Phytomedicines for Prevention and Treatment of Mental Health Disorders." *Psychiatric Clinics of North America, Complementary and Integrative Therapies for Psychiatric Disorders* 36 (1): 37–47. https://doi.org/10.1016/j.psc.2012.12.004.

González-Burgos Elena, Carlos Fernandez-Moriano, and M. Pilar Gómez-Serranillos. 2015. "Potential Neuroprotective Activity of Ginseng in Parkinson's Disease: A Review." *Journal of Neuroimmune Pharmacology* 10 (1): 14–29. https://doi.org/10.1007/s11481-014-9569-6.

Hinge Nikita Subhash, Himanshu Kathuria, and Murali Monohar Pandey. 2022. "Engineering of Structural and Functional Properties of Nanotherapeutics and Nanodiagnostics for Intranasal Brain Targeting in Alzheimer's." *Applied Materials Today* 26 (March): 101303. https://doi.org/10.1016/j.apmt.2021.101303.

Irfan Zainab, Sofia Khanam, Varnita Karmakar, Sayeed Mohammed Firdous, Bothaina Samih Ismail Abou El Khier, Ilyas Khan, Muneeb U. Rehman, and Andleeb Khan. 2022. "Pathogenesis of Huntington's Disease: An Emphasis on Molecular Pathways and Prevention by Natural Remedies." *Brain Sciences* 12 (10): 1389. https://doi.org/10.3390/brainsci12101389.

Jain Nimansha, and Alice S. Chen-Plotkin. 2018. "Genetic Modifiers in Neurodegeneration." *Current Genetic Medicine Reports* 6 (1): 11–19. https://doi.org/10.1007/s40142-018-0133-1.

Johnson Shelby L., Hyun Y. Park, Nicholas A. DaSilva, Dhiraj A. Vattem, Hang Ma, and Navindra P. Seeram. 2018. "Levodopa-Reduced Mucuna Pruriens Seed Extract Shows Neuroprotective Effects against

Parkinson's Disease in Murine Microglia and Human Neuroblastoma Cells, Caenorhabditis Elegans, and Drosophila Melanogaster." *Nutrients* 10 (9): 1139. https://doi.org/10.3390/nu10091139.

Kalia Lorraine V., and Anthony E. Lang. 2015. "Parkinson's Disease." *The Lancet* 386 (9996): 896–912.

Kawamata Hibiki, and Giovanni Manfredi. 2011. "Introduction to Neurodegenerative Diseases and Related Techniques." In *Neurodegeneration: Methods and Protocols*, edited by Giovanni Manfredi and Hibiki Kawamata, 3–8. Totowa, NJ: Humana Press. https://doi.org/10.1007/978-1-61779-328-8_1.

Kay Chris, Michael R. Hayden, and Blair R. Leavitt. 2017. "Chapter 3 – Epidemiology of Huntington Disease." In *Handbook of Clinical Neurology*, edited by Andrew S. Feigin and Karen E. Anderson, 144: 31–46. Huntington Disease. Elsevier. https://doi.org/10.1016/B978-0-12-801893-4.00003-1.

Kumar Puneet, Harikesh Kalonia, and Anil Kumar. 2010. "Huntington's Disease: Pathogenesis to Animal Models." *Pharmacological Reports* 62 (1): 1–14. https://doi.org/10.1016/S1734-1140(10)70238–3.

Kumar Vikas. 2006. "Potential Medicinal Plants for CNS Disorders: An Overview." *Phytotherapy Research* 20 (12): 1023–35. https://doi.org/10.1002/ptr.1970.

Lamptey Richard N. L., Bivek Chaulagain, Riddhi Trivedi, Avinash Gothwal, Buddhadev Layek, and Jagdish Singh. 2022. "A Review of the Common Neurodegenerative Disorders: Current Therapeutic Approaches and the Potential Role of Nanotherapeutics." *International Journal of Molecular Sciences* 23 (3): 1851. https://doi.org/10.3390/ijms23031851.

Li Song, Jie Dong, Cheng Cheng, and Weidong Le. 2016. "Therapies for Parkinson's Diseases: Alternatives to Current Pharmacological Interventions." *Journal of Neural Transmission* 123 (11): 1279–99. https://doi.org/10.1007/s00702-016-1603-9.

Masters Colin L., Randall Bateman, Kaj Blennow, Christopher C. Rowe, Reisa A. Sperling, and Jeffrey L. Cummings. 2015. "Alzheimer's Disease." *Nature Reviews Disease Primers* 1 (1): 1–18. https://doi.org/10.1038/nrdp.2015.56.

Mathur Sunil, and Joseph Sutton. 2017. "Personalized Medicine Could Transform Healthcare." *Biomedical Reports* 7 (1): 3–5. https://doi.org/10.3892/br.2017.922.

Medina Alex, Yasamin Mahjoub, Larry Shaver, and Tamara Pringsheim. 2022. "Prevalence and Incidence of Huntington's Disease: An Updated Systematic Review and Meta-Analysis." *Movement Disorders* 37 (12): 2327–35. https://doi.org/10.1002/mds.29228.

Menard Caroline, Stephane Bastianetto, and Remi Quirion. 2013. "Neuroprotective Effects of Resveratrol and Epigallocatechin Gallate Polyphenols Are Mediated by the Activation of Protein Kinase C Gamma." *Frontiers in Cellular Neuroscience* 7 (December). https://doi.org/10.3389/fncel.2013.00281.

Merelli Amalia, Liliana Czornyj, and Alberto Lazarowski. 2013. "Erythropoietin: A Neuroprotective Agent in Cerebral Hypoxia, Neurodegeneration, and Epilepsy." *Current Pharmaceutical Design* 19 (38): 6791–6801. https://doi.org/10.2174/1381612811319380011.

Mischley Laurie K, Jason Allen, and Ryan Bradley. 2012. "Coenzyme Q10 Deficiency in Patients with Parkinson's Disease." *Journal of the Neurological Sciences* 318 (1–2): 72–75. https://doi.org/10.1016/j.jns.2012.03.023.

Murakami Hidetomo, Tomotaka Shiraishi, Tadashi Umehara, Shusaku Omoto, and Yasuyuki Iguchi. 2023. "Recent Advances in Drug Therapy for Parkinson's Disease." *Internal Medicine* 62 (1): 33–42. https://doi.org/10.2169/internalmedicine.8940-21.

Ondo William G., Nicte I. Mejia, and Christine B. Hunter. 2007. "A Pilot Study of the Clinical Efficacy and Safety of Memantine for Huntington's Disease." *Parkinsonism & Related Disorders* 13 (7): 453–54. https://doi.org/10.1016/j.parkreldis.2006.08.005.

O'Suilleabhain Padraig, and Jr Dewey Richard B. 2003. "A Randomized Trial of Amantadine in Huntington Disease." *Archives of Neurology* 60 (7): 996–98. https://doi.org/10.1001/archneur.60.7.996.

Paleacu Diana. 2007. "Tetrabenazine in the Treatment of Huntington's Disease." *Neuropsychiatric Disease and Treatment* 3 (5): 545–51.

Pang Shirley Yin-Yu, Philip Wing-Lok Ho, Hui-Fang Liu, Chi-Ting Leung, Lingfei Li, Eunice Eun Seo Chang, David Boyer Ramsden, and Shu-Leong Ho. 2019. "The Interplay of Aging, Genetics and Environmental Factors in the Pathogenesis of Parkinson's Disease." *Translational Neurodegeneration* 8 (1): 23. https://doi.org/10.1186/s40035-019-0165-9.

Park Han-A., and Amy C. Ellis. 2020. "Dietary Antioxidants and Parkinson's Disease." *Antioxidants* 9 (7): 570. https://doi.org/10.3390/antiox9070570.

Rahman Md. Habibur, Johny Bajgai, Ailyn Fadriquela, Subham Sharma, Thuy Thi Trinh, Rokeya Akter, Yun Ju Jeong, Seong Hoon Goh, Cheol-Su Kim, and Kyu-Jae Lee. 2021. "Therapeutic Potential of Natural Products in Treating Neurodegenerative Disorders and Their Future Prospects and Challenges." *Molecules* 26 (17): 5327. https://doi.org/10.3390/molecules26175327.

Roos Raymund A. C. 2010. "Huntington's Disease: A Clinical Review." *Orphanet Journal of Rare Diseases* 5 (December): 40. https://doi.org/10.1186/1750-1172-5-40.

Sajini Deepak Vasudevan, Praveen Thaggikuppe Krishnamurthy, and Amritha Chakkittu Kandiyil. 2022. "Neuroprotective Activities of Orientin: A Review." *Current Traditional Medicine* 8 (1): e130921196409. https://doi.org/10.2174/2215083807666210913104505.

Scott Lesley J., and Karen L. Goa. 2000. "Galantamine." *Drugs* 60 (5): 1095–122. https://doi.org/10.2165/00003495-200060050-00008.

Se Thoe Ewen, Ayesha Fauzi, Yin Quan Tang, Sunita Chamyuang, and Adeline Yoke Yin Chia. 2021. "A Review on Advances of Treatment Modalities for Alzheimer's Disease." *Life Sciences* 276 (July): 119129. https://doi.org/10.1016/j.lfs.2021.119129.

Tang Mengxi, and Changiz Taghibiglou. 2017. "The Mechanisms of Action of Curcumin in Alzheimer's Disease." *Journal of Alzheimer's Disease: JAD* 58 (4): 1003–16. https://doi.org/10.3233/JAD–170188.

Tardy Anne-Laure, Etienne Pouteau, Daniel Marquez, Cansu Yilmaz, and Andrew Scholey. 2020. "Vitamins and Minerals for Energy, Fatigue and Cognition: A Narrative Review of the Biochemical and Clinical Evidence." *Nutrients* 12 (1): 228. https://doi.org/10.3390/nu12010228.

Trevino J. T., R. C. Quispe, F. Khan, and V. Novak. 2020. "Non-Invasive Strategies for Nose-to-Brain Drug Delivery." *Journal of Clinical Trials* 10 (7): 439.

Vaddadi K. S., E. Soosai, E. Chiu, and P. Dingjan. 2002. "A Randomised, Placebo-Controlled, Double Blind Study of Treatment of Huntington's Disease with Unsaturated Fatty Acids." *NeuroReport* 13 (1): 29.

Vasudevan Sajini Deepak, Praveen Thaggikuppe Krishnamurthy, Amritha Chakkittukandiyil, and Ravi Naik Mudavath. 2024. "Orientin Modulates Nrf2-ARE, PI3K/Akt, JNK-ERK1/2, and TLR4/NF-kB Pathways to Produce Neuroprotective Benefits in Parkinson's Disease." *Neurochemical Research* 49 (6): 1577–87. https://doi.org/10.1007/s11064-024-04099-8.

Venkatesan Ramu, Eunhee Ji, and Sun Yeou Kim. 2015. "Phytochemicals that Regulate Neurodegenerative Disease by Targeting Neurotrophins: A Comprehensive Review." *BioMed Research International* 2015: 1–22. https://doi.org/10.1155/2015/814068.

Venuto Charles S., Andrew McGarry, Qing Ma, and Karl Kieburtz. 2012. "Pharmacologic Approaches to the Treatment of Huntington's Disease." *Movement Disorders: Official Journal of the Movement Disorder Society* 27 (1): 31–41. https://doi.org/10.1002/mds.23953.

Walker Eric A., and Mark V. Pellegrini. 2024. "Bacopa Monnieri." In *StatPearls*. Treasure Island (FL): StatPearls Publishing. http://www.ncbi.nlm.nih.gov/books/NBK589635/.

Walker Francis O. 2007. "Huntington's Disease." *The Lancet* 369 (9557): 218–28. https://doi.org/10.1016/S0140-6736(07)60111-1.

Wareham Lauren K., Shane A. Liddelow, Sally Temple, Larry I. Benowitz, Adriana Di Polo, Cheryl Wellington, Jeffrey L. Goldberg, et al. 2022. "Solving Neurodegeneration: Common Mechanisms and Strategies for New Treatments." *Molecular Neurodegeneration* 17 (1): 23. https://doi.org/10.1186/s13024-022-00524-0.

Yadav Satyndra Kumar, Jay Prakash, Shikha Chouhan, and Surya Pratap Singh. 2013. "*Mucuna Pruriens* Seed Extract Reduces Oxidative Stress in Nigrostriatal Tissue and Improves Neurobehavioral Activity in Paraquat-Induced Parkinsonian Mouse Model." *Neurochemistry International* 62 (8): 1039–47. https://doi.org/10.1016/j.neuint.2013.03.015.

Yuan Haidan, Qianqian Ma, Li Ye, and Guangchun Piao. 2016. "The Traditional Medicine and Modern Medicine from Natural Products." *Molecules* 21 (5): 559. https://doi.org/10.3390/molecules21050559.

Zr Chen, Huang Jb, Yang Sl, and Hong Ff. 2022. "Role of Cholinergic Signaling in Alzheimer's Disease." *Molecules (Basel, Switzerland)* 27 (6). https://doi.org/10.3390/molecules27061816.

19 Challenges during the Use of Natural Remedies for Neurodegenerative Disorders

K.M. Muhasina,[1] Puja Ghosh,[2] Mehrukh Zehravi,[3] N.P. Deepika,[4] Krishna Veni Nagappan,[5] Md Ekhtiar Uddin,[6] and Dhanabal S. Palaniswamy[2]

[1]Department of Pharmaceutical Analysis, Prime College of Pharmacy, Erattayal, Palakkad, Kerala

[2]Department of Pharmacognosy, JSS College of Pharmacy, Ooty, Tamil Nadu, India

[3]Department of Clinical Pharmacy, College of Dentistry & Pharmacy, Buraydah Private Colleges, Buraydah, Kingdom of Saudi Arabia

[4]Gujarat Grassroots Innovations Augmentation Network (GIAN), Satellite, Ahmedabad, Gujarat

[5]Department of Pharmaceutical Analysis, JSS College of Pharmacy, Ooty, Tamil Nadu, India

[6]Department of Pharmacy, Southeast University, Dhaka, Bangladesh

19.1 INTRODUCTION

The use of herbal remedies for neurological illnesses, including Alzheimer's (1), Parkinson's (2), and Huntington's (3) diseases, has received significant interest in few years from patients and healthcare workers seeking complementary therapies to conventional drugs. Natural remedies, such as dietary modifications, conventional therapy, and herbal supplements, often claim significance, including alleviation of symptoms, enhanced quality of life, and cognitive improvement. Nevertheless, the intense use of these therapies may lead to many problems that prevent their safety, effectiveness, and incorporation into conventional treatments.

A significant problem is the absence of scientific data to prove the effectiveness of several natural remedies in the treatment of neurodegenerative illnesses. Also, early interventions may indicate supportive evidence, and large-scale clinical trials are necessary to validate these points. Furthermore, the quality and standardization of herbal products may vary greatly, making it more difficult to validate appropriate dosages, potential side effects, and long-term safety (4). Then, there is a chance that natural remedies and prescription medications may interact. Individuals with neurological conditions often adhere to complex treatment regimens, and using herbal remedies in combination with prescription medications may have unfavourable effects or reduce therapeutic efficacy. Furthermore, misunderstandings and the widespread availability of unregulated herbal products might lead to overuse, misuse, or worsening of these negative effects.

Ultimately, the socioeconomic and cultural factors influencing the use of natural remedies substantially impact their consumption and effectiveness. The disposition of patients, availability of healthcare resources, and the cost spectrum of different options may influence the effective implementation of these therapies. This chapter meticulously analyses these issues and emphasizes the

DOI: 10.1201/9781003478652-21

need for comprehensive regulation, research, and education on the safe and effective use of natural medicines in the treatment of neurological disorders.

19.2 NATURAL REMEDIES AS TREATMENT ALTERNATIVES

When we discussed the advancements in neurodegenerative disorders, there was a lack of progress in viable treatment modalities. The use of natural products has significant promise for the prevention and treatment of neurological illnesses, and herbal drugs are showing notable efficacy in different aspects of neurological disorders. Recent research has identified several neuroinflammatory mechanisms and pathways significant with the onset and progression of neurological diseases, including the activation of microglial and the NF-κB, MAPK, ROS, NLRP3, TLR4, Nrf2, PI3K, and Akt pathways and also the influence of gut microbiota. In this chapter, we tried to discuss the mechanism of neurological disorders and the activity of natural products on these mechanisms and also provide theoretical pieces of evidence for the support of effective treatment for neurological disorders (5).

Natural remedies have received significant attention as complementary or alternative therapies for neurological disorders, including Parkinson's disease, Alzheimer's disease, multiple sclerosis, and Huntington's disease. These remedies usually include phytoconstituents, essential oils, herbal medication, supplements, and lifestyle modifications like controlled diet, yoga, and exercise, which may provide neuroprotective effects and enhance the health of the brain. The use of herbal medicines in the management of neurological illnesses and problems that might affect their safety and efficacy, as well as their incorporation into standard medication protocols. Up to this time, several concerns have emerged about their role in the management of neurodegenerative disorders, mostly centred on the absence of scientific validation and proving their clinical outcomes. The ambiguities lead to essential inquiries regarding their legal perspectives from a bioethical point, necessitating the decisional autonomy of the patient, the safety of the patient, and the differentiation between beneficial and, non-beneficial treatment (5).

19.3 CHALLENGES IN THE USE OF NATURAL REMEDIES AS TREATMENT MODALITIES

The primary problem is the absence of robust scientific research and data to validate the claims of natural remedies. Although much research indicates significant advantages. These results are either insufficiently strong to advise for wider use (6). The inconsistency in dose, quality, and preparation of natural products affects their standardization/validation for therapeutic use. Also, there are concerns about the safety of natural therapies, especially their interactions with conventional pharmaceutical dosage forms and their possible adverse effects. A significant number of persons with neurological disorders are already undergoing intricate therapies, and the use of herbal supplements or alternative remedies may result in severe side effects or reduced activity of prescribed medications (7). Inadequate monitoring may expose patients to hazardous/dangerous herbals that may jeopardize their health. The next session outlines the ethical, regulatory, and scientific problems associated with patients, healthcare professionals, and researchers in using natural therapies for the treatment of neurodegenerative illnesses. It is important to address these problems to guarantee the safe, responsible, and efficient use of natural remedies in addressing these types of illnesses.

19.4 CHALLENGES ASSOCIATED WITH THE NATURAL REMEDIES FOR THE TREATMENT OF NEUROLOGICAL DISORDERS

The growing significance of natural remedies for the treatment of neurological disorders, including Parkinson's, Alzheimer's, and other age-related problems, leads to a saga of alternative or complementary treatments to traditional medicines (2). However, various therapies, including herbal food

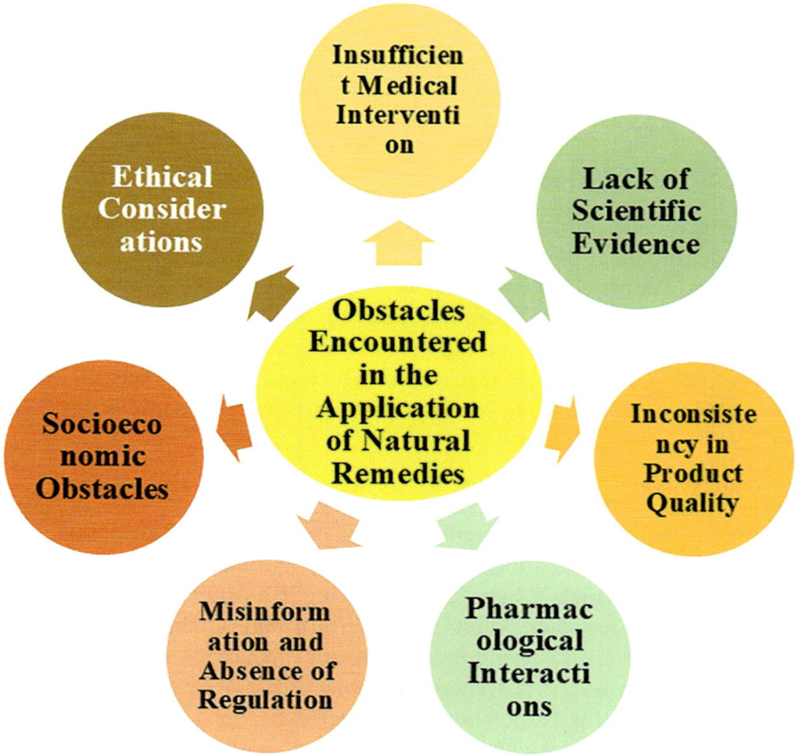

FIGURE 19.1 Obstacles encountered in the use of natural remedies for the treatment of neurodegenerative disorders.

supplements (8), vitamins (9), and lifestyle changes (10), may have significant advantages and also some considerable difficulties. These problems include several dimensions, such as safety, regulation, scientific evidence, and patient outcomes. Comprehending and mitigating these problems are necessary for the efficient and secure integration of natural remedies/treatments in the management of neurodegenerative illnesses (Figure 19.1).

19.4.1 Lack of Scientific Data

A notable difficulty is the insufficiency of scientific data that validates the activity of most of the complementary treatments for neurological disorders. Notwithstanding a few encouraging results in in vivo and limited human trials, a significant deficiency still exists in large-scale and high-quality clinical trials capable of determining the advantages and possible mechanisms of natural treatments. Several studies face criticism because of smaller sample sizes, durations, and methodological errors, hindering the generalization and reproducibility of their findings. Consequently, the healthcare sector may be reluctant to endorse natural treatments in the absence of more significant data to substantiate their therapeutic activity (11).

19.4.2 Inconsistency in the Product Quality and Dose

The efficacy of natural remedies may significantly vary based on some variables such as the growth techniques, the plant's origin, storage conditions, and processing. Against pharmaceutical medications, which undergo rigorous quality control checks, herbal remedies often lack standardization regarding their efficacy, active constituents, and purity. This variation may result in different therapeutic outcomes and may decrease the overall efficacy of a treatment. Also, establishing the specific dose for herbal drugs may be strenuous due to the absence of globally acceptable guidelines or standards, resulting in the risk of overdosing or underdosing (12).

19.4.3 Pharmacological Interactions and Safety Problems Associated

The safety of herbal medicines in connection with conventional drugs is an important issue. A number of people with neurological disorders are now undergoing significant treatments to alleviate their symptoms. Herbal supplements may interact with certain drugs, resulting in certain effects such as decreased therapeutic activity, potential toxicity, or side effects. Some herbs may disrupt the activity of neurologic activators such as cholinesterase inhibitors. The safety profiles of several herbal therapeutics remain inadequately researched, which may result in specific dangers, particularly when the patients are susceptible to over-sensitization towards specific drugs (13).

19.4.4 Misinformation and Absence of Regulation

A notable difficulty in the use of natural therapies is the pervasive misinformation and the absence of rigorous regulation within the natural health sector. A multitude of patients resort to internet resources or unverified testimonials for guidance, resulting in a dependence on anecdotal evidence instead of scientifically substantiated knowledge. Certain natural therapies are promoted with overstated assertions of efficacy, fostering unreasonable expectations among patients and carers. Moreover, the natural health sector is sometimes subject to less stringent regulation than pharmaceutical medications, resulting in potential consumer exposure to contaminated, inaccurately labelled, or counterfeit items. The absence of control may jeopardize patient health and safety (14).

19.4.5 Cultural and Socioeconomic Obstacles

Cultural considerations significantly influence the adoption of natural medicines. In many areas, a robust history of using herbal medicine and alternative remedies exists, with patients often preferring these approaches over conventional treatments. However, challenges may emerge in the amalgamation of natural remedies with modern medicine, especially when there is inadequate communication between healthcare providers and patients on the potential benefits or risks. Furthermore, socioeconomic factors, including the expense and accessibility of natural medicines, may impede their use. A considerable proportion of patients may be unable to afford premium supplements or see practitioners skilled in integrative medicine (15).

19.4.6 Ethical Considerations

The use of natural treatments prompts ethical considerations, especially regarding their accessibility, possible exploitation, and environmental sustainability. Certain plant-based medicines originate from endangered species or need unsustainable harvesting methods, potentially damaging ecosystems. Moreover, there are apprehensions over the commodification of traditional knowledge and the exploitation of indigenous communities that have historically depended on natural cures. Ethical considerations must be meticulously evaluated to guarantee that the use of natural therapies honours both environmental integrity and the cultural legacy linked to them (5).

19.4.7 Postponed or Insufficient Medical Intervention

Ultimately, a concern linked to the use of natural medicines is that patients can postpone or abandon more efficacious conventional treatments in preference for these alternatives. Neurodegenerative illnesses need prompt and evidence-based medical intervention; hence, dependence on unverified alternative cures may exacerbate symptoms or hinder patients from fully benefiting from existing treatments. Although some people may consider natural treatments beneficial as adjunctive choices, they need not be regarded as replacements for established medicines.

19.5 FUTURE PERSPECTIVES ON THE USE OF HERBAL REMEDIES FOR THE TREATMENT OF NEURODEGENERATIVE DISORDERS

The use of natural therapies for neurodegenerative conditions, including Alzheimer's disease, Parkinson's disease, and other types of dementia, has considerable potential but encounters substantial obstacles. These obstacles generally originate from concerns relating to effectiveness, safety, and the complicated molecular pathways underlying these disorders. Nevertheless, other potential strategies and methods may be used to overcome these challenges and facilitate the progress of natural remedies. Various neurodegenerative diseases are driven by complex molecular pathways, including protein aggregation, oxidative stress, mitochondrial dysfunction, and neuroinflammation (16). A comprehensive knowledge of these processes at the cellular and molecular levels is crucial for developing customized natural therapies. Researchers are now using advanced technologies, including genomes, proteomics, and metabolomics, to elucidate these processes. Genetic and epigenetic factors profoundly affect susceptibility to neurodegenerative diseases (17). Tailored treatment strategies that consider an individual's genetic composition may enhance the effectiveness of natural therapies. People with certain genetic markers may benefit more from various phytochemical compounds (18).

A significant problem associated with natural remedies (such as herbal extracts, essential oils, and nutritional supplements) is the inconsistency in the strength and purity of bioactive components across various products (19). Establishing tight criteria for procurement, processing, and quality control of these goods is crucial to maintaining consistent therapeutic results.

Natural substances sometimes lack the stringent regulatory scrutiny applied to commercial pharmaceuticals. Stronger regulatory frameworks are required to ensure the safety and effectiveness of natural remedies, particularly for long-term usage. Regulatory agencies like the FDA and EMA may be essential in defining standards for clinical trials, including natural treatments (20). One of the fundamental constraints of natural chemicals is their capacity to cross the blood–brain barrier (BBB), which shields the brain from dangerous substances but also impedes the delivery of therapeutic agents. Innovative drug delivery systems crossing the BBB may enhance the efficacy of phytomedicines, including nanoparticles, liposomes, and nanocarriers (21). Inadequate absorption or rapid metabolism can occasionally restrict the bioavailability of natural substances. Innovative formulations, such as nanoencapsulation, are currently being investigated by researchers in order to improve the efficacy of bioactive compounds by increasing their solubility and stability (22).

For natural remedies to be widely accepted, it is necessary to conduct additional high-quality, large-scale clinical studies that verify their safety and efficacy. Comprehensive research, including randomized controlled trials and extended follow-up investigations, is necessary to integrate natural products into evidence-based medicine (23).

A prominent area of research is the investigation of the potential synergistic effects of combining natural remedies with conventional pharmaceutical medications. Natural compounds have the potential to enhance the therapeutic efficacy of conventional medications or mitigate their adverse effects, thereby promoting a more integrative therapy approach concentrating on particular neurodegenerative mechanisms. Numerous natural substances have demonstrated the potential to address specific pathogenic pathways associated with neurodegenerative disorders (24).

Investigating these compounds and their molecular targets may provide more effective therapies. Natural compounds with anti-inflammatory properties, including curcumin (derived from turmeric), resveratrol (extracted from grapes), and omega-3 fatty acids (sourced from fish oils), warrant more investigation on their potential effects on neuroinflammation. Oxidative stress is a critical contributor to cerebral injury linked to conditions such as Alzheimer's and Parkinson's diseases. Compounds with substantial antioxidant properties, including flavonoids, polyphenols, and vitamins, have shown potential in reducing oxidative damage. Further examination of their specific mechanisms of action will be essential (25).

The accumulation of misfolded proteins, such as amyloid-beta in AD and alpha-synuclein in PD, is a defining feature of both conditions. Certain natural compounds, such as EGCG (extracted from green tea) and berberine, have shown an influence on autophagy (the cellular waste disposal process) and may aid in the removal of detrimental protein aggregates (26). Various natural compounds, such as ginseng, lion's mane mushroom, and *Bacopa monnieri*, have shown the capacity to enhance neurogenesis and synaptic plasticity, potentially augmenting cognitive function and slowing the progression of neurodegenerative diseases. Many natural treatments for neurodegenerative diseases derive from ancient medicinal traditions, such as Ayurveda, traditional Chinese medicine, and indigenous healing practices (27). Collaborative endeavours between modern scientists and traditional healers may result in the identification of groundbreaking therapeutic compounds. Modern techniques such as high-throughput screening may evaluate the efficacy of these traditional medicines. This interdisciplinary field, combining ethnobotany, pharmacology, and traditional knowledge, has significant potential for discovering new natural chemicals having neuroprotective effects

Increasing evidence suggests that gut health profoundly influences neurodegenerative diseases, with the gut–brain axis playing a crucial role in inflammation and neurodegeneration. Natural medicines that affect the gut microbiota, such as prebiotics, probiotics, and polyphenol-rich diets, may provide a novel approach for managing neurodegenerative diseases. Anti-inflammatory diets, including the Mediterranean diet, and natural compounds with neuroprotective properties may function as preventive or adjunct therapy for neurodegenerative diseases (28). The influence of dietary and natural supplements on reducing neuroinflammation is a progressively important area of study (29). A comprehensive therapeutic approach that combines natural products with lifestyle alterations (including exercise, diet, mindfulness, and cognitive training) may be more effective than other treatments. Natural compounds may be used in cognitive rehabilitation or physical therapy to impede or reverse disease progression. The development of natural medicines for the prevention of neurodegenerative diseases, particularly in early stages or among high-risk individuals, might substantially impact public health. Natural treatments, such as dietary modifications, vitamin supplementation, and cognitive exercises, may aid in delaying or preventing the onset of disease in at-risk populations.

19.6 CONCLUSION

The use of natural therapies for neurodegenerative disorders is promising but poses considerable challenges. The integration of these medicines into clinical practice requires careful assessment owing to scientific uncertainty, quality control issues, safety risks, and cultural factors. Future research, improved regulation, and expanded communication between healthcare providers and patients are essential for resolving these issues and ensuring the effective and safe use of natural therapies in the treatment of neurodegenerative diseases. The prospects for natural treatment in neurodegenerative diseases seem promising, although a multidisciplinary approach is essential. Addressing the challenges of efficacy and safety, as well as comprehending the molecular underpinnings of neurodegeneration, will be crucial for realizing the potential of natural therapies. Ongoing research, collaboration between conventional and modern medicine, and innovative technologies for drug administration and clinical evaluation will facilitate the development of safe and effective natural treatments for neurodegenerative diseases.

REFERENCES

1. Cahlíková, L, et al. Natural compounds (small molecules) as potential and real drugs of Alzheimer's disease: A critical review. *Studies in Natural Products Chemistry* 42 (2014): 153–194.
2. Deepika, NP, et al. Ethnopharmacological validation of Karkataka Taila-An edible crab Rasayana in rotenone-induced in vitro and in vivo models of Parkinson's disease. *Journal of Ethnopharmacology* 335 (2024): 118691.

3. Lum, PT, et al. Protective effect of natural products against Huntington's disease: An overview of scientific evidence and understanding their mechanism of action. *ACS Chemical Neuroscience* 12.3 (2021): 391–418.
4. Solanki, I, Parihar, P, Parihar, MS. Neurodegenerative diseases: From available treatments to prospective herbal therapy. *Neurochemistry International* 95 (2016): 100–108.
5. Di Paolo, M, et al. Natural products in neurodegenerative diseases: A great promise but an ethical challenge. *International Journal of Molecular Sciences* 20.20 (2019): 5170.
6. Zhang, J, et al. Quality of herbal medicines: Challenges and solutions. *Complementary Therapies in Medicine* 20.1–2 (2012): 100–106.
7. Shan, JJ, et al. Challenges in natural health product research: The importance of standardization. *Proceedings-Western Pharmacology Society* 50. [Western Pharmacology Society]; 1998, 2007.
8. da Costa, IM, et al. Supplementation with herbal extracts to promote behavioral and neuroprotective effects in experimental models of Parkinson's disease: A systematic review. *Phytotherapy Research* 31.7 (2017): 959–970.
9. Zhao, X, et al. Benefits of vitamins in the treatment of parkinson's disease. *Oxidative Medicine and Cellular Longevity* 2019.1 (2019): 9426867.
10. Santiago, JA, Potashkin, JA. Physical activity and lifestyle modifications in the treatment of neurodegenerative diseases. *Frontiers in Aging Neuroscience* 15 (2023): 1185671.
11. Rahman, MH, et al. Therapeutic potential of natural products in treating neurodegenerative disorders and their future prospects and challenges. *Molecules* 26.17 (2021): 5327.
12. Rahman, MM, et al. Multifunctional role of natural products for the treatment of Parkinson's disease: At a glance. *Frontiers in Pharmacology* 13 (2022): 976385.
13. Houghton, PJ, Melanie-Jayne, H. Natural products and derivatives affecting neurotransmission relevant to Alzheimer's and Parkinson's disease. *Neurosignals* 14.1–2 (2005): 6–22.
14. Micallef, J, et al. What should be done to combat misinformation about health products?. *Therapies* 79.1 (2024): 87–98.
15. Abe-Inge, V, et al. Plant-based dietary shift: Current trends, barriers, and carriers. *Trends in Food Science & Technology* 143 (2024): 104292.
16. Rehman, MU, et al. Mitochondrial dysfunctions, oxidative stress and neuroinflammation as therapeutic targets for neurodegenerative diseases: An update on current advances and impediments. *Neuroscience & Biobehavioral Reviews* 144 (2023): 104961.
17. Migliore, L, Fabio C. Genetics, environmental factors and the emerging role of epigenetics in neurodegenerative diseases. *Mutation Research/Fundamental and Molecular Mechanisms of Mutagenesis* 667.1–2 (2009): 82–97.
18. Zhang, Y-J, et al. Antioxidant phytochemicals for the prevention and treatment of chronic diseases. *Molecules* 20.12 (2015): 21138–21156.
19. Osman, A, et al. Quality consistency of herbal products: Chemical evaluation. *Progress in the Chemistry of Organic Natural Products 122: Botanical Dietary Supplements and Herbal Medicines*. Cham, Switzerland: Springer Nature, 2023, pp. 163–219.
20. Alqahtani, S, et al. Priority review drugs approved by the FDA and the EMA: Time for international regulatory harmonization of pharmaceuticals?. *Pharmacoepidemiology and Drug Safety* 24.7 (2015): 709–715.
21. Mahomoodally, MF, et al. The potential application of novel drug delivery systems for phytopharmaceuticals and natural extracts–current status and future perspectives. *Mini Reviews in Medicinal Chemistry* 21.18 (2021): 2731–2746.
22. Alfei S. Nanotechnology applications to improve solubility of bioactive constituents of foods for health-promoting purposes. *Nano-Food Engineering: Volume One*. Cham: Springer International Publishing, 2020, pp. 189–257.
23. Khorsan, R, et al. Systematic review of integrative health care research: Randomized control trials, clinical controlled trials, and meta-analysis. *Evidence-Based Complementary and Alternative Medicine* 2011.1 (2011): 636134.
24. Vasudevan Sajini, Deepak, et al. Orientin modulates Nrf2-ARE, PI3K/Akt, JNK-ERK1/2, and TLR4/NF-kB Pathways to produce neuroprotective benefits in Parkinson's Disease. *Neurochemical Research* 49.6 (2024): 1577–1587.
25. Tumilaar, SG, et al. A comprehensive review of free radicals, oxidative stress, and antioxidants: Overview, clinical applications, global perspectives, future directions, and mechanisms of antioxidant activity of flavonoid compounds. *Journal of Chemistry* 2024.1 (2024): 5594386.
26. Payne, A, et al. Epigallocatechin-3-Gallate (EGCG): New therapeutic perspectives for neuroprotection, aging, and neuroinflammation for the modern age. *Biomolecules* 12.3 (2022): 371.

27. Ven Murthy, MR, et al. Scientific basis for the use of Indian ayurvedic medicinal plants in the treatment of neurodegenerative disorders: 1. Ashwagandha. *Central Nervous System Agents in Medicinal Chemistry (Formerly Current Medicinal Chemistry-Central Nervous System Agents)* 10.3 (2010): 238–246.
28. Muhasina, KM, et al. From gut microbiomes to infectious pathogens: Neurological disease game changers. *Molecular Neurobiology* 62.1 (2025): 1184–1204.
29. Deepika, NP, et al. Neuroprotective role of coconut oil for the prevention and treatment of Parkinson's disease: Potential mechanisms of action. *Biotechnology and Genetic Engineering Reviews* 40.4 (2024): 3346–3378.

20 Conclusion and Future Aspects during Treatment of Neurodegenerative Disorders

Mirunalini Gobinath,[1] *L. Priyanka Dwarampudi,*[1] *R. Shanmugam,*[2] *Parikshit Roychowdhury,*[3] *and Subhajit Dutta*[2]

[1]Department of Pharmacognosy & Phytopharmacy, JSS College of Pharmacy, JSS Academy of Higher Education & Research, Ooty, The Nilgiris, Tamil Nadu, India

[2]TIFAC CORE HD, JSS College of Pharmacy, JSS Academy of Higher Education & Research, Ooty, The Nilgiris, Tamil Nadu, India

[3]Department of Pharmaceutics, JSS College of Pharmacy, JSS Academy of Higher Education & Research, Ooty, The Nilgiris, Tamil Nadu, India

20.1 INTRODUCTION

Neurodegenerative disorders refer to conditions in which nerve cells or neurons in the brain or spinal cord gradually deteriorate or are lost, leading to a decline in neurological function. As a result of this degeneration, neurological function declines, affecting many areas of neurological activity, including movement and cognition. These conditions usually deteriorate with time as they are chronic and progressive (1). Damage to the nervous system can be due to synaptic dysfunction, disruption of neuronal communication, or deposition of abnormal proteins. Impaired cognitive and motor symptoms are commonly observed in patients with disrupted neuronal signaling due to synaptic dysfunction. These neurons in the brain show a maze-like organization, which permits them to execute assorted functions with great efficiency (2). Nevertheless, due to neurodegeneration, which causes abnormalities in neuronal connections and communication, these organized networks cannot perform their normal activities. Thus, neurodegeneration is a multifaceted process involving complex interactions between cellular and molecular mechanisms that leads to a gradual loss of neuronal structure and function. To develop therapies that have potential benefits in delaying or halting these debilitating diseases, it is crucial to understand the basic fundamental processes that drive neurodegeneration. These neurodegenerative disorders include Alzheimer's disease, amyotrophic lateral sclerosis (ALS), frontotemporal dementia, Huntington's disease (HD), Lewy body dementia, multiple sclerosis, and Parkinson's disease (3) (Table 20.1).

Alzheimer's disease generally occurs due to the excessive buildup of neurofibrillary tangles of tau proteins and amyloid beta plaques. The clinical presentation shows gradual dementia and other neuropsychiatric symptoms. The etiology of Alzheimer's is diverse. They are also observed to determine if the function of cholinergic neurotransmitters is diminished. Consequently, this results in cognitive deficits due to cholinergic neuronal damage. Various epigenetic factors, such as non-coding RNA, histone modifications, and DNA methylation, play key roles in the onset of the disease (3, 4).

ALS, previously known as Lou Gehrig's disease, is a neurodegenerative disorder that affects motor neurons. Abnormal RNA processing and protein aggregation, excitotoxicity, mitochondrial dysfunction, and defects in neurofilaments represent a few possible underlying root causes of the

DOI: 10.1201/9781003478652-22

disease. Considering the emerging evidence of sophisticated interplay between molecular and genetic pathways, the pathophysiological mechanisms of ALS are complex and multifaceted. It is believed that excessive stimulation of glutamate neurotransmitters might incite the disease via the activation of calcium ions and free radicals, which are bounced to cause neurodegeneration by cellular damage to mitochondria and other organelles, which activates the inflammatory cascade (5).

Frontotemporal dementia, as the name suggests, is caused by degeneration of the frontotemporal lobes. This is substantially caused by gene mutations in the proteins in the brain. Early clinical manifestations, including inappropriate social etiquette, changes in personality, and other neuropsychological symptoms, were observed. As the disease progresses, changes in cognitive function and motor deficits are observed, which act as a merger between Alzheimer's and Parkinson's symptoms (6).

HD is a movement-based neurodegenerative disease characterized by irregular and involuntary movements of the head, neck, limbs, or face (often referred to as chorea). It is a genetic disorder that involves mutations in the huntingtin gene, which subsequently leads to illness. Numerous repeats in CAG on chromosome 4 can cause damage to the caudate nucleus, which in turn results in decreased acetylcholine and GABA levels and increased glutamate neurotransmitters (7).

Parkinson's disease involves progressive depletion of dopaminergic neurons in the substantia nigra of the basal ganglia, which is characterized by dyskinesia, tremors, and postural instability. Neurotransmitters, such as serotonin and noradrenaline, are diminished and lead to depression-like symptoms, whereas acetylcholine is elevated, which in turn causes motor problems. Pathological manifestations are due to the aggregation of misfolded proteins, such as α-synuclein, ubiquitin, and other neurofilaments. These changes have also been observed in patients with Lewy body dementia (8, 9).

Multiple sclerosis is a chronic neurodegenerative disease that affects the axons of the brain and spinal cord. This causes demyelination, neuronal degeneration, and autoimmune inflammation. The mechanism involved is unclear; however, after the trigger activates autoreactive T-lymphocytes, the inflammatory cascade begins to demyelinate axons and form plaques. Thus, as a result, atrophy of oligodendrocytes and gliosis occurs. The primary manifestations include retrobulbar neuritis, which causes impaired vision and headaches (10).

TABLE 20.1
Overview of Neurodegenerative Disorders

Neurodegenerative Disease	Pathophysiology	Clinical Features
Alzheimer disease	Neuritic plaques embedded as β-amyloid deposits and intracellular neurofibrillary tangles, which cause gross atrophy and loss of neuronal communication	Loss of short-term memory, language dysfunction, errors in speech, impaired reasoning, ability to handle complex tasks, inability to recognize faces, agitation, and other behavioral problems
Parkinson's disease	Affects the substantia nigra of the basal ganglia due to chronic depletion of neurons that innervate dopamine	Bradykinesia, postural instability, along with resting tremor or rigidity, apraxia, and agnosia
Huntington disease	Mutations in huntingtin gene cause an increased number of CAG trinucleotide repeats	Dysfunction of subcortical motor circuits, irregular and involuntary movements of limbs, face, or head and neck
Multiple sclerosis	Immune-mediated disease that damages axons and the myelin sheath of neurons in the brain and spinal cord	Affects all nerves innervated by the spinal cord and brain due to autoimmunity
Amyotrophic lateral sclerosis	Abnormal RNA	Twitching, muscle cramps, gradual loss of muscle function, spasticity, and muscle weakness

20.2 CURRENT TREATMENT STRATEGIES AND CHALLENGES FOR NEURODEGENERATIVE DISEASES

Treating neurodegenerative disorders is often associated with disease-specific conditions. Currently, there is an abundance of recognized management approaches that either focus on the etiology of the illness or make an effort to alleviate its symptoms.

Although there isn't a cure for Alzheimer's at this time, there are therapies that can help control the symptoms and enhance quality of life. Current treatment strategies include symptomatic therapies that attempt to reduce Alzheimer's symptoms without affecting the disease's fundamental course (11). In the brain, acetylcholine levels are raised by medications such as galantamine, rivastigmine, and donepezil, which aid in the treatment of memory loss and other cognitive impairments. Acetylcholine is a chemical messenger essential to thought and memory that is increased in the brain by these drugs. Memantine functions by controlling glutamate, another crucial neurotransmitter in the brain. It can decrease excessive glutamate activity, which may have a role in the neurotoxicity of Alzheimer's. The behavioral treatments include antidepressants that are used to control emotional illnesses, including anxiety and sadness, and antipsychotics, which are used cautiously to treat serious behavioral problems such as psychosis or aggressiveness, anxiolytics, and sleep aids that are used to treat sleep issues and anxiety (4, 12).

Other disease-modifying management is by focusing on the underlying molecular mechanisms that underlie Alzheimer's disease; these therapies aim to interfere with the molecular pathological process. Alzheimer's disease presently has no FDA-approved disease-modifying medications; however, some interesting treatments are being developed. Anti-amyloid therapies that focus on a protein called amyloid beta cause brain plaques that are characteristic of Alzheimer's disease. The first amyloid beta-targeting medication authorized by the FDA is aducanumab; however, opinions on its efficacy vary. Tau therapies that target tau, an additional protein that creates brain tangles linked to Alzheimer's disease. NSAIDs and immunomodulators could decrease the progression of Alzheimer's disease by reducing inflammation. Some antioxidant and neuroprotective medications are used to encourage the growth of new neurons or protect existing ones from harm (13).

Even though progress has been achieved, researchers, doctors, and families still face several challenges due to the disease's complexity. A significant number of available therapies concentrate more on symptom management than on healing or appreciably delaying the course of the illness. Current symptomatic therapies, such as memantine and cholinesterase inhibitors, have limited effects and frequently lose their effectiveness over time. While there are new disease-modifying medications on the market, like monoclonal antibodies that target beta-amyloid plaques, their effectiveness varies, and they are not always beneficial. Some monoclonal antibodies cause serious adverse effects, such as brain swelling and hemorrhage, whereas other current treatments have modest side effects, including nausea and dizziness. Treatment techniques are complicated by genetic predispositions that impact Alzheimer's disease risk and development, like the APOE ε4 allele. Some patients may not be able to afford modern diagnostics and disease-modifying medicines (3, 14).

Symptomatic treatments for Parkinson's disease work as symptomatic therapy to enhance quality of life by controlling both motor and non-motor symptoms. Parkinson's disease symptoms, such as tremor, bradykinesia, stiffness, and postural instability, are treated using symptomatic approaches. Levodopa is the most often prescribed drug for Parkinson's disease. It helps make up for the brain's decreased synthesis of dopamine since it is an initial precursor to dopamine. Benserazide or carbidopa are frequently used with levodopa to minimize adverse effects and boost effectiveness. As an alternative to levodopa, medications such as ropinirole, rotigotine, and pramipexole directly affect dopamine receptors. Levodopa can be administered with them or without them to treat motor problems. MAO-B inhibitors, for example, rasagiline and selegiline, can prevent the breakdown of dopamine by an enzyme and boost the neurotransmitter's availability. They are commonly used in conjunction with levodopa in early-stage Parkinson's disease. By preventing levodopa's breakdown and lowering motor fluctuations, COMT inhibitors like tolcapone and entacapone can prolong the drug's duration of effect. Anticholinergics have the potential to have cognitive adverse effects, which limit their usage even if they can help treat tremors. Although its precise mode of action is unknown, amantadine is occasionally used to treat dyskinesias. It is believed to work via influencing the glutamate and dopamine pathways. The surgical approach, like deep brain stimulation, entails implanting electrodes to modify aberrant neural activity in certain brain areas, such as the globus pallidus interna or subthalamic nucleus. Lesions such as thalamotomy and pallidotomy, which are less prevalent than DBS, generate tiny lesions in the brain to lessen motor symptoms. The goal of physical therapy is to assist patients in retaining their physical function by enhancing their mobility, balance, and flexibility. Parkinson's disease-related speech and swallowing issues are treated with speech therapy (15, 16).

Disease-modifying treatments for this disease are treated with the goal of slowing or stopping its development by addressing its underlying causes. In this management, sometimes gene therapy and stem cell therapy are used. Gene therapy is a medical procedure in which genetic material is introduced into cells to modify their function. Research is being done on gene therapy as a possible Parkinson's disease. Introducing genes that could aid in dopamine neuron protection or stimulate the development of new dopamine neurons is the aim for gene therapy. Another kind of treatment for Parkinson's disease that is being researched is stem cell therapy. Stem cells are undeveloped cells with the capacity to differentiate into a wide variety of cell types. To treat Parkinson's disease, stem cells would be transplanted into the brain with the expectation that they will differentiate into dopamine neurons. The goal of immunotherapies against alpha-synuclein, a protein linked to Parkinson's disease, is to lessen the protein's brain distribution and aggregation (15, 17).

The most widely faced challenges in the treatment of Parkinson's disease are levodopa, which loses its effectiveness as the condition worsens. This leads to "off" times when symptoms resurface, seriously interfering with day-to-day activity. Levodopa usage over an extended length of time can result in dyskinesias, or uncontrollable movements, especially during "on" times when the drug is acting as intended. It's still difficult to strike the correct balance between controlling symptoms and reducing adverse effects. Numerous non-motor symptoms, like as cognitive decline, anxiety, autonomic dysfunction, depression and sleep difficulties sensory abnormalities, can be brought on by Parkinson's disease. It can be difficult to coordinate the multidisciplinary approach needed to treat them. The current therapies are symptomatic, which means they just treat the symptoms without addressing the underlying cause of the illness. Disease-modifying treatments are desperately needed in order to stop or delay the course of Parkinson's disease and maybe provide patients with a long-term cure. There are inherent dangers associated with procedures, including deep brain surgery, such as bleeding, infection, or hardware issues. To reduce these hazards, thorough monitoring and patient selection are essential (12, 18).

Lou Gehrig's disease, often known as ALS, is a progressive neurological illness that affects motor neurons. Many patients begin with lower motor neuron symptoms, which inevitably advance to encompass bulbar and lower motor neuron symptoms. Managing ALS symptoms and enhancing the patient's quality of life are the goals of symptomatic therapy. The drugs used, like benzodiazepines, baclofen, and tizanidine, are useful for cramps and muscle spasticity. NSAIDs and opioids help reduce pain. Some drugs, such as glycopyrrolate, atropine, or scopolamine patches, contribute to reducing the overproduction of saliva. Physical therapy improves in preserving the range of motion and lessening spasticity. Exercises for strengthening and stretching are commonplace. As the illness worsens, speech therapists frequently suggest assistive communication devices to help with communication problems. For people who have trouble breathing, non-invasive ventilation devices like BiPAP machines might be helpful. Invasive ventilation may be necessary in more serious situations (5).

The primary objective of disease-modifying treatments is to impede the development of ALS. An FDA-approved medication called riluzole can somewhat increase lifespan by lowering glutamate excitotoxicity. Another FDA-approved drug, Edaravone, is an antioxidant drug which can slow the decline in their functional abilities in some ALS patients. Stem cell therapy can repair and replace the damaged neurons.

Most of the challenges faced in the treatment are due to ALS's symptoms overlapping with those of other neuromuscular illnesses; ALS diagnosis is sometimes delayed. While it might be demanding to attain, early intervention is vital. It is obscure to find a treatment plan that works for every patient with ALS since the disease exhibits considerable variation in terms of progression pace, symptom presentation, and hereditary variables (19). The research and development of treatments for ALS is complicated by the involvement of several disease pathways, including neuroinflammation, protein aggregation, mitochondrial dysfunction, and oxidative stress. There are only two FDA-accepted preventive medications with low effectiveness, despite substantial research, underscoring the need for new potent therapies (20).

A hereditary alteration of the huntingtin (HTT) gene causes HD, which is a progressive neurodegenerative disorder that causes aberrant movements, symptoms of mental illness, and cognitive impairment. Treatment includes FDA-approved drugs like tetrabenazine and deutetrabenazine are used in this disease by depleting dopamine. Risperidone, olanzapine, and quetiapine may help in controlling mental symptoms and lowering chorea. Myoclonus and chorea can be treated with clonazepam and valproic acid. Lithium and valproic acid are the drugs that can help control mood swings and agitation (7).

Disease-modifying therapies like tominersen are one of the antisense oligonucleotides that bind to the altered HTT mRNA to decrease the amount of poisonous huntingtin protein produced. The HTT gene is being directly edited using methods like CRISPR/Cas9, which may be able to fix the genetic mutation causing HD. RNA interference treatments try to break down mutant HTT mRNA in order to reduce huntingtin protein levels (21).

The precise processes by which the mutated huntingtin protein damages neurons remain incompletely known, which makes the creation of tailored treatments more difficult. Although HD is frequently identified after symptoms appear, early management may be more beneficial. One of the main challenges is finding biomarkers for early diagnosis. Psychiatric symptoms, including psychosis, depression, and anxiety, may be very difficult to treat and have a big influence on a patient's quality of life (22).

20.3 EMERGING THERAPEUTIC APPROACHES

Although neurodegenerative disorders encompass a wide range of conditions, they all have similarities in their characteristic features. Common hallmarks of the disease that can be observed are progressive degeneration of neurons and a decline in their capacity, activation of microglia, induction of pro-inflammatory mediators, and increased production of reactive oxygen species, which in turn contribute to oxidative stress.

One of the highly promising therapies for neurodegenerative illnesses may be gene therapy. Developing successful treatment strategies against neurodegenerative disorders requires understanding the underlying mechanisms of disease through the manipulation of gene expression. The process of introducing therapeutic genes to target specific cells or tissues to treat genetic abnormalities and to provide beneficial effects or alter biological processes. For instance, if there is a presence of a defective functional gene to be corrected, current advances in the development of gene-based therapy can be helpful (23, 24).

20.3.1 Gene Therapy

Gene therapy is usually carried out by delivering the genes that encode restorative or protective molecules to the target area of the brain for neurodegenerative diseases. The main objective is to restore the function of neurons by promoting synapse formation and the regeneration of axons. Neuronal growth factors are used in the current approaches of the research. These predominant molecules, when delivered to the targeted area of the brain which has neuronal degradation, ameliorate the survival of neurons and reverse the process of atrophy. The main objective of gene therapy is to treat the disorder from its root cause gene by replacing the defective gene that triggers the condition (25, 26).

In neurodegenerative diseases such as Alzheimer's and Parkinson's, various evidence from preclinical studies suggests the potential activity of neurotrophic factors which totally inhibit the process of neuronal degeneration, resulting in improved cognitive and motor performances. Sometimes, vector viruses are used to deliver the transgene, for instance, AAV of parvovirus (ssDNA) type, adenovirus of dsDNA, HSV of herpesvirus (DsDNA) type, and lentivirus of retrovirus (ssRNA) type (27, 28). Some examples of gene therapy are shown in Table 20.2.

20.3.2 Oligonucleotide Therapeutics

This involves modulating gene expression to modify the malfunctioning gene that causes the disease. There are different types of oligonucleotide therapeutics, these include the following:

1. **Antisense Oligonucleotides (ASOs)**
   These are short, single-stranded, synthetic nucleotides that bind to complementary sequences of mRNA. They use mechanisms like RNase H-mediated degradation, splicing modulation, and translation inhibition, which interrupt protein synthesis. After the discovery of non-coding microRNAs, antisense therapy has become overrated (29). They can reduce the disease-causing proteins by either encoding the mRNA or inhibiting the translation process. In certain motor neuron diseases, RNA splicing approaches can be designed to restore some essential protein functions. ASOs can correct mutations which impacts dysfunctional proteins (30).
2. **Small interfering RNA (siRNA)**
   These are double-stranded RNA molecules that instigate RNA interference (RNAi), which is a specific natural defense mechanism against exogenous genes. They have the potential to silence disease-related genes in a sequence-specific manner. They have

TABLE 20.2
Examples of Gene Therapy That Are Used in Neurodegenerative Diseases

S. No	Name of the Disease	Gene Therapy	Mechanism
1	Spinal muscular dystrophy	Zolgensma (onasemnogene abeparvovec)	Delivers functional copy of the SMN1 gene, which is crucial for the survival of motor neurons, with the aid of adeno-associated virus vector (AAV9)
2	Amyotrophic lateral sclerosis	SOD1 gene silencing	Gene silencing of the mutated gene linked to SOD1 by developing antisense oligonucleotides (ASO) or CRISPR-Cas9 technology
3	Parkinson disease	AAV2 GDNF/AAV2 neurturin	Improving motor functions by delivering neuronal growth factors like glial cell line-derived neurotrophic factor (GDNF) and neurturin that enhance the regeneration of dopaminergic neurons and protect them
4	Huntington disease	HTT gene silencing	Downregulating the mutant huntingtin gene with the help of CRISPR-Cas9, ASOs, or RNAi techniques
5	Alzheimer disease	AAV-NGF therapy	Targets the cholinergic neurons that affect Alzheimer's disease

the tendency to mediate targeted mRNA degradation or mRNA translation repression. However, if siRNA gets incorporated into RNA-induced silencing complex (RISC), it unwinds the double-strand, which is divided into a single-stranded molecule and leaves the guide strand. RISC also protects them from degradation. When the guide strand finds the complementary sequences of mRNA molecules at a specific point, it cleaves the mRNA with the help of enzymes. Thus, they silence the gene by halting their protein synthesis (31).

3. **microRNA (miRNA)**
   These are small non-coding RNA that binds to the 3′-untranslated region of proteins that are 20–24 nucleotides long, which are involved in posttranscriptional regulation of gene expression. They can either be inhibitors or cause degradation of the translation step in protein synthesis. In recent years, they've identified miRNA expression, which is abundant in neuronal cells taking part in the development of neuronal axons and appendages, neurotransmission, and plasticity of synapses. Hence, they might also play a crucial role in dopaminergic neurons and regions of the brain. miRNAs also have an important role in autophagy regulation. Thus, there can be an opportunity to implement gene-based therapy or oligonucleotide therapeutics using miRNA as gene regulation (32).

20.3.3 Stem Cell Therapy

Stem cells have the capacity to renew themselves, proliferate continuously, and differentiate into various types of cells. These include embryonic stem cells (ESCs), induced pluripotent stem cells (iPSCs), and mesenchymal stem cells, which have their respective mechanisms. In the case of neurodegeneration, they have the potential to replace damaged cells by integrating into damaged brain tissue. They also secrete neurotrophic factors and molecules with anti-inflammatory properties, which further promote survival of neuronal cells and decrease inflammation (33).

A study by Barker et al. states that some investigations from clinical trials have shown promising results using fetal mesencephalic tissue transplantation, ESC-derived dopaminergic neurons, and iPSC-derived dopaminergic neurons in Parkinson's disease (34, 35). It is also mentioned in other diseases like Alzheimer's disease and HD. But the challenges that might be faced are properties such as immunogenicity, safety, ethical considerations, and functionality. Research in stem cell therapy is also known as regenerative medicine, which aims to restore lost function and to slow the disease progression (36).

20.3.4 Traditional Therapy

Complementary and alternative therapies, including traditional medicine, play a significant role in treating neurodegenerative disorders. These extensive theoretical and practical frameworks have not changed since they were created by ancient cultures. Some examples of alternative medical practices include homeopathy, naturopathy, Unani medicine, traditional Chinese medicine, and Indian Ayurvedic medicine. Several alternative medical approaches place a strong focus on using internal natural energies to achieve a harmonious state of mind and body (37, 38).

Complementary and alternative treatments, such as natural products and herbal medicines, have been used to enhance neurodegenerative disease therapy because there are no disease-modifying medications for the disease. Herb-based traditional oriental remedies from East Asia have been utilized for the treatment of dementia for a very long time. Research into herbal medicines and the standardization of natural goods is yielding new insights. Combination treatments are also valued for their comprehensive approaches to treating many issues and enhancing treatment results overall. Several herbs have shown encouraging outcomes in early studies. Integrating herbal medicine with neurodegenerative disease treatment offers a possible new avenue for treatment. It provides a range of approaches to deal with the complex pathophysiological processes of the illness, making it a potentially difficult therapy in the future to replace current care (39).

The awareness of herbal treatments and their capacity to delay the beginning of certain ailments is becoming more widely acknowledged as scientific knowledge expands. Clinical trials, molecular evaluations, and pharmacological studies are examples of research methodologies that have advanced and provide promise for future, precise, and effective harnessing of the medicinal potential of herbal substances. Furthermore, they can address the underlying disease to enhance synaptic plasticity, cognitive resilience, and neurogenesis. Herbal therapy has the potential to provide customized, holistic treatments that are tailored to the various requirements of people with neurological issues, as it blends conventional knowledge with research-based findings (40).

Subsequently, for Alzheimer's and Parkinson's, evidence-based research toward herbal medicines is currently increasing (Table 20.3). For other neurological diseases, complementary or alternative therapies with holistic practices like acupuncture, mindfulness and meditation, aromatherapy, yoga and tai chi, chiropractic care, and massage therapy are encouraged apart from herbal medicine. These approaches have the benefit of neuroprotective effects, alleviate symptoms, promote emotional and physical well-being, improve mobility and quality of life, and reduce stress by promoting relaxation and sensory stimulation (41).

TABLE 20.3
Summary of Herbal Medicine Used in Neurodegenerative Diseases

Medicinal Plant	Neurodegenerative Disease	Mechanism of Action
Ginkgo biloba	Alzheimer's disease	The active metabolite, ginkgolides, possesses anti-inflammatory, antioxidant, and free radical scavenging action and improves cerebral blood flow and enhances cognitive function.
Curcumin	Alzheimer's disease	Neuroprotection and enhances cognitive function by interfering with β-amyloid plaques and toxic oligomers aggregation, and clearance of tau phosphorylation.
Bacopa monnieri	Alzheimer's disease	Neuroprotection and nootropic properties enhance cognitive function; certain studies showed that within 6 months there's a drastic improvement.
Huperzia serrata	Alzheimer's disease	Acetylcholinesterase inhibition by huperzine A and cognitive enhancement by enhancing the availability of acetylcholine.
Ashwagandha	Parkinson disease	Neuroprotective effects due to the presence of withanoside, antioxidant, and anti-inflammatory potential for motor function improvement promote synaptic and neuronal healing.
Ginseng	Parkinson disease	Bioactive molecule ginsenosides possess neuroprotective, anti-inflammatory properties by stimulating PI3K/Akt pathway, improving motor symptoms and quality of life.
Phyllanthus	Alzheimer's disease	Diminishes oxidative stress, acetylcholinesterase activity, and lipid peroxidation, and improves antioxidant levels in the brain. These help in the neuroprotective function.
Green tea	Parkinson disease	Polyphenolic content present in green tea has beneficial neuroprotective, antioxidant, and metal chelation activity, which also regulates protein kinase C due to the presence of active moieties like epigallocatechin gallate.

20.3.5 Precision Medicine

Personalized medicine or precision medicine is an umbrella term commonly used to describe customized therapy or tailored medical treatment according to their individuality, variability in genes and environment, and lifestyle of each patient. The specific steps that must make sure to ensure a precise medicinal strategy are to perform genetic profile studies. This can be done by identifying specific gene mutations or variability in the diseased condition, which can predict risk of the disease, progression, and response to certain treatments. Precision medicine also helps in determining the measurable indicators, like biomarkers, which track the disease progression or treatment response at an early stage. Some other methods, including predictive modeling, patient stratification, targeted therapies, and personalized treatment plans, can overcome the challenges that are faced through conventional or regular therapeutic strategies. With the help of improved technologies such as artificial intelligence and data analysis, implementing precision medicine into routine practices holds the promise for transforming diagnosis and management (42, 43).

20.4 FUTURE PERSPECTIVES AND CHALLENGES

Predicting the future of the treatment perspectives for neurodegenerative disorders is crucial. Research on ageing has garnered more attention in recent years, especially when employing models that can accurately replicate the multifaceted pathophysiology of age-related disorders in situ. Some of the new treatment protocols, like senolytics, have been overrated in the case of chronic diseases, as age is the primary reason for many neurological degeneration and diseases like Alzheimer's and Parkinson's. Upregulation of factors such as cell cycle arrest, mitochondrial damage, telomere shortening, prone to DNA damage, expression of lytic enzymes, as well as growth factors contribute to age-related inflammatory diseases. Thus, senolytics have the tendency to inhibit the senescent cells, which halts the above factors. Other therapeutic approaches, including gene therapy, regenerative medicine, oligonucleotide therapeutics, precision medicine, and herbal medicine, are blooming into the realms of neurodegenerative treatment. However, having the ability to apply our newly acquired information to create more effective instruments and creative strategies for treatment will be a challenge going forward (44, 45).

20.5 CONCLUSION

In conclusion, the future aspect of treatment for neurodegenerative diseases holds great promise with tremendous advancements into several domains. Ongoing research and technological innovations might provide a deeper understanding of the intricate pathophysiology and mechanisms of the disease to improve further direction in treatment using precise strategies. Emerging therapeutic approaches, such as regenerative medicine, and gene and RNA-based therapeutics might offer hope in preserving cognitive function. The future holds possibilities for revolutionary advancements in our knowledge, ability to identify, and treat neurodegenerative disorders via sustained innovation and coordinated efforts across disciplines.

REFERENCES

1. Heemels MT. Neurodegenerative diseases. *Nature*. 2016 Nov 1;539(7628):179.
2. Kampmann M. Molecular and cellular mechanisms of selective vulnerability in neurodegenerative diseases. *Nat Rev Neurosci*. 2024 May;25(5):351–71.
3. Scheltens P, Strooper BD, Kivipelto M, Holstege H, Chételat G, Teunissen CE, et al. Alzheimer's disease. *The Lancet*. 2021 Apr 24;397(10284):1577–90.
4. Nahar L, Charoensup R, Kalieva K, Habibi E, Guo M, Wang D, Kvasnica M, Onder A, Sarker SD. Natural products in neurodegenerative diseases: Recent advances and future outlook. *Front Pharmacol*. 2025 Mar 19;16:1529194.

5. Moens TG, Da Cruz S, Neumann M, Shelkovnikova TA, Shneider NA, Van Den Bosch L. Amyotrophic lateral sclerosis caused by FUS mutations: Advances with broad implications. *Lancet Neurol*. 2025 Feb 1;24(2):166–78.
6. Bang J, Spina S, Miller BL. Frontotemporal dementia. *The Lancet*. 2015 Oct 24;386(10004):1672–82.
7. Walker FO. Huntington's disease. *The Lancet*. 2007 Jan 20;369(9557):218–28.
8. Bloem BR, Okun MS, Klein C. Parkinson's disease. *The Lancet*. 2021 Jun 12;397(10291):2284–303.
9. Walker Z, Possin KL, Boeve BF, Aarsland D. Lewy body dementias. *The Lancet*. 2015 Oct 24;386(10004):1683–97.
10. Jakimovski D, Bittner S, Zivadinov R, Morrow SA, Benedict RH, Zipp F, et al. Multiple sclerosis. *The Lancet*. 2024 Jan 13;403(10422):183–202.
11. Breijyeh Z, Karaman R. Comprehensive review on Alzheimer's disease: Causes and treatment. *Molecules*. 2020 Dec 8;25(24):5789.
12. Bellenguez C, Küçükali F, Jansen IE, Kleineidam L, Moreno-Grau S, Amin N, et al. New insights into the genetic etiology of Alzheimer's disease and related dementias. *Nat Genet*. 2022 Apr;54(4):412–36.
13. Belder CRS, Schott JM, Fox NC. Preparing for disease-modifying therapies in Alzheimer's disease. *Lancet Neurol*. 2023 Sep 1;22(9):782–3.
14. Kim J, Basak JM, Holtzman DM. The role of apolipoprotein E in Alzheimer's disease. *Neuron*. 2009 Aug 13;63(3):287–303.
15. Allain H, Bentué-Ferrer D, Akwa Y. Disease-modifying drugs and Parkinson's disease. *Prog Neurobiol*. 2008 Jan 1;84(1):25–39.
16. Colombo D, Pnevmatikou P, Melloni E, Keywood C. Therapeutic innovation in Parkinson's disease: A 2020 update on disease-modifying approaches. *Expert Rev Neurother*. 2020 Oct 2;20(10):1047–64.
17. Lindholm D, Mäkelä J, Di Liberto V, Mudò G, Belluardo N, Eriksson O, et al. Current disease modifying approaches to treat Parkinson's disease. *Cell Mol Life Sci*. 2016 Apr 1;73(7):1365–79.
18. Vijiaratnam N, Simuni T, Bandmann O, Morris HR, Foltynie T. Progress towards therapies for disease modification in Parkinson's disease. *Lancet Neurol*. 2021 Jul 1;20(7):559–72.
19. Hardiman O, Al-Chalabi A, Chio A, Corr EM, Logroscino G, Robberecht W, et al. Amyotrophic lateral sclerosis. *Nat Rev Dis Primer*. 2017 Oct 5;3(1):1–19.
20. Masrori P, Van Damme P. Amyotrophic lateral sclerosis: A clinical review. *Eur J Neurol*. 2020; 27(10):1918–29.
21. Caron NS, Dorsey ER, Hayden MR. Therapeutic approaches to Huntington disease: From the bench to the clinic. *Nat Rev Drug Discov*. 2018 Oct;17(10):729–50.
22. Tabrizi SJ, Flower MD, Ross CA, Wild EJ. Huntington disease: New insights into molecular pathogenesis and therapeutic opportunities. *Nat Rev Neurol*. 2020 Oct;16(10):529–46.
23. Choong CJ, Baba K, Mochizuki H. Gene therapy for neurological disorders. *Expert Opin Biol Ther*. 2016 Feb 1;16(2):143–59.
24. Butterfield DA, Halliwell B. Oxidative stress, dysfunctional glucose metabolism and Alzheimer disease. *Nat Rev Neurosci*. 2019 Mar;20(3):148–60.
25. Deverman BE, Ravina BM, Bankiewicz KS, Paul SM, Sah DWY. Gene therapy for neurological disorders: Progress and prospects. *Nat Rev Drug Discov*. 2018 Sep;17(9):641–59.
26. Simonato M, Bennett J, Boulis NM, Castro MG, Fink DJ, Goins WF, et al. Progress in gene therapy for neurological disorders. *Nat Rev Neurol*. 2013 May;9(5):277–91.
27. Weinberg MS, Samulski RJ, McCown TJ. Adeno-associated virus (AAV) gene therapy for neurological disease. *Neuropharmacology*. 2013 Jun 1;69:82–8.
28. Scoles DR, Pulst SM. Oligonucleotide therapeutics in neurodegenerative diseases. *RNA Biol*. 2018 Jun 3;15(6):707–14.
29. Testa CM. Antisense oligonucleotide therapeutics for neurodegenerative disorders. *Curr Geriatr Rep*. 2022 Sep 1;11(3):19–32.
30. Grabowska-Pyrzewicz W, Want A, Leszek J, Wojda U. Antisense oligonucleotides for Alzheimer's disease therapy: From the mRNA to miRNA paradigm. *eBioMedicine*. 2021 Dec 1;74.
31. Bortolozzi A, Manashirov S, Chen A, Artigas F. Oligonucleotides as therapeutic tools for brain disorders: Focus on major depressive disorder and Parkinson's disease. *Pharmacol Ther*. 2021 Nov 1;227:107873.
32. Ma YM, Zhao L. Mechanism and therapeutic prospect of miRNAs in neurodegenerative diseases. *Behav Neurol*. 2023;2023(1):8537296.
33. Kim SU, de Vellis J. Stem cell-based cell therapy in neurological diseases: A review. *J Neurosci Res*. 2009;87(10):2183–200.
34. Barker RA, Parmar M, Studer L, Takahashi J. Human trials of stem cell-derived dopamine neurons for Parkinson's disease: Dawn of a new era. *Cell Stem Cell*. 2017 Nov 2;21(5):569–73.

35. Zhang P, Kishimoto Y, Grammatikakis Ioannis, Gottimukkala K, Cutler RG, Zhang S, et al. Senolytic therapy alleviates Aβ-associated oligodendrocyte progenitor cell senescence and cognitive deficits in an Alzheimer's disease model. *Nat Neurosci.* 2019 May;22(5):719–28.
36. Ashraf SS, Hosseinpour Sarmadi V, Larijani G, Naderi Garahgheshlagh S, Ramezani S, Moghadamifar S, et al. Regenerative medicine improve neurodegenerative diseases. *Cell Tissue Bank.* 2023 Sep 1;24(3):639–50.
37. Jeon SG, Song EJ, Lee D, Park J, Nam Y, Kim J il, et al. Traditional oriental medicines and Alzheimer's disease. *Aging Dis.* 2019 Apr 1;10(2):307–28.
38. Kushwah S, Maurya NS, Kushwaha S, Scotti L, Chawade A, Mani A. Herbal therapeutics for Alzheimer's disease: Ancient Indian medicine system from the modern viewpoint. *Curr Neuropharmacol.* 2023 Mar 30;21(4):764–76.
39. Luthra R, Roy A. Role of medicinal plants against neurodegenerative diseases. *Curr Pharm Biotechnol.* 2022;23(1):123–39.
40. Ratheesh G, Tian L, Venugopal JR, Ezhilarasu H, Sadiq A, Fan TP, et al. Role of medicinal plants in neurodegenerative diseases. *Biomanufacturing Rev.* 2017 Oct 24;2(1):2.
41. Cai C, Fang J, Ke H. Editorial: Traditional medicine and phytochemicals for neurodegenerative diseases treatment: Application of interdisciplinary technologies in novel therapeutic target and drug discovery. *Front Neurosci.* 2023 Aug 8;17.
42. Leonard HL, Nalls MA, Day-Williams A, Esmaeeli S, Jarreau P, Bandres-Ciga S, et al. Open science in precision medicine for neurodegenerative diseases. *Nat Rev Drug Discov.* 2024 Apr;23(4):233–4.
43. Rossi SL, Subramanian P, Bovenkamp DE. The future is precision medicine-guided diagnoses, preventions and treatments for neurodegenerative diseases. *Front Aging Neurosci.* 2023 Mar 17;15:1128619.
44. Pramotton FM, Spitz S, Kamm RD. Challenges and future perspectives in modeling neurodegenerative diseases using organ-on-a-chip technology. *Adv Sci.* 2024 Aug;11(32):2403892.
45. Noble W, Burns MP. Challenges in neurodegeneration research. *Front Psychiatry.* 2010 Mar 10;1:7.

Index

Note: Page locators in *italic* indicate a figure, and page locators in **bold** indicate a table on the corresponding page.

A

acetylcholinesterase inhibitors, 16, 18, 115, 117, 159, 161, 262, 286
adamantane derivatives, 114, 115
ALS (amyotrophic lateral sclerosis), 3, 27, 28, 40, 114, 120, 160, 178, 243
amyloid-β (Aβ), 8, 11, 23, 78, 117, 119, 148
amyloid-β aggregation, 122, 148
anticholinergics, 91, 244, 245, 282
antioxidant activity, 16, 31, **93**, 207, 214, 230, 233, 277
apoptosis, 5–9, 47, 89, 109, 143, 154
autonomic dysfunction, **75**, 77, 260, 282
axonal transport, 6, 228

B

BACE-1 (β-secretase), 115
Bacopa monnieri, 16, 23, 66, 96
baicalein, **92**, **239**
BBB (blood–brain barrier), 164, 192, 203, 205
benzothiazoles, 116, 120, 121
bioavailability, 19, 21, 23, 33, 47, 102, 229, 230, 237, 238
brain-derived neurotrophic factor (BDNF), 88, 104, 216, 236

C

carbidopa, 81, 91, 245, 264
cardamonin, 119
cardiovascular side effects, 248, 249
catalase, 31, 62, 92, 105, 232
catechin, 29, 35, 41, **48**, 50–52, 54, 68, 233
chalcones, 118, 120
cholinesterase inhibitors, 7, **12**, 15, 81, 84, 135, 187, 193
chronic neuroinflammation, 231
citalopram, 82
constipation, **62**, 80, 81, 91, **136**, 245, 246, 255, 256
curcumin, 16, 19, 27, 91, 103, 202, 206, 256, 264

D

deep brain stimulation, 94, **136**, 243, 244
dementia, 3, 11, 18, 22, 61, 80, 85, 224
dendrimers, 137, 195, 200, 207
dendritic polyglycerol (DPG), 201
depression, 11, 25, 65, 248, 251
dizziness, 247, 249, 250
donepezil, 19, 81, 114, 118, 244, 245
dopamine, 10, **12**, 54, 59, 78, 125, 206
droxidopa, 82
duloxetine, 112, 113

E

electron spin resonance (ESR), 170
entacapone, 91, 114, 264
epicatechin, 29, 47, 51, 54, 64, 219, 232, 233, 236
excitotoxicity, 5, 28, **63**, 175, 227, 228, 231
extracellular vesicles (EVS), 10, 102, 149, 161, 165

F

flavonoids, 18, 32, 33, 50, **63**, 87, 90, 102, 119, 188, 205, 262
frataxin deficiency, 45, 53
free radical scavenging, 286
Friedreich's ataxia, 7, 45, 47, 50, 54

G

galantamine, 12, 15, 81, 114, 159, 262, 263
genistein, 65, 234
Ginkgo biloba (EGb761), 16, 21, 257, 258, 262, 264, 286
glutamate excitotoxicity, 230, 231, 282
glutathione, 8, 31, 62, 89, 103, 232, 234

H

hallucinations, 60, 62, 74–79, 81, 91, 249, 250
heat shock proteins (Hsp27, Hsp70), 34
hesperidin, 30, 12, 66, 92, 118, 234, 239
hesperidin methyl chalcone, 118
hippocampus, 8, 15, 161, 225, 236, 261, 262
huntingtin protein, 64, 144, 227, 231, 283
Huntington's disease, 56, 58, 62, 63, 69
hyperphosphorylation, 11, 230, 262

I

in vitro studies, 51, 234, 237
in vivo studies, 19, 78, 114, 137, 163, 167, 264
indole derivatives, 117, 120
insomnia, 15, **62**, 82, 83, 247, 254–259, 264
integrative medicine, 123, 131, 132
interleukin-6 (IL-6), 106
ion suppression, 179, **180**
iron accumulation, 7

K

kaempferol, 92, 232, 233, 239, 266
kale, 217, 232, 239

L

levodopa, 12, 61, 81, 83, 88, 90, 92–94
Lewy bodies, 73–77, 85, 114, 225–227
Lewy body dementia, 11, 74, 75, 279, 280
lifestyle modifications, 131, 249, 252, 254, 256, 259, 272
lipid peroxidation, 8, 10, 38, 49, 50, 64, 89, 92, 105, 118
lipopolysaccharide (LPS), 118
liposomes, 119, 137–139, 148, 162, 196

M

malondialdehyde (MDA), 62, 230, 234
MAO-B inhibitors, 91, **93**, 244, 247, 263, 281
memantine, 12, 15, 19, 62, 114, 115, 159, 244, 262, 265
metabolomics, 172, 179, **183**, 275
midodrine, 82, 247
mirtazapine, 82
mitochondrial dysfunction, 5, 6, 9, 11, 28, 32, 36, 38, 104, 214
modafinil, 83, 252
monoamine oxidase-B (MAO-B) inhibitors, 264
motor neuron degeneration, 27, 34, 36, 98, 103, 105, 106, 108, 111, 226
motor symptoms, 59, 61, 74, 79, 84, 243, 244, 247
Mucuna pruriens, 90, 92, 264
multiple sclerosis (MS), 197, 202

N

nanogels, 139, 195, 199, 200, 207
nanomedicine, 140, 144, 146, 148
nanoparticles, 21, 69, 111, 119, 137, **138**, 139–148, 162, 163, 191, 195, 196, 205, 206, 230, 238, 266, 275
nanotechnology, 21, 119, 126, 128, 137, 140, 144, 148, 162, 163, 195–196, *196*, 206, 207
naringenin, 30, 234, 239
natural AChE inhibitor, 263
neurodegenerative diseases (overview), 12, 224, 281, **284**, **286**
neurofibrillary tangles, 4, 6, 76, 77, 143, 225, 230, 279, 280
neurogenesis, **12**, 17, 22, 32, 104, 109, 163, 216, 219, 229, 236, 240, 241, 263, 276, 286
neuroinflammation, 5, *7*, 8–9, 15, 20, 28, 29, 32, 33, 34, 35, 36, 37, 38, 57, 84, 87, 89–91, **92–93**, 106, 108, 135, 143–144, 160, 185, 192, 203, 213–215, 217–221, 227, 230–232, 235–236, 240–241, 243, 263, 275–276, 283
neuroprotection, 5, 17, 18, 22, 32, 37, 47, 90, **93**, **94**, 103, 105, 107–108, 139, 163, 193, 212–214, 217, 219, 221, 232, 235, 237, 240, **286**
neuroprotective agents, 90, 143, 147, 205
nicotinic acetylcholine receptors, 89
NMDA antagonists, **136**, 248
NMDA receptor, **12**, 15, **62**, **91**, **136**, 159–161, 227, 236, 244, 247, 248, 250, 262, 264, 265
nortriptyline, 82
Nrf2 pathway, 31, 89
Nusinersen (Spinraza), 98

O

OEA (oleoylethanolamide), 162
omega-3 fatty acids, 35, 37, 212, 214, 216, 218, 220, 221, 240, 258
onasemnogene abeparvovec (Zolgensma), 98, 110, 284
organic farming, 130, 131
orthostatic hypotension, 62, 80, 82, 247
oxidative DNA damage, 233, 234
oxidative injury, 104
oxidative stress, 104–110, 114, 127, 135, 143
oxidative stress markers, 34, 104, 176, 177, 234
oxybutynin, 82

P

paramagnetic species, 172, 173, 176, 178
Parkinson's disease (PD), 73, 87
peptides, 15–17, 116, 138, 140, 141, 164, 171, 172, 198, 200, 205, 219
permaculture, 130
personalized medicine, 84, 181, 192, 239–241, 287
pharmacognosy, 125, 129, 132
pharmacokinetics, 110, 144, 146, 197, 198, 200, 202
phytoconstituents, 62, 65, 67, 69, 87, 88, 90–94, 127
pramipexole, 81, 83, 91, 114, 116, 264, 281
protein aggregation, 10, 11, 32, 38, **63**, 114, 160, 177
protein misfolding and aggregation, 227, 259
proteomics, 172, 179, **183**, 275
pyrazolines, 117, 118

Q

quantum computing, 182, **183**, 184
quercetin, 21, 30, 33, 66, 92, 102, 111, 218, 219, 232, 234, 235, 239, 240

R

rasagiline, 81, 91, 114, 264, 281
reactive oxygen species (ROS), 6, 16, 31, 49, 173, 215, 227
regulatory approval, 162, 207
REM sleep behaviour disorder (RBD), 79
Risdiplam (Evrysdi), 98
rivastigmine, 15, 81, 84, 114, 119
ropinirole, 267
rotigotine, 91, 264, 281

S

safinamide, 91
scalability, 110, 145, 146, 198, 200, 202, 204, 205, 238
selegiline, 81, 91, 114, 264, 281
sildenafil, 83
single-cell analysis, 182
SiNPs (silica nanoparticles), 196, 205, 206
SIRT1 activation, 65, 104, 105
sleep hygiene, 248, 255
SMN protein, 99
Sofalcone, 97–100, 103
spatialomics, 181
stress management, 5, 249, 254, 255, 259
superoxide dismutase (SOD), 103

T

tadalafil, 83
tamsulosin, 82
tardive dyskinesia, **62**
thioflavin T, 116
TNF-α, 89, 92, 109, 118
tolcapone, 91, 281
tremors, 56, 60, 76, 79, 90, 156, 226, 247, 248, 255, 263, 280, 282
TrkB signaling, 236

U

urinary incontinence, 80

V

venlafaxine, 82
vitamin E, 35, 36, 47–49, 216

W

weight gain/loss, 251, 252

For Product Safety Concerns and Information please contact our
EU representative GPSR@taylorandfrancis.com Taylor & Francis
Verlag GmbH, Kaufingerstraße 24, 80331 München, Germany